Forsthoffer's Rotating Equipment Handbooks

Forsthoffer's Rotating Equipment Handbooks

Volume 3: Compressors

W E (Bill) Forsthoffer

President, Forsthoffer Associates
Washington Crossing, PA, USA

ELSEVIER

UK Elsevier Ltd, The Boulevard, Langford Lane, Kidlington, Oxford OX5 1GB, UK
USA Elsevier Inc, 360 Park Avenue South, New York, NY 10010-1710, USA
JAPAN Elsevier Japan, Tsunashima Building Annex, 3-20-12 Yushima, Bunkyo-ku, Tokyo 113, Japan

ISBN 1-85617-469-7

British Library Cataloguing in Publication Data
A CIP Catalogue record for this book is available from the British Library

Published by
Elsevier Advanced Technology,
The Boulevard, Langford Lane, Kidlington, Oxford OX5 1GB, UK
Tel: +44(0) 1865 843000
Fax: +44(0) 1865 843971

Typeset by Land & Unwin Ltd., Northamptonshire
Transferred to digital print 2007
Printed and bound by CPI Antony Rowe, Eastbourne

Contents

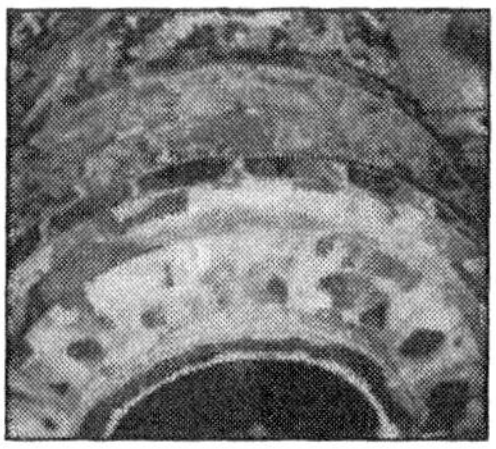

Preface

This Series has evolved from my personal experience over the last 40 years with the design, selection, testing, start-up and condition monitoring of Rotating Equipment. Most of the concept figures were originally written on a blackboard or whiteboard during a training session and on a spare piece of paper or I beam during a start-up or a problem solving plant visit.

My entire career has been devoted to this interesting and important field. Then and now more than ever, the cost of rotating equipment downtime can severely limit revenue and profits. A large process unit today can produce daily revenues in excess of US$5 million. And yet, the Operators, Millwrights and Engineers responsible for the safety and reliability of this equipment have not been afforded the opportunity to learn the design basis for this equipment in practical terms. I have also observed in the last ten years or so, that the number of experienced personnel in this field is diminishing rapidly.

Therefore the series objective is to present, in User friendly (easy to access), practical terms (using familiar analogies), the key facts concerning rotating equipment design basis, operation, maintenance, installation and condition monitoring to enable the reader (Engineer, Operator and Millwright) to:

- Understand the effect of process and environmental changes on equipment operation, maintenance and reliability
- Condition Monitor equipment on a component basis to optimize up-time, mean time between failure (MTBF) and mean time to repair (MTTR)
- Select, audit and test the Equipment that will produce highest safety and reliability in the field for the lowest life cycle cost.

The hope is that the knowledge contained in this series will enable

Plant Operations, Maintenance and Engineering Personnel to easily access the material that will allow them to present their recommendations to management to solve existing costly problems and produce new projects of optimum reliability.

This volume, Compressors, provides details of the various types of machine characteristics, operation in a process system (using the concept of required and produced gas head), performance relationships, selection and what determines the turbo compressor curve shape. It also covers surge/ stall/ stonewall, the effects of fouling, the design basis of journal and thrust bearings, balance drums, seals, critical speeds, control and protection guidelines, series and parallel operation, component condition monitoring, trouble-shooting and many other aspects.

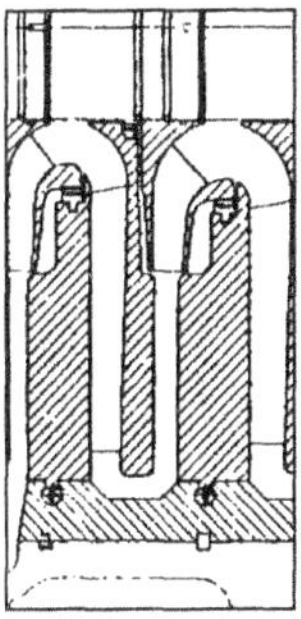

Acknowledgements

This series is a result of interactions with literally thousands of dedicated engineers, machinists, operators, vendors, contractors and students who have been an integral part of my career.

I consider myself very fortunate to have been associated with the best of the best mentors, business associates and dear friends throughout my career. Most especially, in chronological order Dick Salzmann, Bob Aimone, Merle Crane, Walt Neibel, the late Murray Rost, Mike Sweeney and Jimmy Trice. Bob, Merle, Murray and Mike have contributed specifically to the material in this series while Dick, Walt and Jimmy have tactfully kept me on track when necessary.

Special thanks to all of the global machinery vendors who have allowed me to use their material for my training manuals and now this publication.

Last but certainly least; my career would not have been possible without the support, encouragement and assistance from my wife Doris and our children Jennifer, Brian, Eric, Michael and Dara.

A special additional note of thanks to Michael who helped assemble the material, and hopefully learned some in doing so, since he has elected to pursue a career in rotating machinery.

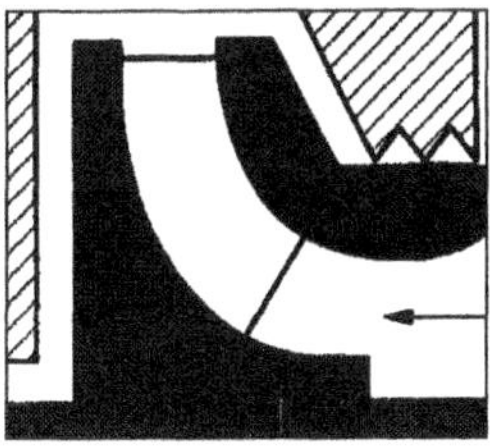

About the author

Bill Forsthoffer began his life-time career in rotating machinery in 1962 with De Laval Turbine Inc. as a summer trainee. After obtaining a Bachelor of Arts degree in Mathematics and Bachelor of Science degree in Mechanical Engineering, during which time he worked for De Laval part time in the Test, Compressor and Steam Turbine Departments, he joined De Laval full time in the Compressor Engineering Department in 1968. He was responsible for all phases of centrifugal compressor component and auxiliary design and also made many site visits to provide field engineering assistance for start up and problem resolution.

Bill joined Mobil Oil Corporate Engineering in 1974 and was responsible for all aspects of rotating equipment specification, technical procurement, design audits, test, field construction, commissioning, start-up and troubleshooting.

After 15 years at Mobil, Bill founded his own consulting company in 1990 and has provided rotating equipment consulting services to over 100 companies. Services include: project reliability assurance, training (over 7,000 people trained) and troubleshooting.

Bill is active in the industry as President of Forsthoffer Associates Inc., frequently writes articles for Turbo Machinery International Magazine and conducts many site specific and public workshops each year.

He can be contacted at bill@forsthofferassociates.com

DEDICATION

IN MEMORY OF MY PARENTS
FOR THEIR CONSTANT INSPIRATION

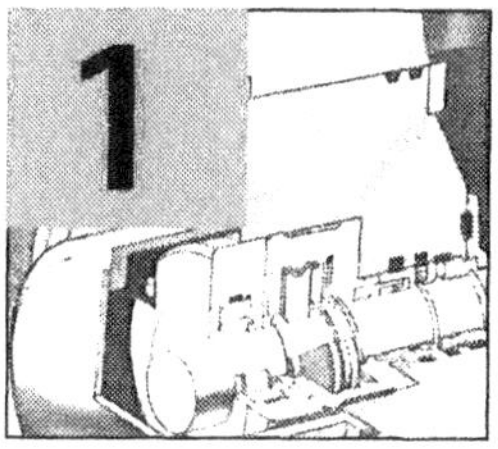

1 Rotating equipment overview

- Introduction
- Definition of rotating equipment
- Classifications of rotating equipment
- Site equipment examples
- Performance and mechanical design similarities
- The equipment 'train' or 'unit'
- Important fundamentals

Introduction

Take a minute and list all the different types and kinds of rotating equipment you can think of. Even if you have not been involved with rotating equipment for a long time, when you consider the types of equipment that you come in contact with every day, your list will be sizeable. Imagine if we pursued our objectives by looking at each individual piece of equipment. You would never remember all the aspects and the book would be long and very boring. We will not attempt this approach. Rather, this section will divide all types of rotating equipment into four major classifications. The function of each individual classification will be defined. Throughout this book we will cover many types of rotating equipment.

One good thing to remember is to always ask yourself what the function of this particular type is, what does it do? We will find that many aspects covered in this book will have the same common function.

Our approach therefore, will be to observe the similarities in both performance and mechanical aspects of various types of equipment. We will see that many of these relationships apply regardless of the type of equipment that is considered.

Now think of any rotating equipment unit that you have come in contact with and review that unit considering the different components that comprise it. You will find that every unit of rotating equipment consists of a driven machine, driver, transmission device and is supported by auxiliary equipment. That is, each unit is made up of all the classifications of rotating equipment that are described in this section. This is an important fact to remember in troubleshooting equipment. In essence then, each unit is a system.

Definition of rotating equipment

DEFINITION
Rotating equipment moves products
Products = solids, liquids, gases

Figure 1.1 Definition

Figure 1.1 presents a basic definition of rotating equipment.

As we shall see, there are different classifications of rotating equipment. Regardless of the classification, rotating equipment moves product. More properly stated:

Rotating equipment moves money!!

Stop the equipment and the source of revenue stops! This is a very important fact to remember. If you want management to approve your recommendation, you must be able to justify it economically! The form of any recommendation to management should be as shown in Figure 1.2.

A successful recommendation
Must state clearly and simply:
■ The problem
■ The problems' cost to the company (parts, labor and loss revenue)
■ The recommended solution
■ Its' cost
■ Proof that the solution will work (has it worked somewhere else?)
■ Savings to company by implementing proposed solution

Figure 1.2 A successful recommendation

If you proceed as shown in Figure 1.2 you will be able to obtain and maintain management support. Remember, you can learn a great deal in this book. However, if you cannot implement what you have learned the information is totally useless to the company. If you cannot obtain management support, you will never implement any action plan.

Classifications of rotating equipment

There are four (4) basic function classifications of rotating equipment. Refer to Figure 1.3 which defines the classifications of rotating equipment.

Classifications of rotating equipment
■ Driven
■ Drivers or prime movers (provide power)
■ Transmission devices
■ Auxiliary equipment

Figure 1.3 Classifications of rotating equipment

Figure 1.4 is a partial listing of some rotating equipment types grouped according to function.

Major types of rotating equipment

I. Driven equipment

- A. Compressors
  - 1. Dynamic
    - Centrifugal
    - Axial
    - Integral Gear
  - 2. Positive displacement
    - Screw
    - Rotary lobe
    - Reciprocating
    - Diaphragm
    - Liquid Ring
- B. Pumps
  - 1. Dynamic
    - Centrifugal
    - Axial
    - Slurry
    - Integral Gear
  - 2. Positive displacement
    - Plunger
    - Diaphragm
    - Gear
    - Screw
    - Progressive cavity
- C. Extruders
- D. Mixers
- E. Fans

II. Drivers – prime movers

- A. Steam turbines
- B. Gas turbines
- C. Motors
  - Induction
  - Synchronous
  - Vari-speed
- D. Engines
  - Internal combustion
  - Diesel
  - Gas

III. Transmission devices

- A. Gears
  - Helical
  - Double helical
- B. Clutches
- C. Couplings

IV. Auxiliary equipment

- A. Lube and seal systems
- B. Buffer gas systems
- C. Cooling systems

Figure 1.4 Major types of rotating equipment

Site equipment examples

Following is an example of typical site rotating equipment.

Figures 1.5, 1.6, 1.7, 1.8 show examples of each rotating equipment classification.

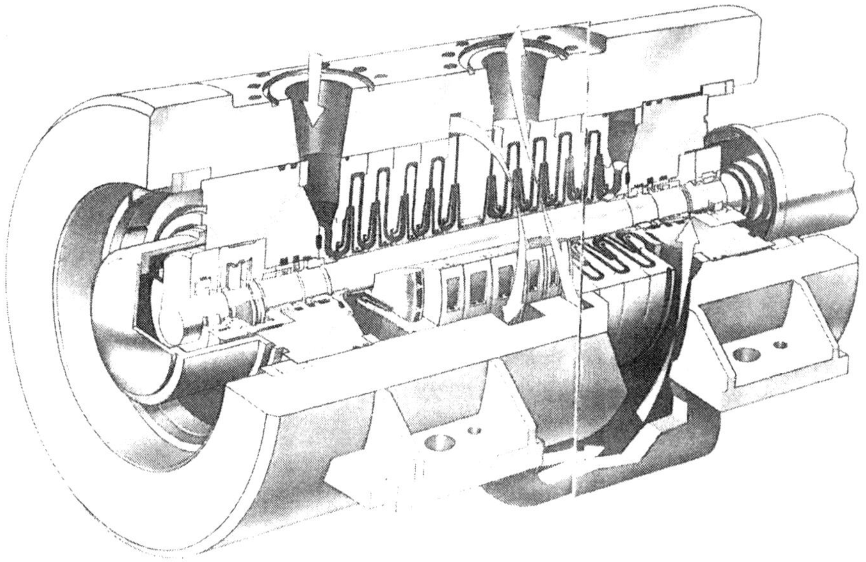

Figure 1.5 High pressure centrifugal compressor (Courtesy of Dresser Rand)

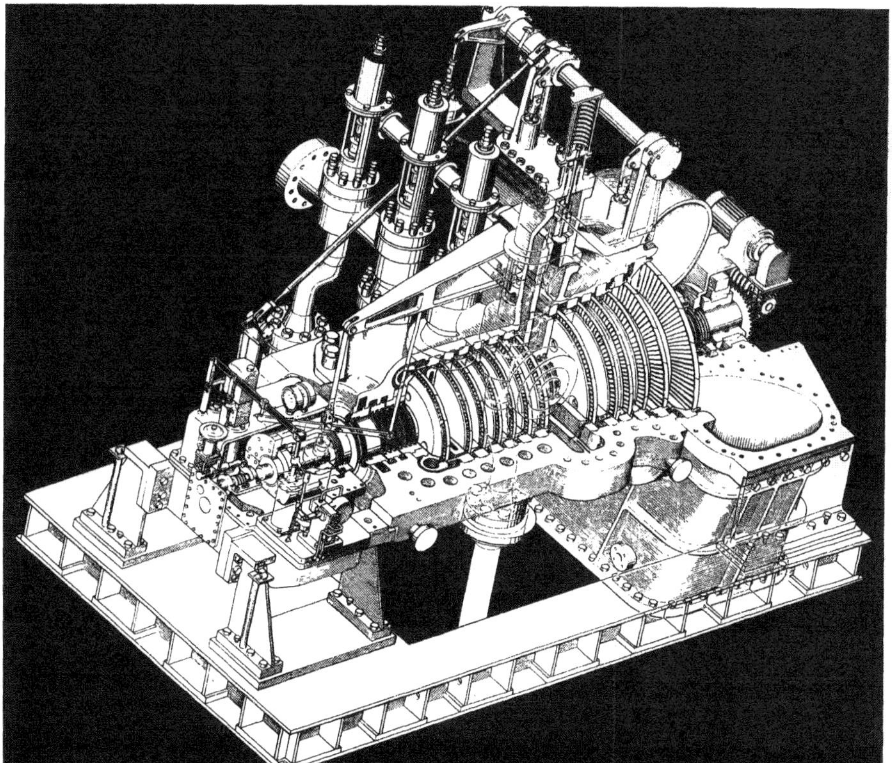

Figure 1.6 Extraction – condensing steam turbine (Courtesy of MHI)

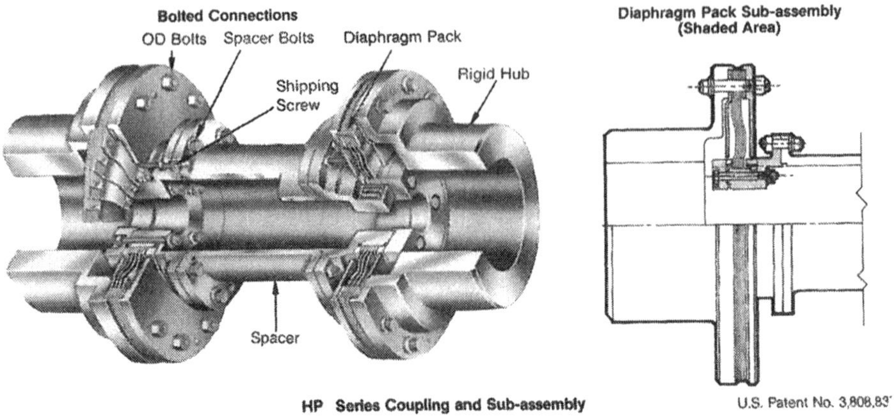

Figure 1.7 Multiple, convoluted diaphragm-spacer coupling (Courtesy of Zurn Industries)

Figure 1.8 Horizontal oil console arrangement (Courtesy of Oltechnique)

Performance and mechanical design similarities

During this book we will be examining many different types of rotating equipment. However, the task will be a lot easier if we begin our study by first focusing on the similarities of the equipment and then the specific differences.

As an example, we have chosen to first present pumps then compressors as topics. As was just discussed, both pumps and compressors are driven types of equipment and move product. Regardless of the product phase or state, their functions are identical. Refer to Figure 1.9 which compares dynamic pump and compressor performance.

Pumps and compressors

- **Both pumps and compressors** move a fluid from one energy level to another.
- **A pump** moves an incompressible fluid - a liquid. For our purposes, the volume of a liquid does not change with pressure and temperature.
- **A compressor** moves a compressible fluid - a gas. The volume of a gas changes with pressure, temperature and gas composition.
- The principles of dynamic machines apply both to pumps and compressors. However, since gases are compressible, the volume flow rate and hence the gas velocity in a passage is affected.

Figure 1.9 Pumps and compressors

The same comments can be made concerning mechanical components. Refer back to Figures 1.5 and 1.6 and ask are the functions of the casings, internal seals, shaft end seals and bearings the same? Absolutely! A bearing performs the same function whether it is in a pump, compressor, gearbox, etc.

Figure 1.10 shows how both performance and mechanical functions are similar regardless of the classification or type of equipment.

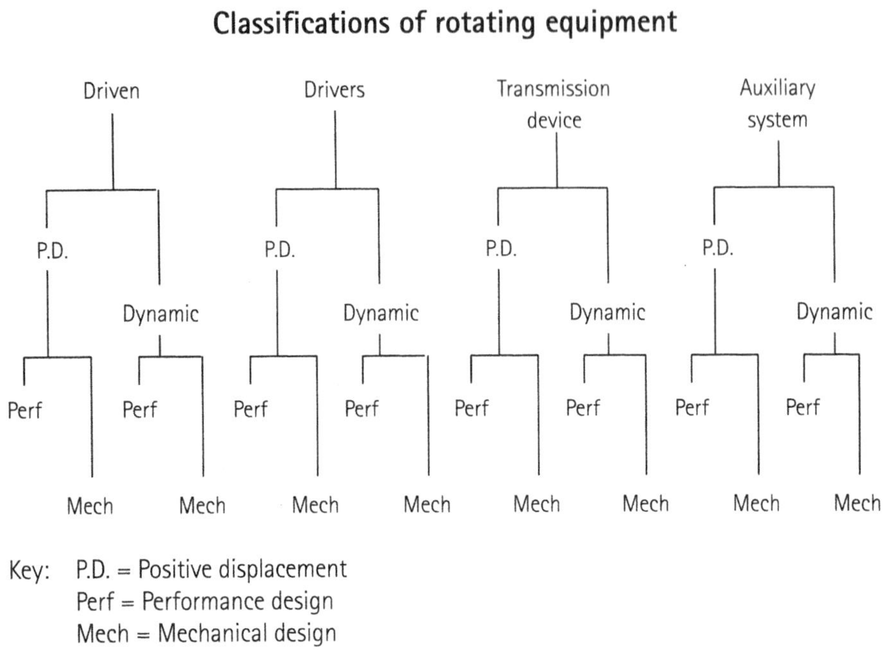

Figure 1.10 Classifications of rotating equipment

As we proceed through this book, we will discover that positive displacement or dynamic performance principles will be the same regardless of the type of equipment (pump, compressor or turbine). Also, the mechanical principles presented for bearings, seals, etc. will apply whether the component is in a pump, turbine, gear, etc.

The equipment 'train' or 'unit'

As stated, the objective is to learn the functions of equipment and major components so that they can be effectively condition-monitored to maximize site safety and reliability. Having defined the four (4) classifications of equipment, how many of these classifications are present in a pump unit or compressor train? ... All four (4)!!

Regardless of the type of unit or train, a driven, driver, transmission(s) and auxiliary system(s) must always be present. When you are asked to inspect G-301 or K-101 you are actually inspecting G-301 pump unit or K-101 compressor train. Failure to recognize this fact will severely limit your troubleshooting scope and ability.

As an example, a call from the unit shift manager may state that G-301 discharge pressure is zero – what are possible causes? A few are:

- Process change
- Pump wear
- Coupling failure
- Pump shaft failure
- Driver shaft failure
- Pump or driver shaft seizure (no oil)
- Pump seal failure
- Process valve closed
- Steam inlet valve closed (if driver is a steam turbine)

Do you get the point! ... The entire unit or train, all four machinery classifications must always be considered in rotating equipment design, revamps and troubleshooting.

Important fundamentals

Before discussing specific facts concerning all the rotating equipment on site, some important fundamentals need to be presented. The environment or surroundings for any piece of rotating equipment play an important part in determining the availability of that particular item (Refer to Figure 1.11).

This figure shows that the rotating equipment environment is the process unit in which the equipment is installed. The surroundings of the equipment will be defined early in the project. Proper design of process conditions, piping and foundations, selection of other components (drivers, transmission devices and auxiliaries) and proper specification of ambient conditions all must be considered. If any of these items are not taken into account, the end user of the equipment will be faced with a history of an unreliable process and will pay dearly in terms of lost product revenue.

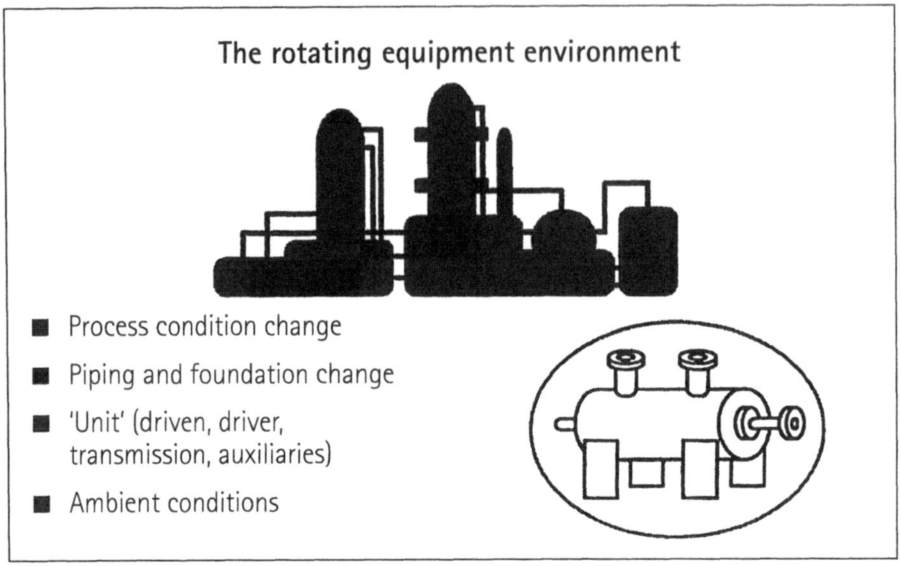

Figure 1.11 The rotating equipment environment

It is important to understand that the life span of rotating equipment is extremely long compared to the specification, design and installation phase. Refer to Figure 1.12. A typical installation will have a specification design and installation phase of only approximately 10% of the total life of the process unit. Improper specification, design or installation will significantly impact the maintenance requirements, maintenance cost and availability of a particular piece of machinery.

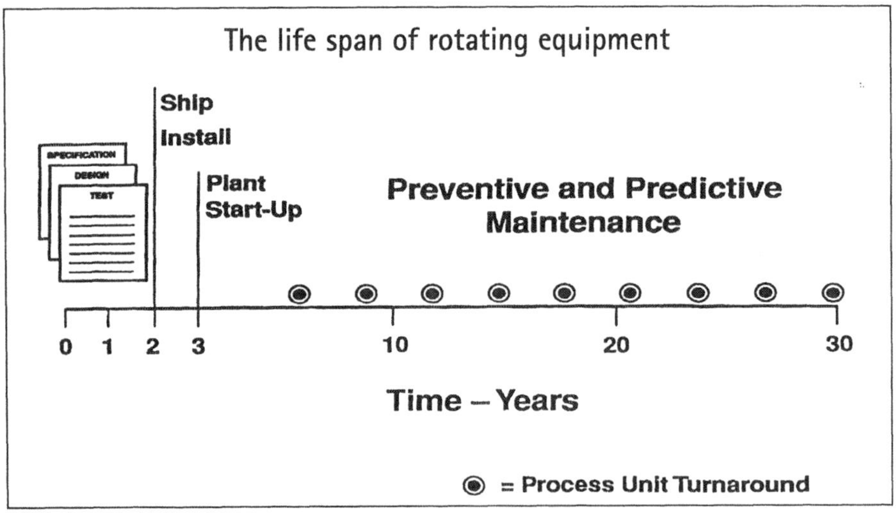

Figure 1.12 The life span of rotating equipment

The objectives of the end user are shown in Figure 1.13.

The objectives

- Maximum reliability (on stream time)
- Maximum product throughput
- Minimum operating costs

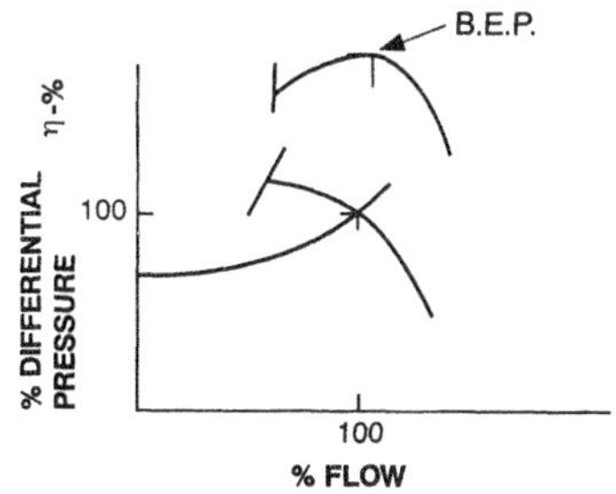

Figure 1.13 The objectives

In order to maximize the profit, a piece of machinery must have maximum reliability, maximum product throughput and minimum operating cost (maximum efficiency). In order to achieve these objectives, the end user must play a significant part in the project during the specification and design phase and not only after the installation of the equipment in the field. Effective field maintenance starts with the specification phase of a project. Inadequate specifications in terms of equipment operating conditions, mechanical design, instrumentation and the location of the instrumentations will reduce equipment availability.

The definitions of reliability and availability are shown in Figure 1.14.

Definitions: Reliability and availability

Reliability — The amount of time equipment operates in one (1) year

$$\text{Reliability (\%)} = \frac{\text{Operating hours per year}}{\text{8700 hours}}$$

Availability — The amount of time equipment operates in one (1) year – the planned downtime

$$\text{Availability (\%)} = \frac{\text{Operating hours per year}}{\text{8700 – planned downtime (hours)}}$$

Note: The values should be greater than 95%.

Figure 1.14 Definitions: Reliability and availability

Reliability does not take into account planned down time. Availability considers planned downtime (turnarounds, etc.). Both values are stated in percentage. Typical equipment availabilities when properly specified are 97% +.

Availability directly affects the product revenue. Product revenue is the value obtained from the product produced in one day. For refineries, the loss of revenue can exceed a million U.S. Dollars a day. This will occur if a critical piece of equipment (unspared) is shut down. Remember, even though the unit may be down for a short period, the time necessary to bring the process back within specification may be significantly longer. It is very valuable to obtain the product revenue figure for the unit in which you are working. As will be shown, this value can significantly influence management in terms of decision making.

It is important to understand that upset conditions with any piece of rotating equipment can occur in a very short period of time. In addition, the requirements for reliable operation (a minimum of three (3) years continuous run) require enormous amounts of equipment revolutions all on a very thin film of lubrication oil. Failure to maintain this film is one of the major reasons for reduced equipment availability. Figure 1.15 presents a typical fact sheet concerning a turbo-compressor.

Interesting facts

- A typical turbo-compressor that runs at 6,000 RPM rotates at 100 times per second.
- It also rotates 8,640,000 times per day.
- And ... 3,153,600,000 times per year.
- All on an oil film of less than one thousandth of an inch.

Figure 1.15 Interesting facts

As an exercise you may want to determine how far this shaft would travel in one year if its diameter were four inches. There are 5,280 feet in one mile and one revolution around the earth equals 25,000 miles.

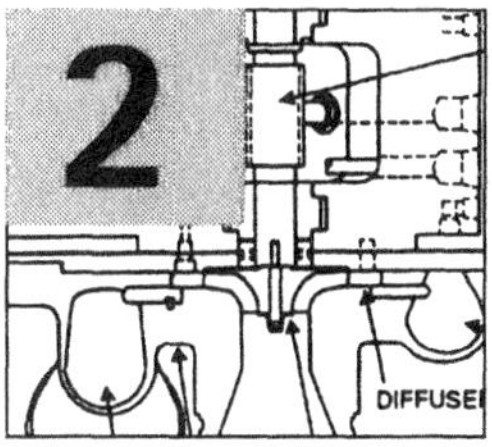

2 Compressor types and applications

- Introduction
- Positive displacement compressors
- Dynamic compressors

Introduction

In this chapter we will overview compressor types and their typical applications. The two basic classifications of compressors are positive displacement and dynamic compressors.

Positive displacement compressors are constant volume, variable energy (head) machines that are not affected by gas characteristics.

Dynamic compressors are variable volume, constant energy (head) machines that are significantly affected by gas characteristics.

The type of compressor that will be used for a specific application therefore depends on the flow rate and pressure required and the characteristic of the gas to be compressed.

In general, dynamic compressors are the first choice since their maintenance requirements are the lowest. The next choice are rotary type positive displacement compressors since they do not contain valves and are gas pulsation free. The last choice are reciprocating compressors since they are the highest maintenance compressor type and produce gas pulsations. However, the final selection depends upon the application requirements as discussed below.

Figure 2.1 presents a flow range chart showing the various types of compressor applications as a function of flow (ACFM) and discharge pressure (PSIG).

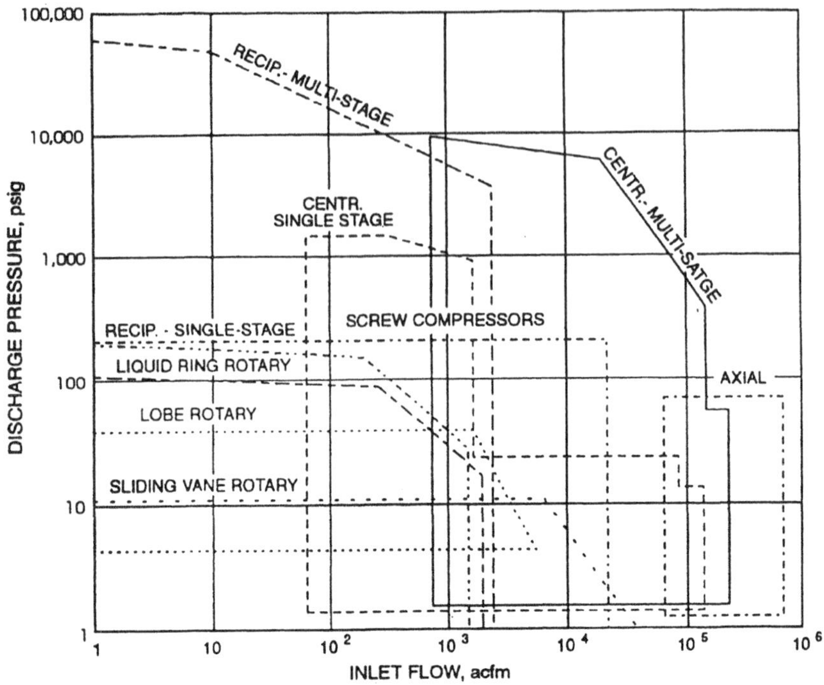

Figure 2.1 Compressor application range chart

Table 2.1 shows the typical operating ranges for the various types of compressors used in the refining, chemical and gas processing industries.

Table 2.1 Typical operating range of various types of gas compressors

Machine Type	Capacity ICFM Min Max	T_2 Max °F	P_1 Max PSIA	P_2 Max PSIA	P/R MIN	P/R MAX
Rotary lobe	1 to 40,000	350	35	55	1.0+	2.4
Rotary vane	45 to 3,300	350	45	65	1.3	3.2
Rotary screw	50 to 20,000	350	150	615	2.0	6.0
Recip	1 to 10,000	800	1000	10,000	3.0	50.0
Liquid ring	10 to 10,000	N/A	100	140	1.0+	10.0
Centrifugal Single stage	700 to 150,000	500	1000	1,400	1.0+	3.4
Centrifugal Multi stage	300 to 150,000	800	2000	6,000	2.0	10.0
Axial	75,000 to 350,000	800	30	150	1.0	10.0

Table 2.2 describes the typical applications for the various types of compressors presented in Figure 2.1 and Table 2.1.

Table 2.2 Typical compressor applications

Compressor type	Application
Rotary lobe	Conveying – powder, polyethylene
Rotary vane air	Air blowers (low volume). Also used as gas turbine starters.
Rotary screw	Plant and instrument air, low flow process – off gas, recycle, sulphur blowers
Rotary liquid ring	Crude unit vacuum, various saturated gas applications
Reciprocating	Plant and instrument air off gas (low flow) recycle (low flow) H_2 make-up, gas reinjection (low flow)
Centrifugal single stage	Air blowers, recycle
Centrifugal single stage – Integral gear	Low flow recycle, off gas, plant air (can replace a recip in low flow, medium to high molecular weight applications).
Centrifugal multi-stage side load (Horizontal split or barrel)	(Propane, propylene, ethylene, Freon, mixed gas) refrigeration
Centrifugal multi-stage barrel	Recycle, reinjection, syn gas
Centrifugal multi-stage – integral gear	Plant and instrument air, *Process applications require proven field experience
Axial compressor	FCC blower, MTBE effluent, gas turbine air

Positive displacement compressors

Positive displacement compressors are used for low flow and/or low molecular weight (hydrogen mixture) applications. The various types are presented below.

Rotary lobe

A rotary lobe compressor consists of identically synchronized rotors. The rotors are synchronized through use of an external, oil-lubricated, timing gear, which positively prevents rotor contact and which minimizes meshing rotor clearance to optimize efficiency. This feature also allows the compressor to be oil free in the gas path. The rotors of the two-lobe compressor each have two lobes. When the rotor rotates, gas is trapped between the rotor lobes and the compressor casing. The rotating rotor forces the gas from the gas inlet port, along the casing, to the gas discharge port. Discharge begins as the edge of the leading

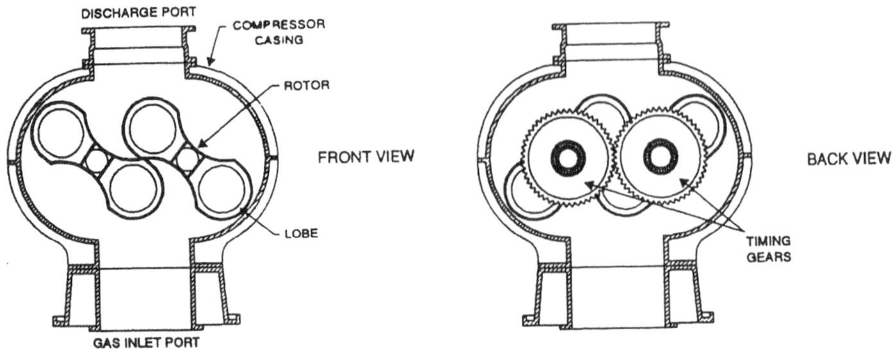

Figure 2.2 Rotary lobe

lobe passes the edge of the discharge port. The trailing lobe pushes the entrapped gas into the discharge port, which compresses the gas against the backpressure of the system. Rotary lobe compressors are usually supplied with noise enclosures or silencers to reduce their characteristic high noise level. A schematic of a two-lobe rotary compressor is shown in Figure 2.2.

Rotary vane

A sliding-vane rotary compressor uses a series of vanes that slide freely in longitudinal slots that are cut into the rotor. Centrifugal force causes the vanes to move outward against the casing wall. The chamber that is formed between the rotor, between any two vanes, and the casing is referred to as a cell. As the rotor turns, an individual vane passes the inlet port to form a cell between itself and the vane that precedes it. As an individual vane rotates toward the end of the inlet port, the volume

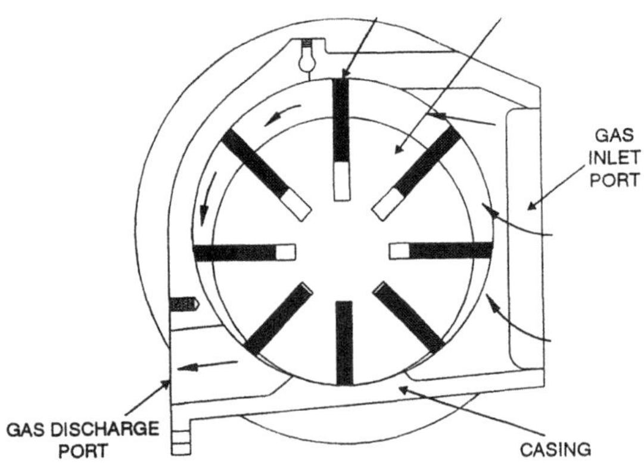

Figure 2.3 Rotary vane

of the cell increases. The increase in the cell volume draws a partial vacuum in the cell. The vacuum draws the gas in through the inlet port. When a vane passes the inlet port, the cell is closed, and the gas is trapped between the two vanes, the rotor and the casing. As rotation continues toward the discharge port, the volume of the cell decreases. The vanes ride against the casing and slide back into the rotor. The decrease in volume increases the gas pressure. The high pressure gas is discharged out of the compressor through the gas discharge port. Sliding-vane rotary compressors are characterized by a high noise level that results from the vane motion. A schematic of a sliding vane rotary compressor is shown in Figure 2.3.

Rotary screw

The single-stage design consists of a pair of rotors that mesh in a one-piece, dual-bore cylinder. The male rotor usually consists of four helical threads that are spaced 90 degrees apart. The female rotor usually consists of six helical grooves that are spaced 60 degrees apart. The rotor speed ratio is inversely proportional to the thread-groove ratio. In the four-thread, six-groove, screw compressor, when the male rotor rotates at 1800 rpm, the female rotor rotates at 1200 rpm.

The male rotor is usually the driven rotor, and the female rotor is usually driven by the male rotor. A film of foil is normally injected between the rotors to provide a seal between the rotor and to prevent metal-to-metal contact. An oil-mist eliminator, installed immediately downstream of the compressor, is required for plant and instrument air service. However, designs are available that do not require lubrication.

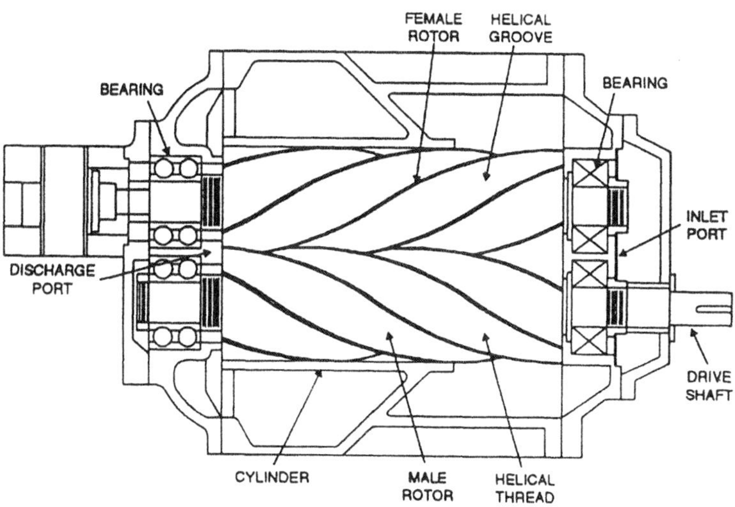

Figure 2.4 Rotary screw

Screw compressors that do not require lubrication are commonly referred to as "dry screw-type compressors".

The inlet port is located at the drive-shaft end of the cylinder. The discharge port is located at the opposite end of the cylinder. Compression begins as the rotors enmesh at the inlet port. Gas is drawn into the cavity between the male rotor threads and female rotor grooves. As rotation continues, the rotor threads pass the edges of the inlet ports and trap the gas in a cell that is formed by the rotor cavities and the cylinder wall. Further rotation causes the male rotor thread to roll into the female rotor groove and to decrease the volume of the cell. The decrease in the volume increases the cell pressure. Oil is normally injected after the cell is closed to the inlet port. The oil seals the clearances between the threads and the grooves, and it absorbs the heat of compression. Compression continues until the rotor threads pass the edge of the discharge port and release the compressed gas and oil mixture. A typical single stage screw compressor is shown in Figure 2.4.

Rotary liquid ring

Liquid ring rotary compressors consist of a round, multi-blade rotor that revolves in an elliptical casing. The elliptical casing is partially filled with a liquid, which is usually water. As the rotor turns, the blades form a series of buckets. As the rotor turns, the buckets carry the liquid

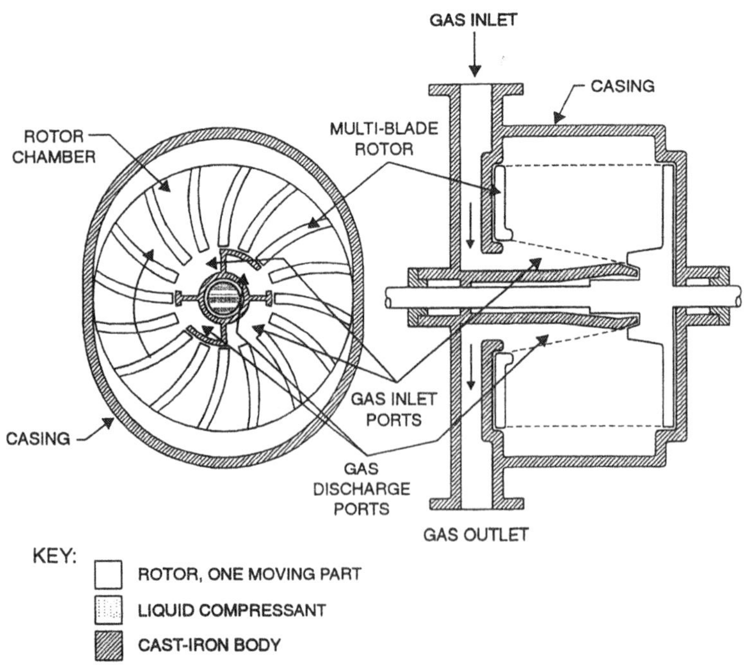

Figure 2.5 Rotary liquid ring (Courtesy of Nash Engineering Company)

around with the rotor. Because the liquid follows the contour of the casing, the liquid alternately leaves and returns to the space between the blades. The space between the blades serves as a rotor chamber. The gas inlet and discharge ports are located at the inner diameter of the rotor chamber. As the liquid leaves the rotor chamber, gas is drawn into the rotor chamber through the inlet ports. As the rotor continues to rotate, the liquid returns to the rotor chamber and decreases the volume in the chamber. As the volume decreases, the gas pressure increases. As the rotor chamber passes the discharge port, the compressed gas is discharged into a gas/liquid separator and then to the process. A typical liquid ring rotary compressor is shown in Figure 2.5.

Reciprocating

The basic components of a reciprocating compressor are a crankshaft, crossheads, piston rod packing, cylinders, pistons, suction valves, and discharge valves. Figure 2.6 is an illustration of a three-stage reciprocating compressor. Note that the third stage piston and cylinder are mounted on top of the second stage piston and cylinder. A prime mover (not shown) rotates the crankshaft. The crankshaft converts the rotary motion of the prime mover into reciprocating motion of the pistons.

The compression cycle of a reciprocating compressor consists of two strokes of the piston: the suction stroke and the compression stroke. The suction stroke begins when the piston moves away from the inlet

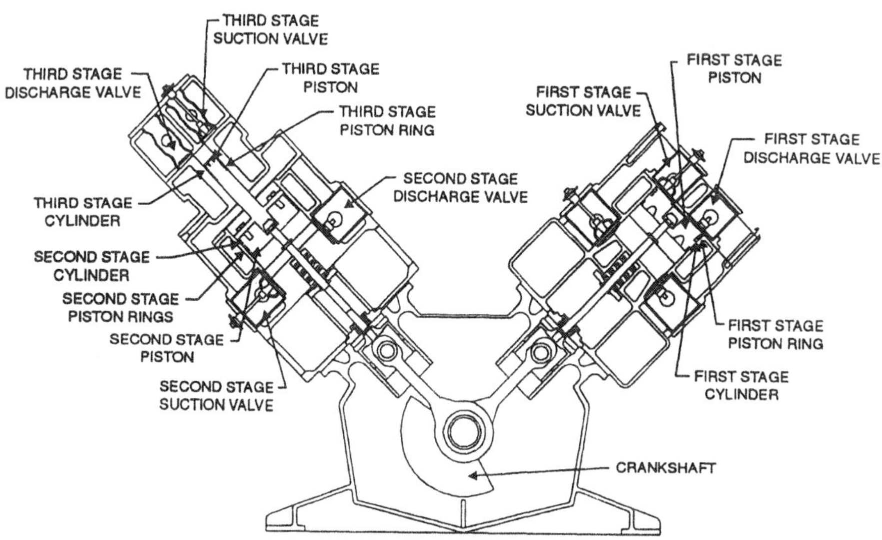

Figure 2.6 Reciprocating

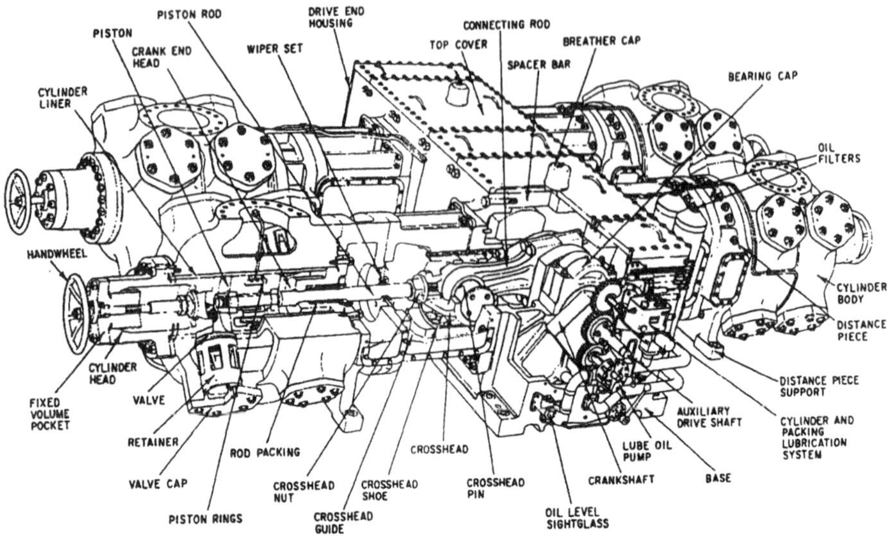

Figure 2.7 Reciprocating

port of the cylinder. The gas in the space between the piston and the inlet port expands rapidly until the pressure decreases below the pressure on the opposite side of the suction valve. The pressure difference across the suction valve causes the suction valve to open and admit gas into the cylinder. The gas flows into the cylinder until the piston reaches the end of its stroke. The compressor stroke starts when the piston starts its return movement. When the pressure in the cylinder increases above the pressure on the opposite side of the suction valve, the suction valve closes to trap the gas inside the cylinder. As the piston continues to move toward the end of the cylinder, the volume of the cylinder decreases and the pressure of the gas increases. When the pressure inside the cylinder reaches the design pressure of the stage, the discharge valve opens and discharges the contents of the cylinder to the suction of the second stage. The second stage takes a suction on the discharge of the first stage, further compresses the gas and discharges to the third stage. The third stage takes a suction on the discharge of the second stage and compresses the gas to the final discharge pressure. Please refer to Figure 2.6.

Figure 2.7 shows a balanced opposed, four throw reciprocating compressor typical of the type that would be used in refinery hydrogen make-up service.

Dynamic compressors

Dynamic compressors are used wherever possible s a result of their low maintenance requirements. The single stage integral gear centrifugal compressor allows the use of a dynamic compressor in many applications where positive displacement compressors have previously been used. The two types of dynamic compressors are centrifugal and axial.

Centrifugal compressors – general principles of operation

Centrifugal compressors increase the energy of a gas by increasing the tangential velocity (V_T) of a gas, as shown in Figure 2.8. The principle of operation of a centrifugal compressor is very similar to the principle of operation of a centrifugal pump. The gas enters the compressor through the inlet nozzle that is proportioned so the gas enters the impeller with a minimum of shock or turbulence. The impeller, which consists of a hub and blades, is mounted on a rotating shaft. The impeller receives the gas from the inlet nozzle and dynamically compresses it by increasing the energy of gas proportional to the product of the blade tip speed velocity (U_T) and the gas tangential velocity change (V_T) in the impeller. V_R represents the velocity of the gas relative to the blade. The resultant velocity (V) is the vector sum of the relative velocity (V_R) and the blade tip speed velocity (U_T).

A diffuser surrounds the impeller, and it serves to gradually reduce the velocity of the gas as the gas leaves the impeller. The diffuser converts the velocity energy to a higher pressure level. In a single stage compressor, the gas exits the diffuser through a volute casing that surrounds the diffuser. The volute casing collets the gas to further reduce the velocity of the gas and to recover additional velocity energy.

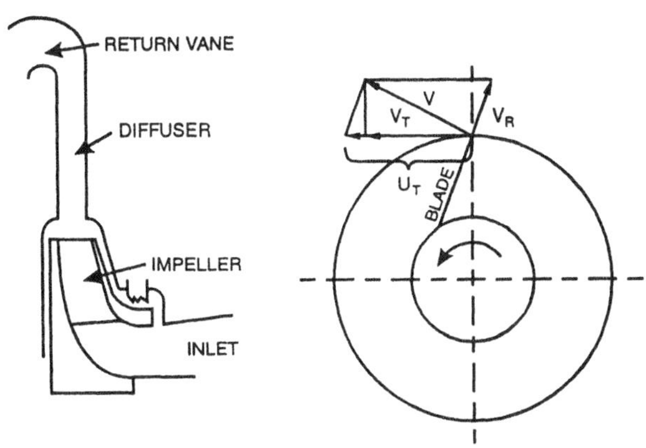

Figure 2.8 Centrifugal compressor

The gas exits through the discharge nozzle. In a multi-stage compressor, the gas exits the diffuser and enters return vanes. The return vanes direct the gas into the impeller of the next stage.

Centrifugal single stage (low ratio)

A typical single stage centrifugal low ratio compressor is shown in Figure 2.9. These types are known as single stage overhung compressors since the impeller is outboard of the radial bearings the cases are radially split.

Centrifugal single stage integral gear

Figure 2.10 shows a "Sundyne" single stage integral gear compressor section. This type of compressor is an in line type (similar to a pump) usually driven by motor through an integrally mounted gear box (not shown). These compressors are used for low flow high energy (head) applications and are being used in many applications previously serviced by positive displacement compressors. These compressors operate at high speeds, 8,000 – 34,000 rpm and are limited to approximately 400 horsepower.

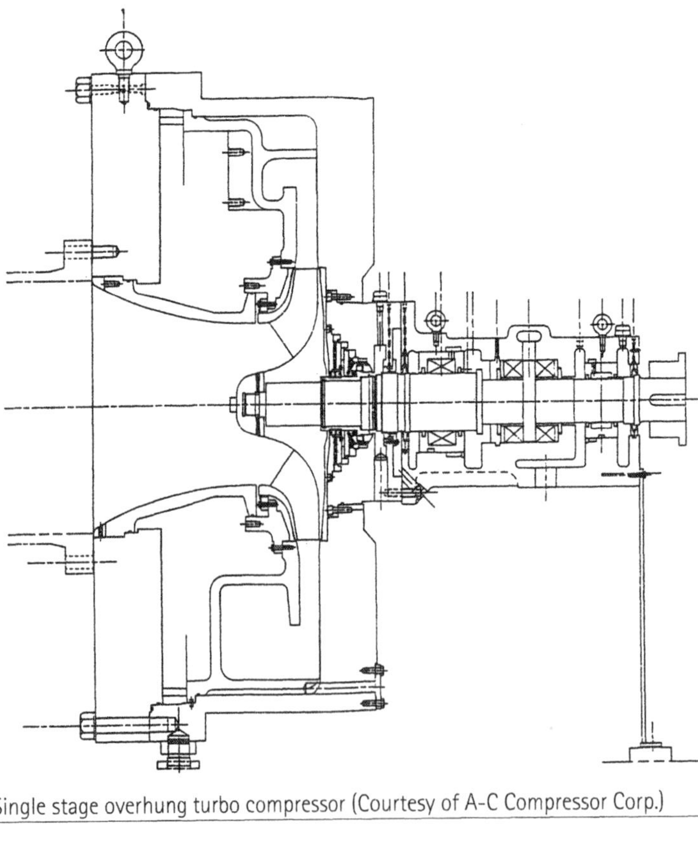

Figure 2.9 Single stage overhung turbo compressor (Courtesy of A-C Compressor Corp.)

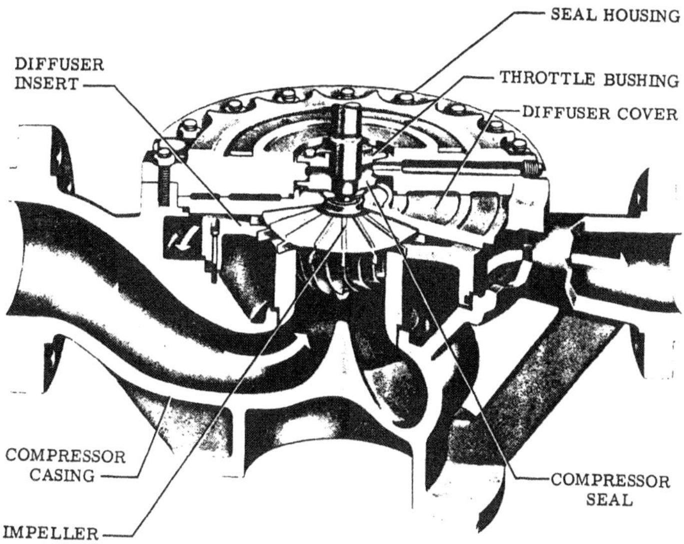

Figure 2.10 Single stage high speed compressor (Courtesy of Sundstrand Corp.)

Centrifugal multi-stage horizontal split

A typical multi-stage horizontally split centrifugal compressor is shown in Figure 2.11. The casing is divided into upper and lower halves along the horizontal centerline of the compressor. The horizontal split casing allows access to the internal components of the compressor without disturbing the rotor to casing clearances or bearing alignment. If possible, piping nozzles should be mounted on the lower half of the compressor casing to allow disassembly of the compressor without removal of the process piping.

Centrifugal multi-stage with side loads

This type of compressor is used exclusively for refrigeration services. The only difference from the compressor shown in Figure 2.11 is that gas is induced or removed from the compressor via side load nozzles. A typical refrigeration compressor is shown in Figure 2.12. Note that this type of compressor can be either horizontally or radially split.

Centrifugal multi-stage (barrel)

A typical multi-stage, radially-split, centrifugal compressor in shown in Figure 2.13. The compressor casing is constructed as a complete cylinder with one end of the compressor removable to allow access to the internal components. Multi-stage, radially-slit centrifugal compressors are commonly called barrel compressors.

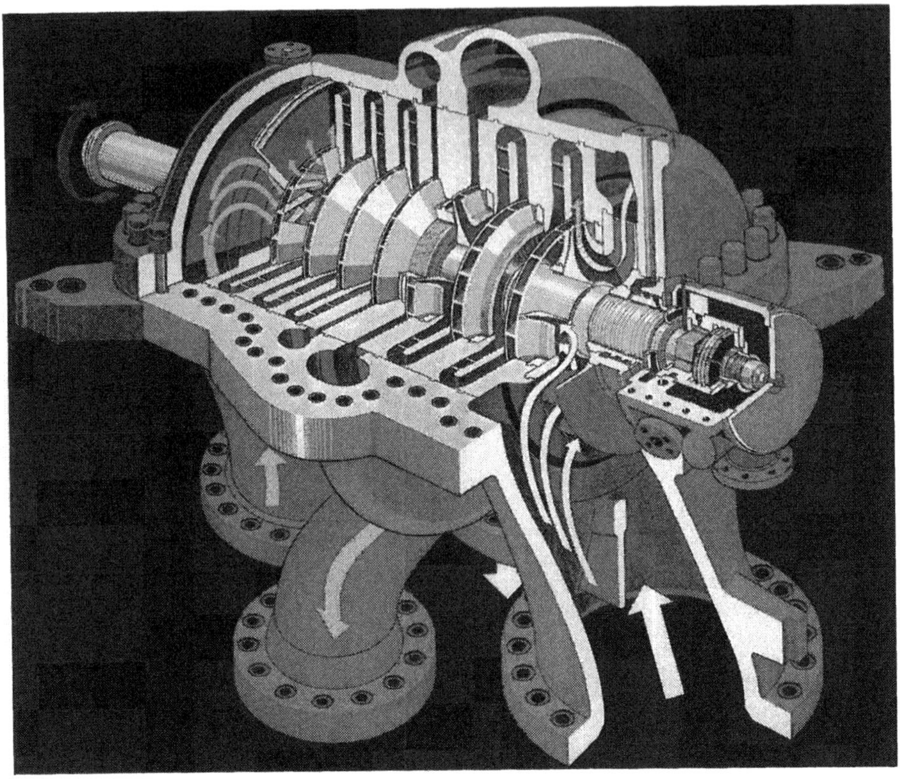

Figure 2.11 Centrifugal multi-stage horizontal split (Courtesy of Mannesmann Demag)

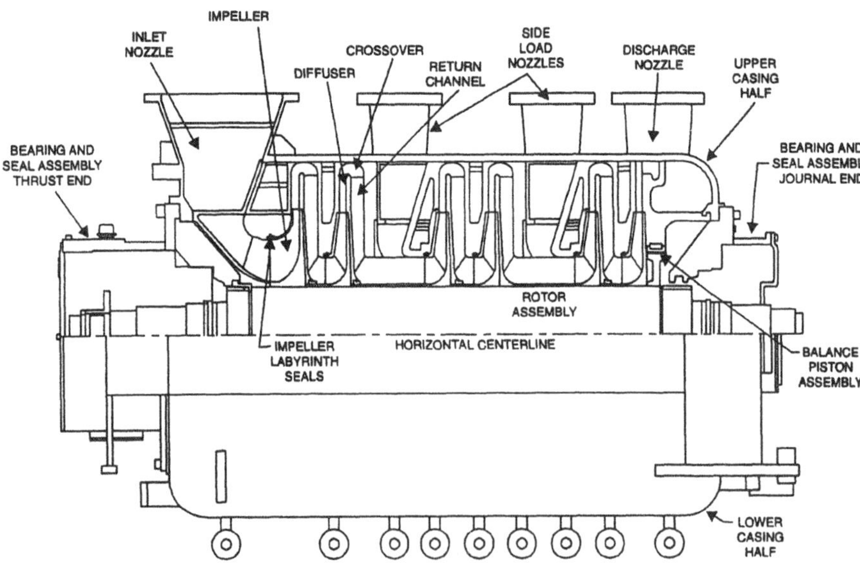

Figure 2.12 Typical multi-stage refrigeration compressor

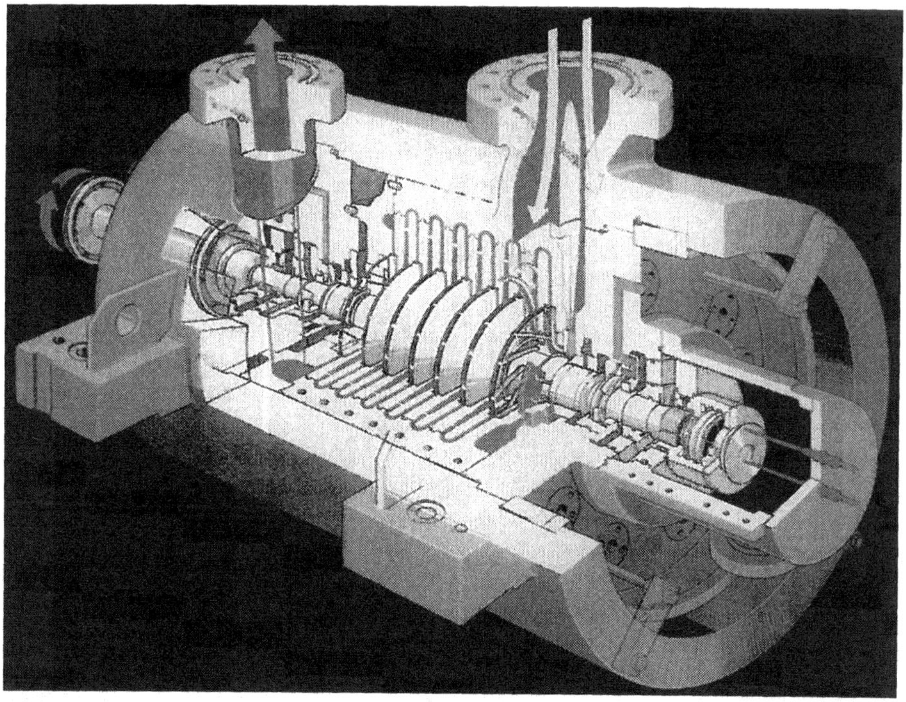

Figure 2.13 Typical multi-stage, radially-split centrifugal compressor (Courtesy of Mannesmann Demag)

Barrel compressors are used for the same types of applications as the multi-stage, horizontally-split centrifugal compressors. Because of the barrel design, however, barrel compressors are normally selected for higher pressure applications or certain low mol gas compositions (hydrogen gas mixtures).

Centrifugal multi-stage integral gear

A typical four-stage, integrally-geared, centrifugal compressor is shown in Figure 2.14. Integrally-geared, centrifugal compressors have a low-speed (bull) gear that drives two or more high-speed gears (pinions). Impellers are mounted at one end or both ends of each pinion. Each impeller has its own casing that is bolted to the gear casing. The gear casing may be horizontally split to allow access to the gears.

The gas enters the compressor through the first stage inlet nozzle to the impeller. The impeller receives the gas from the inlet nozzle, dynamically compresses it, and discharges to the diffuser. The diffuser surrounds the impeller, and it serves to gradually reduce the velocity of the gas as the gas leaves the impeller. The gas exits the diffuser through a volute casing. The volute casing collects the gas, further reduces the velocity of the gas, and recovers additional velocity energy. The gas exits the first stage through the first stage discharge nozzle, enters an

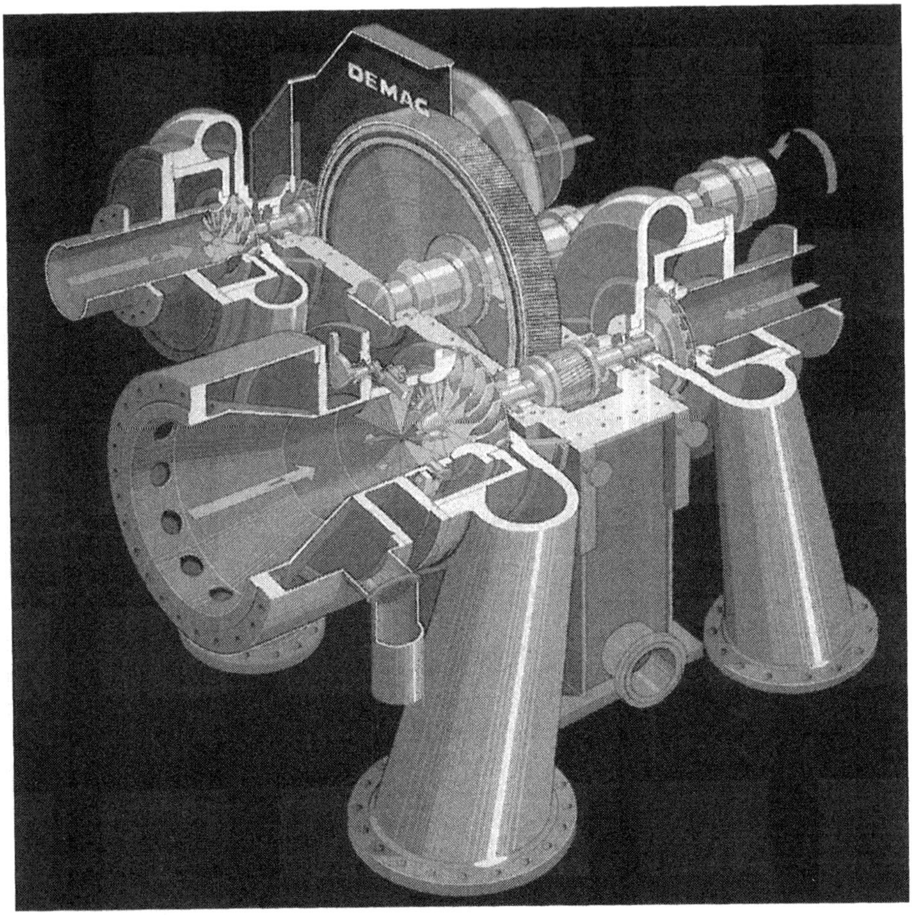

Figure 2.14 Typical integrally-geared, centrifugal compressor (Courtesy of Mannesmann Demag)

intercooler, and then it is piped to the second stage. The discharge from the second stage enters an intercooler, and then it is piped to the third stage.

Axial horizontal split

A typical axial compressor is shown in Figure 2.15. An axial compressor consists of a rotor shaft with a series of rotating blades and a tapered cylindrical casing with fixed stator vanes. Each set of blades is followed by a set of stator vanes. The gas enters the inlet nozzle, which guides the gas to the inlet volute. The inlet volute guides and accelerates the gas stream into the stator vanes. The stator vanes turn the gas stream to properly align the gas with the blades. The blades increase the energy of the gas by increasing the velocity of the gas. The stator vanes act as diffusers to provide resistance to the gas flow, and they cause the gas stream to decrease in velocity and to increase in pressure. The stator vanes also properly orient the gas stream for the next row of blades. Since blades and stator vanes alternate down the length of the casing, the gas is both accelerated and decelerated several times before it leaves the compressor. Pressure is increased each time the gas flow meets a set of stator vanes. The gas exits the compressor through the discharge volute and discharge nozzle.

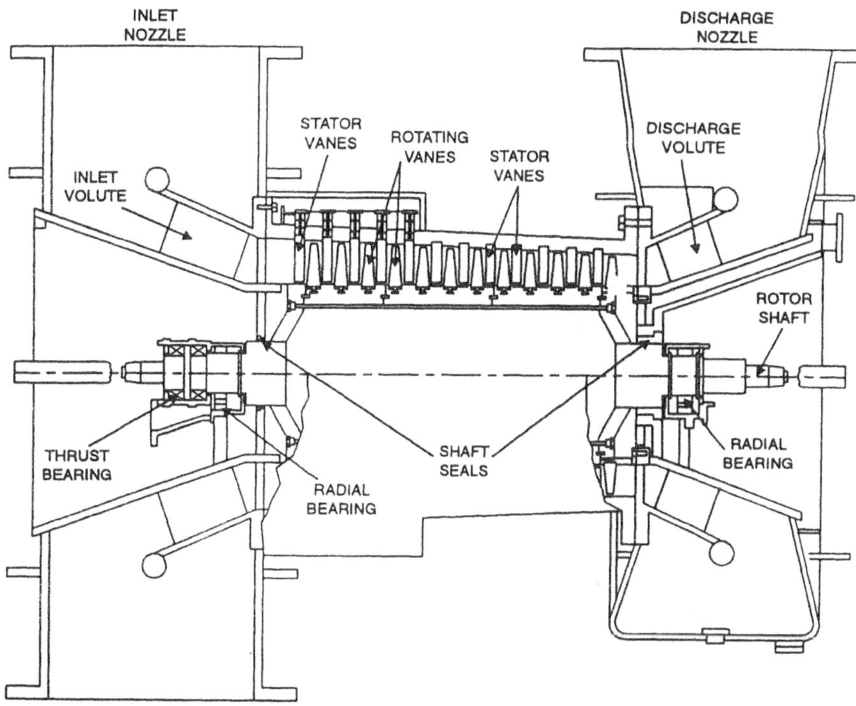

Figure 2.15 Typical axial compressor

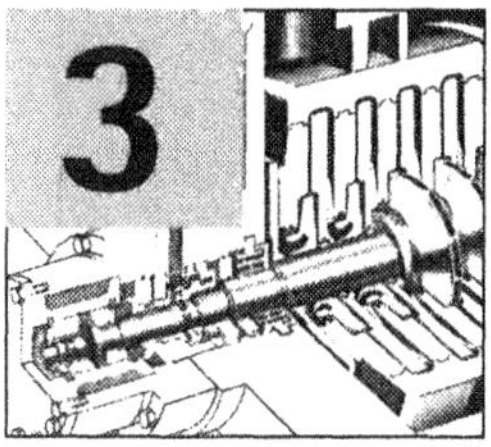

3 Compressor characteristics

(Positive displacement vs dynamic)

- Introduction
- Positive displacement compressors
- Actual volume, standard volume and mass flow
- Dynamic compressors

Introduction

In this section we will discuss the two principle compressor characteristics, positive displacement and dynamic compression. In addition, the concepts of volume flow, mass flow and standard flow will be covered. Although this chapter covers compressors, the characteristics of positive displacement and dynamic are equally applicable to pumps.

Positive displacement compression is defined as the increase of the pressure of a gas caused by operating on a fixed volume in a confined space. Types include; reciprocating, rotary liquid piston, rotary lobe and screw compressors. This concept can best be envisioned by using a simple syringe. As one moves the plunger into the syringe, the volume inside is displaced. It will be displaced regardless of the resistance in which the compressor operates, provided sufficient power is available and the design of the compressor can meet the pressure requirements. A schematic of a positive displacement reciprocating compressor is viewed and one can see that gas will not enter the cylinder until the pressure inside the cylinder is lower than the suction pressure. Conversely, gas will not exit the cylinder until the pressure inside the cylinder is greater than the discharge pressure. The valves shown in this figure are merely check valves. The suction valves act as check valves preventing the compressed gas from escaping back into the suction line. The characteristics then, of a positive displacement compressor are fixed volume, variable pressure capability (energy or head) and not self limiting. By this we mean the compressor will stall or damage itself unless a pressure or power limiting device is included in the compressor scheme. This is usually achieved by using a relief valve.

Before proceeding to the concept of dynamic compression, actual flow, mass flow and standard flow will be discussed as follows.

In the design of any compressor actual volume flow must be used. This is necessary since the design is based on an optimal gas velocity. Gas velocity is the result of a given volume flow acting in a specific area. Think of any compressor, dynamic or positive displacement, compressing a volume of one actual cubic foot per minute and let us assume that the temperature of compression remains constant. If the particular compressor in question has a compression ratio of two, one actual cubic foot per minute entering the compressor will be compressed to a discharge volume of exactly one half cubic foot per minute assuming that the gas is dry.

Standard volume is defined as one volume always referenced to the same pressure and temperature conditions. In customary units, standard pressure is defined as 14.7 pounds per square inch and standard temperature is defined as 60°F. Standard cubic feet then is the ratio of the actual pressure to the referenced standard pressure and the referenced standard temperature to the actual temperature multiplied by the actual volume. Referring back to the previous example of a compressor with a compression ratio of two and no compression temperature increase, one can see that the standard cubic feet per minute in this compressor would remain the same assuming a dry gas. This is because even though the actual volume of the gas does decrease by one half, the discharge standard volume is the ratio of the discharge pressure to the standard atmospheric pressure multiplied by the discharge volume. This will result in the same exit standard volume as the inlet.

Mass flow is the product of the actual volume flow and the density of the specific gas. The concept of mass flow and standard volume flow are the same. That is, the mass flow into the compressor example cited above will be exactly equal to the exit mass flow provided the gas is dry.

Both standard volume flow and mass flow are used to describe process capacities and are used in power calculations. After all, it would be very difficult to charge customers for produced products (gas) unless either a standard volume measure or weight measure were used.

The next concept to be covered is that of dynamic compressors. Dynamic compressors increase the pressure of a gas using rotating blades to increase the velocity of the gas. Types include axial and centrifugal compressors. As we will see later, there are basically two velocities which determine the performance of a dynamic compressor.

Both of these velocities occur at the exit of the moving blade. They are tip speed (the product of blade tip diameter and RPM) and relative velocity (velocity of the gas between two blades). The curve of a centrifugal compressor is significantly different from that of a positive displacement compressor. The centrifugal compressor is a machine of variable capacity, fixed energy or head for a given flow and is self limiting. That is, there is a maximum pressure which the compressor can produce and a maximum horsepower which it requires. Observing the curve of a typical dynamic compressor, one can see that the only place that a dynamic compressor can develop higher energy or head is at a lower flow for a given speed. We will see in the next chapter how the system combined with the compressor operating curve will determine the operating point. It should be noted that the operating curve of an axial compressor at times can approach that of a positive displacement compressor in that its volume range is small and almost approximates the positive displacement curve.

The concept of an equivalent orifice is introduced in this section. Any typical axial compressor blade row or radial compressor impeller can be reduced to a series of equivalent orifices. The inlet of the blade row or impeller being one orifice, the exit being another and the eye and hub seals being additional equivalent orifices.

It should also be noted that both the suction and discharge process system for a given point in time and flow rate can be thought of as equivalent orifices placed at the inlet and discharge of the compressor flanges respectively.

In summary, the characteristics of positive displacement and dynamic compressors are as follows: Flow fixed for positive displacement, variable for dynamic. Head, variable for positive displacement, fixed for dynamic. Limiting characteristics for dynamic compressors, not self limiting for positive displacement (requires relieving device).

Positive displacement compressors

The definition and types of positive displacement compressors is given in Figure 3.1. Figure 3.2 shows a typical process reciprocating compressor. Reciprocating compressors are generally used for the following applications:

- Low flow
- High pressure
- Low molecular weight
- Varying molecular weight

Positive displacement compressors
Increase the pressure of a gas by operating on a fixed volume in a confined space

Types include:

- Reciprocating
- Rotary-liquid piston
- Screw
- Sliding-vane
- Rotary lobe

Figure 3.1 Positive displacement compressors

Figure 3.2 Penn process compressors (Courtesy of Cooper Industries)

The internal construction of a typical reciprocating compressor cylinder is shown in Figure 3.3. As the piston moves from left to right, a constant volume of gas will be displaced. The discharge pressure developed will be the pressure at the discharge flange caused by the process pressure and system resistance. That is, the gas in the cylinder will not exit until the pressure developed in the cylinder is greater than the pressure at the compressor discharge flange.

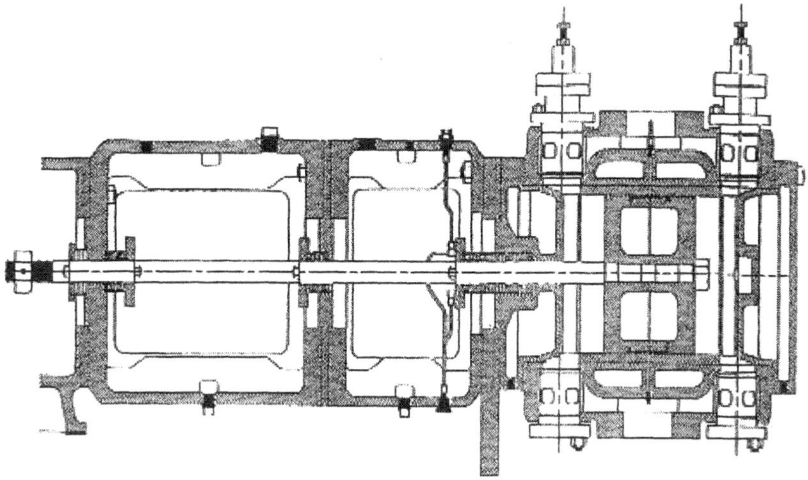

Figure 3.3 Penn process compressors (Courtesy of Cooper Industries)

Actual volume, standard volume and mass flow

Every compressor is designed on the principle of volume flow. Since compressor performance must consider gas velocity, actual flow is always used for design. Figure 3.4 shows the concept for actual volume flow. The volume flow changes directly with temperature and inversely with pressure.

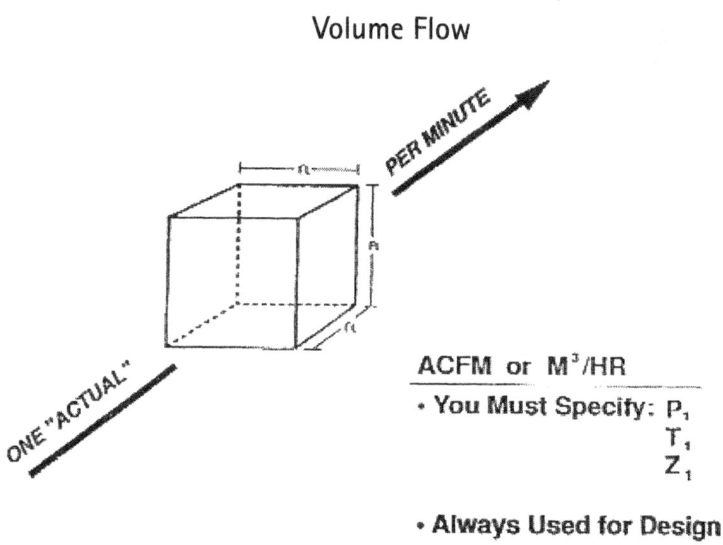

Figure 3.4 Volume Flow

Standard volume and mass flow are used to describe process conditions. If there is not any condensate formed within a compressor section, the standard and mass flow through a compressor remains unchanged. The relationships between standard volume and mass flow are shown in Figure 3.5.

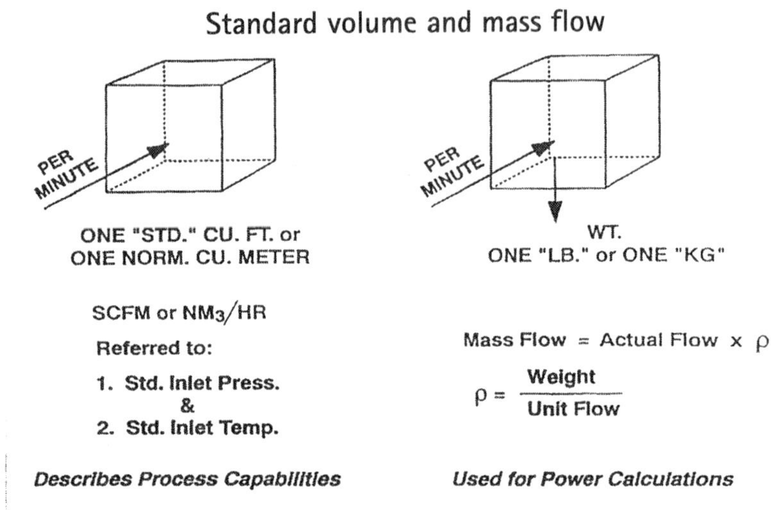

Figure 3.5 Standard volume and mass flow

Figure 3.6 depicts a simple process system.

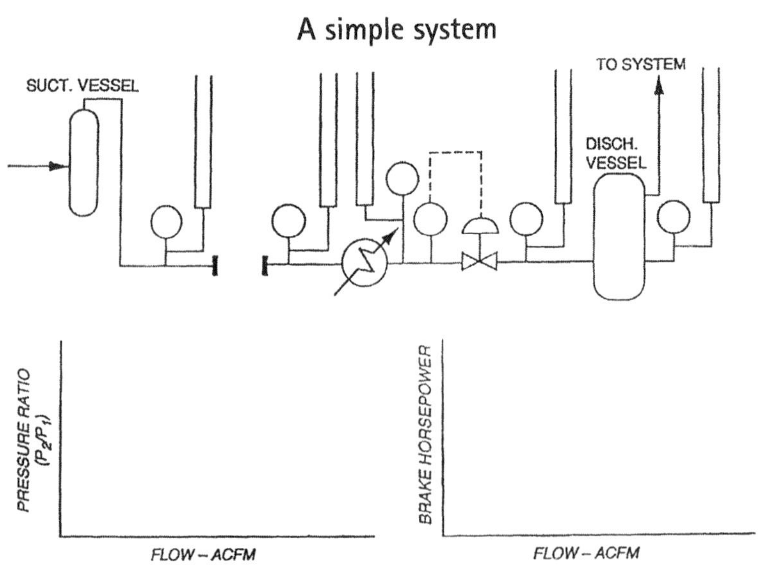

Figure 3.6 A simple system

To demonstrate the characteristics of a positive displacement compressor, insert the symbol for a positive displacement compressor into the space provided in Figure 3.6. The circles shown represent pressure gauges and the vertical tubes represent manometers. For this example pencil in the following set of initial pressures.

- P suction = 20 psig
- P downstream valve = 110 psig
- P disch. vessel = 100 psig
- P upstream valve = 135 psig
- P discharge = 140 psig

Pressures can also be indicated by coloring in the manometers to indicate the relative pressure according to height of the column in the manometer. Therefore the manometer heights will steadily decrease from the discharge to the discharge vessel.

At the conditions noted above, assume that the compressor flow is 500 ACFM and the brake horsepower is 250 BHP. The compression ratio based on the suction and discharge pressure is 4.43. Plot ACFM vs pressure ratio and ACFM vs brake horsepower on the appropriate graphs in Figure 3.6.

Now assume the control valve in the discharge process system is slightly closed. With this action assume the pressure upstream of the control valve increases to 175 psig and the discharge pressure increases to 180 psig. The horsepower also increases 300 BHP. What has happened to the ACFM in this case? Plot the new pressure ratio and brake horsepower in Figure 3.6. Notice that both pressure ratio and brake horsepower curves are essentially a vertical line.

Observing these plots, the characteristics of any positive compressor can be defined as shown in Figure 3.7.

Compressor characteristics

Positive displacement

- constant volume delivery
- variable head capacity
- not self limiting

Figure 3.7 Compressor characteristics

Note that head (energy) is proportional to pressure ratio. 'Not Self Limiting' means that the pressure ratio (head) and brake horsepower will continue to rise until the case pressure is exceeded and/or the driver maximum horsepower is exceeded. To 'limit' these parameters, a pressure relief valve is always required in a system incorporating any type of positive displacement compressor.

Dynamic compressors

The definition and types of a dynamic compressor is presented in Figure 3.8.

Dynamic compressors

Increase the pressure of a gas by using rotating blades to increase the velocity of the gas

Types include:

- Axial
- Centrifugal

Figure 3.8 Dynamic compressors

Figure 3.9 is the assembly drawing of an axial compressor. Axial compressors are used for high flow, low head applications usually above 100,000 ACFM inlet flow. The most common applications for axial compressors are F.C.C. air blowers, MTBE compressors and gas turbine air compressors.

A typical curve for an axial compressor, % volume vs % compression ratio, is shown in Figure 3.10. Note that the pressure rise, from 100% flow to surge flow, is always greater than the corresponding pressure rise for a centrifugal compressor.

An assembly drawing of a multistage horizontal split centrifugal compressor is shown in Figure 3.11.

A centrifugal compressor can also be called a radial compressor because the exit flow from the impellers is radial as opposed to axial. A typical percentage flow vs head centrifugal compressor performance curve is shown in Figure 3.12.

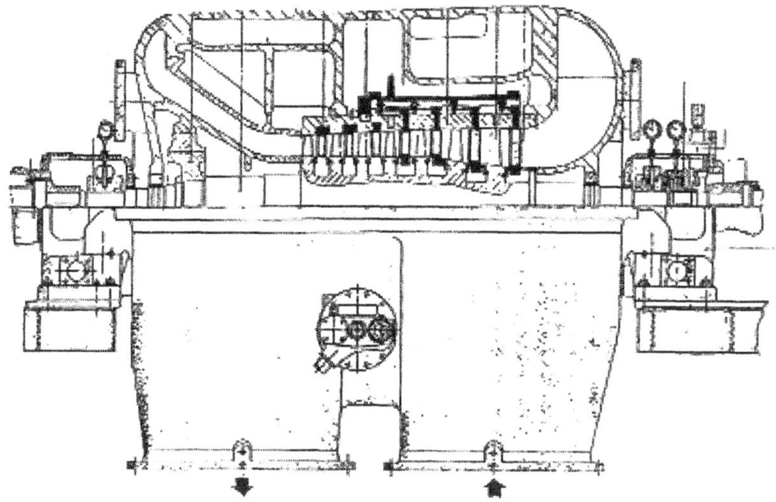

Figure 3.9 Axial compressor (Courtesy of Man Turbo)

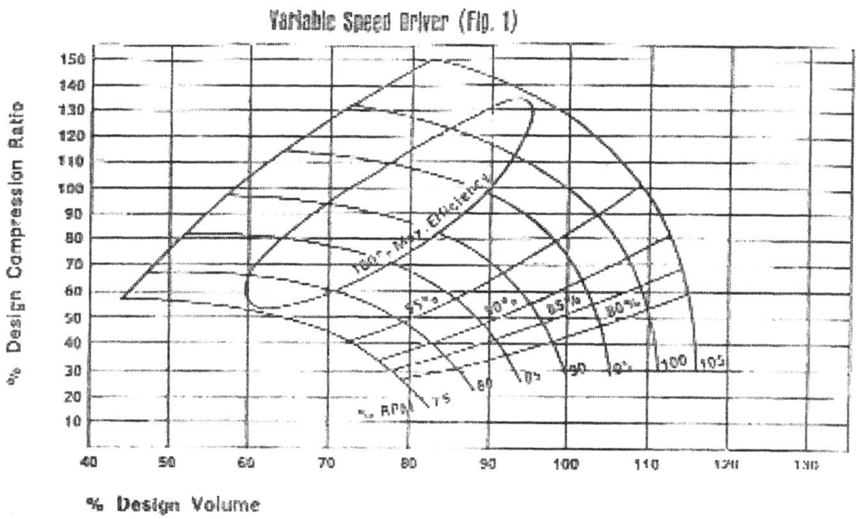

Figure 3.10 Typical Axial curve (Courtesy of Man Turbo)

As a comparison, the head (or pressure) rise shown in Figure 3.12 for the centrifugal compressor is approximately 10% as compared to 30% for the axial compressor shown in Figure 3.10.

The principle of operation of a dynamic compressor always seems more difficult to understand than that of a positive displacement compressor. As stated in the dynamic compressor definition in Figure 3.8, dynamic

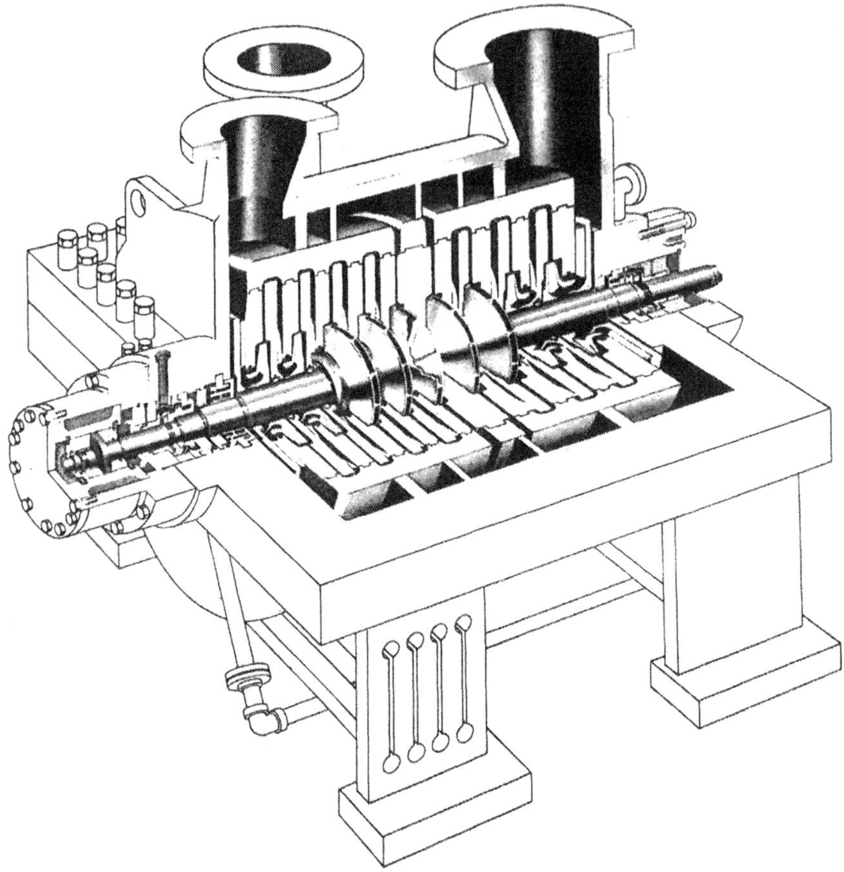

Figure 3.11 Cutaway of horizontally split fabricated compressor (Courtesy of IMO Industries Inc)

compressors increase pressure by using rotating blades to increase the velocity of the gas. Figure 3.13 shows the upper half of a centrifugal compressor impeller with the side plate removed. The velocity vectors 'U' (blade discharge tip speed) and 'V' (discharge gas – 'relative' velocity) are shown. Simply stated, the characteristic shape of any dynamic performance curve (rising head vs decreasing flow rate) is a direct result of the relative velocity through the impeller or blades. The higher the relative velocity, the less energy input from the impeller blades. The lower the relative velocity, the greater the energy input from the impeller blades. The relative gas velocity in any impeller or blade row is determined by the gas passage areas.

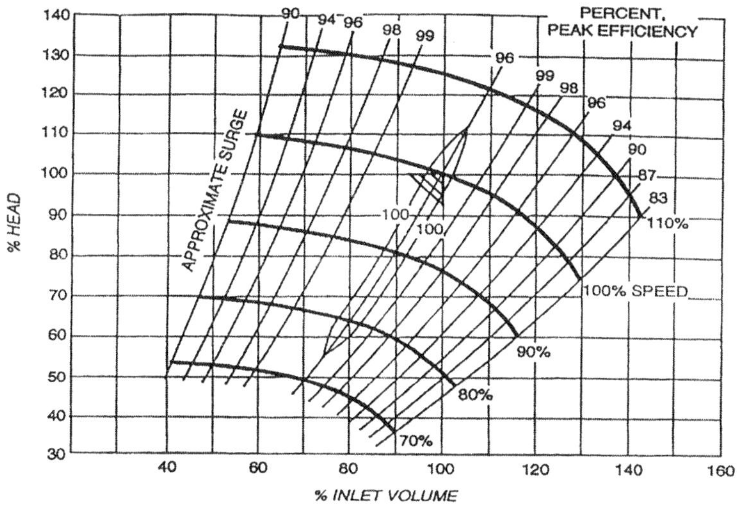

Figure 3.12 Typical multistage compressor curve

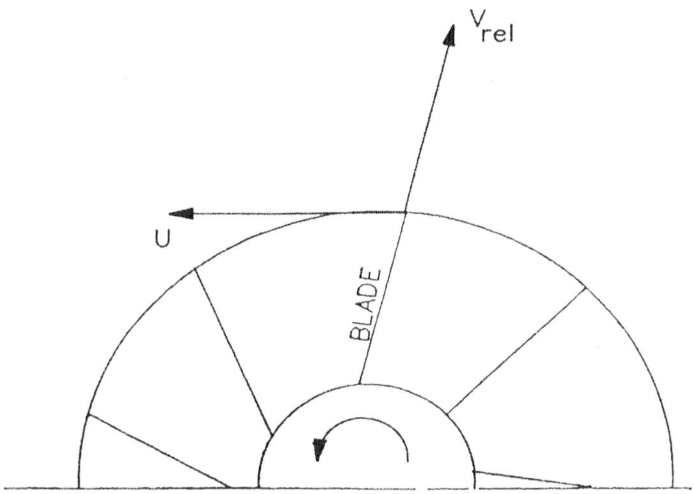

Figure 3.13 Impeller with side plate removed

Figure 3.14 shows how the various sections of any impeller (inlet and discharge) can be interpreted as 'equivalent orifices'. The designer selects the flow areas to reach the optimum gas velocities that will result in the highest impeller efficiencies based on the design flow rate.

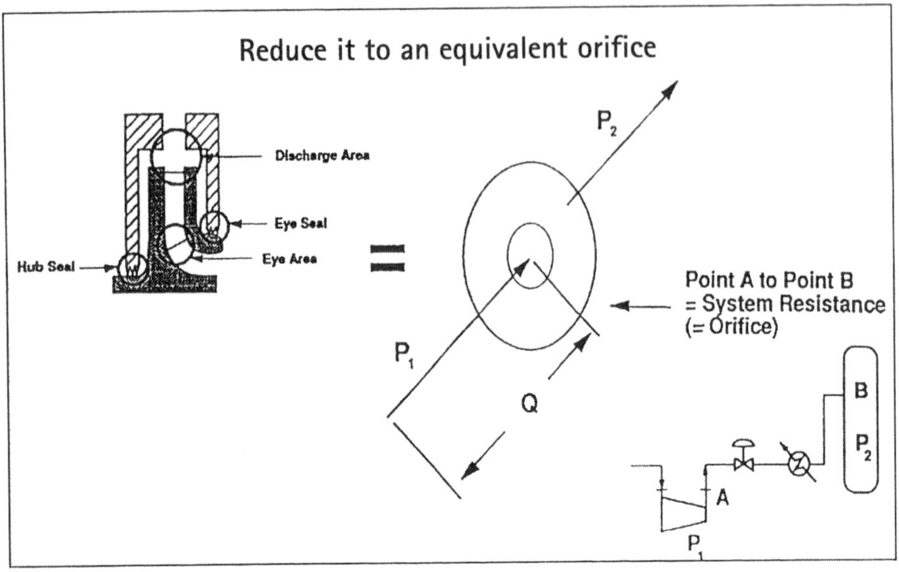

Figure 3.14 Reduce it to an equivalent orifice

Figure 3.15 presents the same simple process system used in Figure 3.6. However, for this exercise insert a dynamic compressor in the space provided. Use the same pressures as in Figure 3.6. The flow rate and horsepower will be larger than that for the positive displacement

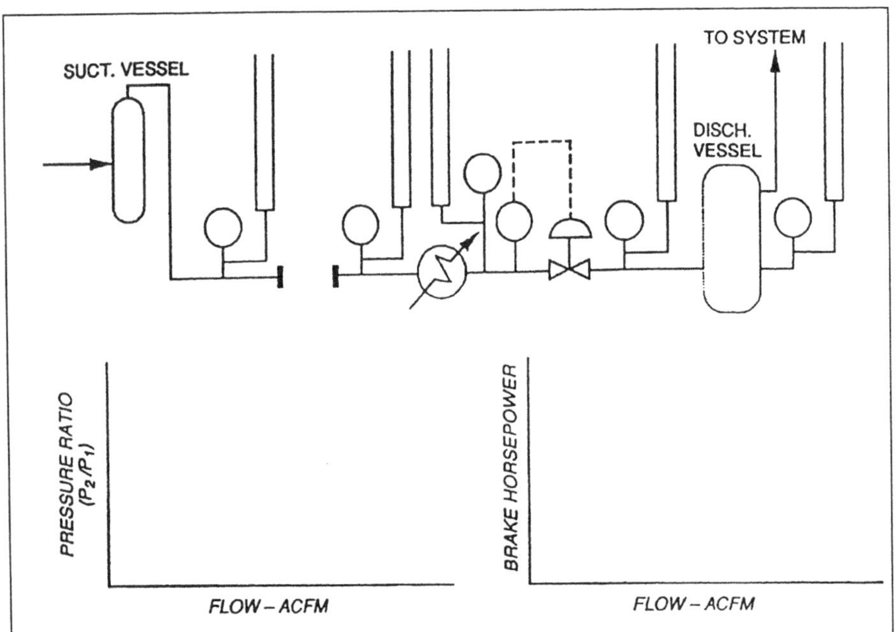

Figure 3.15 A simple system

compressor of Figure 3.6 since dynamic compressors usually handle inlet flow rates above 5000 ACFM. Assume the inlet flow is 20,000 ACFM and the horsepower is 10,000 BHP. Plot ACFM vs pressure ratio and BHP on the appropriate graphs of Figure 3.15.

Again, assume the control valve in the discharge process system is closed slightly. This action causes the compressor discharge pressure to rise to 180 psig. The horsepower however decreases to 6000 BHP. Why? The answer of course, is that the characteristic performance curve of any dynamic compressor is different from a positive displacement compressor. A positive displacement curve is essentially vertical since the actual volume flow is constant. Therefore, the brake horsepower will increase as the pressure ratio does. In the case of a dynamic compressor, increasing the pressure ratio will result in a lower compressor flow rate since the only way a dynamic compressor can develop a higher pressure ratio is at a lower gas velocity (lower flow rate). Simply stated, the longer the gas is in contact with the energy producer (blades), the higher the pressure ratio. For a dynamic compressor, this means the flow rate must decrease. Therefore, plotting the higher-pressure ratio and BHP point on Figure 3.15 will yield a higher pressure ratio, lower flow point.

The characteristics of any dynamic compressor are presented in Figure 3.16.

Compressor characteristics

Dynamic

- Variable volume delivery
- Fixed head capacity (for a certain flow)
- Self limiting

Figure 3.16 Compressor characteristics

As opposed to a positive displacement compressor, the dynamic compressor is self-limiting and does not require a relief valve to protect the compressor and driver.

Figure 3.17 summarizes the characteristics of positive displacement and dynamic compressors by presenting their typical performance curves. Note that the vendors usually do not issue positive displacement curves since volume flow is constant.

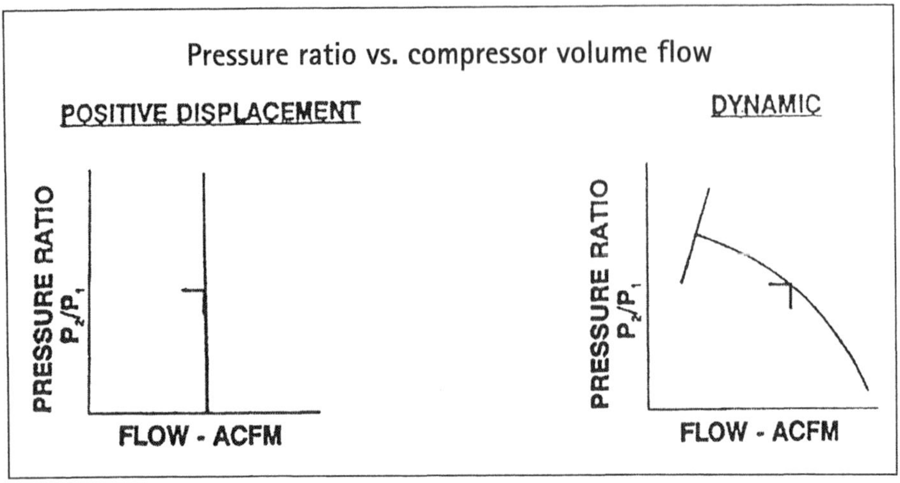

Figure 3.17 Pressure ratio vs. compressor volume flow

4 Operation of a compressor in a system

- Introduction
- The definition of a process system
- System resistance curves
- The operating point
- A positive displacement compressor in the process system
- A dynamic compressor in the process system

Introduction

In this chapter we will define a system, examine system resistance curves (head required by the process) and define the operating point of a compressor. Again, the principles discussed can be applied to a pump operating in a system.

A system is a set of connected things or parts that work together. In a typical process a set of components (vessels, exchangers, furnaces, control valves, etc. and piping) work together to produce a resistance to flow at the turbo-compressor flanges. Examining a simple system one can see that any process system relative to the compressor is comprised of two parts, the suction system and the discharge system. The objective of any compressor in a system is to remove the flow from the suction vessel at the same rate that the flow enters that vessel. In doing so, the pressure is reduced from the suction vessel to the compressor flanges.

In the discharge system the objective is to push the required amount of flow through the system resistance to achieve the final discharge system terminal pressure. This pressure may be in a vessel, may be in a ship or could be in a pipe line. The discharge pressure at the compressor flange then is the accumulation of the system resistance (pipe, valve, exchanger, etc.) from the terminal point (delivery point) back to the compressor discharge flange. An important point to remember is that the compressor differential head or energy required is the net effect of the discharge system resistance and suction system resistance. Increasing discharge system resistance with a constant suction resistance will result in higher compressor energy required. Also increasing suction system resistance with a constant discharge system resistance will result in increasing compressor energy required. Many times the concept of increasing suction system resistance is misunderstood (suction throttling).

If we review the previous chapter and examine the operation of both positive displacement and dynamic compressors in a simple system, two important facts concerning the operation of these machines in a system become evident.

A positive displacement compressor, since it increases the energy of the gas by operating on the gas in a confined space, will always increase the energy, provided sufficient power is available and the machine is designed to meet this objective. Therefore a positive displacement compressor will be relatively insensitive to gas composition changes. In addition, since the positive displacement compressor as previously defined has variable head capability, it will be relatively insensitive to system changes. As a result, a positive displacement compressor will operate at approximately the same delivered volume flow regardless of system resistance or gas composition.

On the other hand, since the characteristics of a dynamic compressor are to increase gas energy (a function of mass and velocity) by working on the gas with blades. Velocities and gas mass (density) play an important part in the make up of this compressor's characteristics. Anything that will result in a velocity and/or density change at the tip of the dynamic compressor blade will result in a different produced differential pressure and a corresponding flow change. Therefore, the dynamic compressor will be extremely sensitive to gas composition changes since these changes will produce mass and velocity changes within the compressor blades or impellers. In addition, this compressor type will also be sensitive to system resistance changes since an increased system resistance requirement will force the compressor to operate at a lower volume throughput. This is because the only way this compressor can produce higher delivered energy is at a lower velocity throughput.

Therefore one can readily see that the operating point of a dynamic compressor will be much more sensitive to both changes in system resistance and the velocities inside the impeller. Both system resistance and the dynamic compressor curve can change when plotted on flow vs pressure coordinates. As will be explained later flow vs head coordinates reflect a dynamic compressor which changes very little provided that the gas composition changes within the limits of approximately 20%. The compressor operating point can be defined as the equilibrium between the required net process system energy and the compressor produced energy. Remember, that the positive displacement compressor characteristic results in relatively constant flow regardless of system required energy, whereas the dynamic compressor characteristics results in significantly large flow changes for changes in system resistance. One additional point to understand is that the process system characteristic curve can also change.

The system resistance curve (head required by the process) is a result of the terminal pressure, that is the discharge pressure at zero flow and the system resistance (square function) as the flow changes. A recycle loop for instance is comprised of friction loss only and will have a relatively steep system resistance curve, whereas an instrument air compressor pumping into a receiver vessel with a relatively small system resistance will have a system resistance curve that will approach a horizontal line. One can see that the combination of a flat dynamic compressor curve and a flat system resistance system curve can lead to a very unstable situation. Remember that the operating point is determined by the intersection of a compressor and system resistance curves. For turbo-compressors both of these curves can change and certainly do.

The Definition of a Process System

The definition of a generic and process system are presented in Figure 4.1. The process system characteristics and the gas composition determine the head required by the process system.

System – Definition

'A system is a set of connected things or parts that work together'

In our case, a set of components (vessels), exchangers, furnaces, control valves, etc. and piping) work together to produce a resistance to flow at the turbo-compressor flanges.

Figure 4.1 System – definition

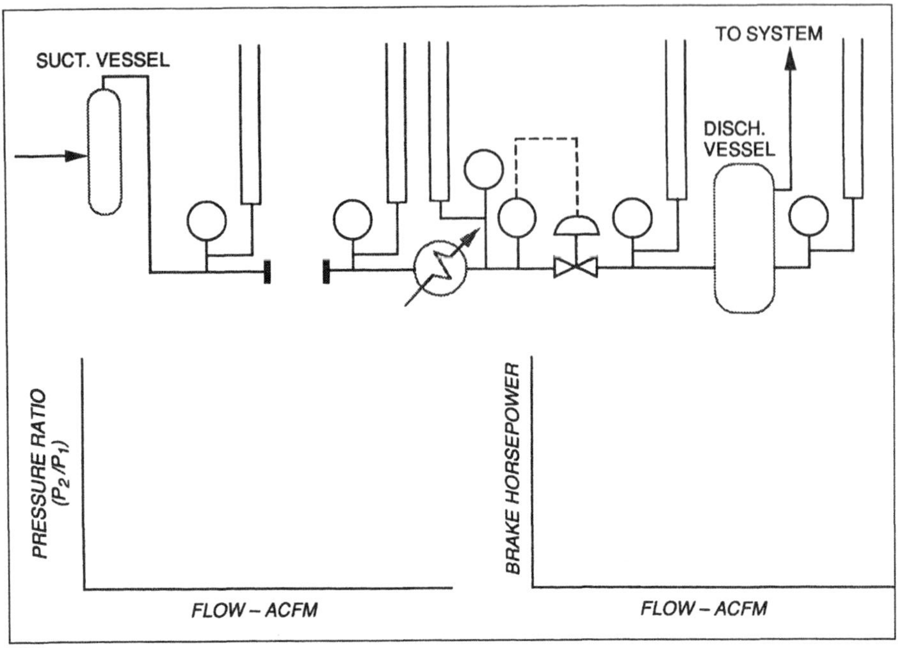

Figure 4.2 A simple system

A simple process system schematic is presented in Figure 4.2. Based on the characteristics of positive displacement and dynamic compressors presented in the previous chapter, draw both the pressure ratio vs flow and brake horsepower vs flow on the appropriate graphs.

System resistance curves

Figure 4.3 shows a system where the system resistance curve (head required by the process) intersects the compressor curve (head produced) at the design point (100% flow).

In Figure 4.4, two different system resistance curves representing two different process systems are shown. The curve starting at zero differential pressure represents a process system where only system resistance is present. When the compressor is shut down, the pressure in the system is zero. This would be the system resistance curve that would describe a recycle process. The curve that starts at approximately 50 PSI differential pressure shows that a pressure differential of approximately 50 PSI exists when the compressor is shutdown in this system. This process system would be typical of a system in which both vessel pressure and system resistance are present. One can see from this figure that the higher the percent of vessel pressure in the process system, the flatter the system resistance curve is.

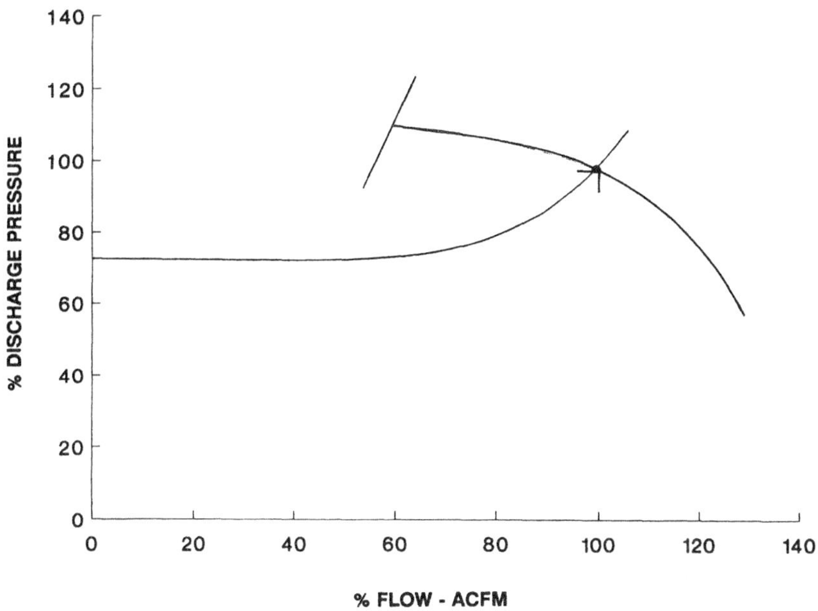

Figure 4.3 % Discharge pressure vs. % flow

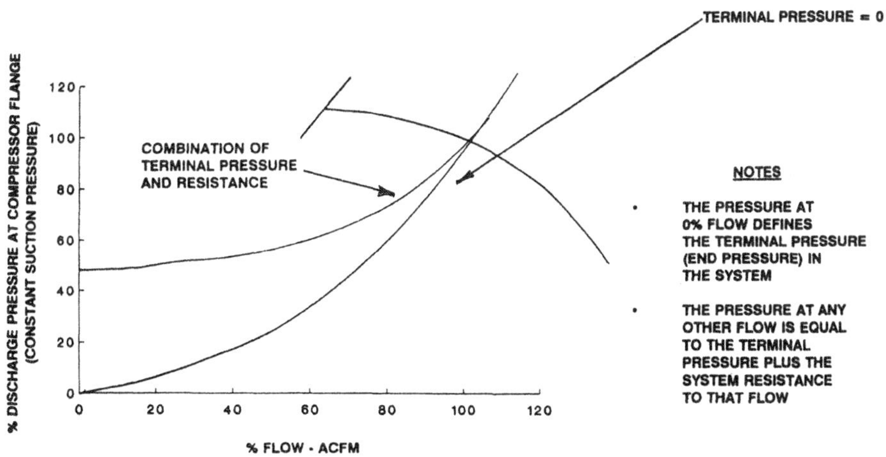

Figure 4.4 The system resistance curve for the discharge system

System resistance characteristics for the three basic types of process systems are shown in Figure 4.5. Note that the steepest curve, the recycle loop, will result in the most stable operation regardless of the compressor curve shape. The instrument air compressor process system curve on the other hand tends to be very flat due to high terminal pressure in the discharge (receiver) vessel. This system will experience instability if the compressor characteristic curve is flat.

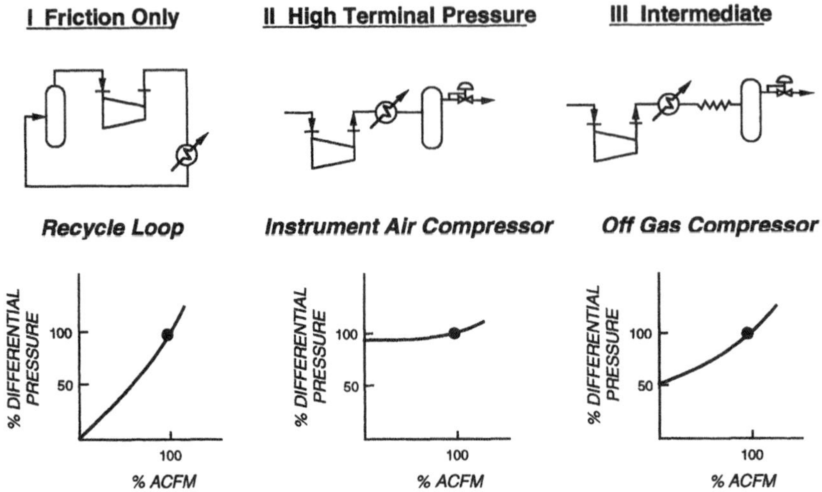

Figure 4.5 System resistance characteristics

Many times, a compressor has not been considered technically acceptable for purchase because of a relatively flat head curve (less than 5% head rise to surge). If the process system that the compressor will operate in is either Type I or Type III in Figure 4.5, this compressor should be considered acceptable based on the fact that the process system steadily rising head characteristic curve will provide a stable

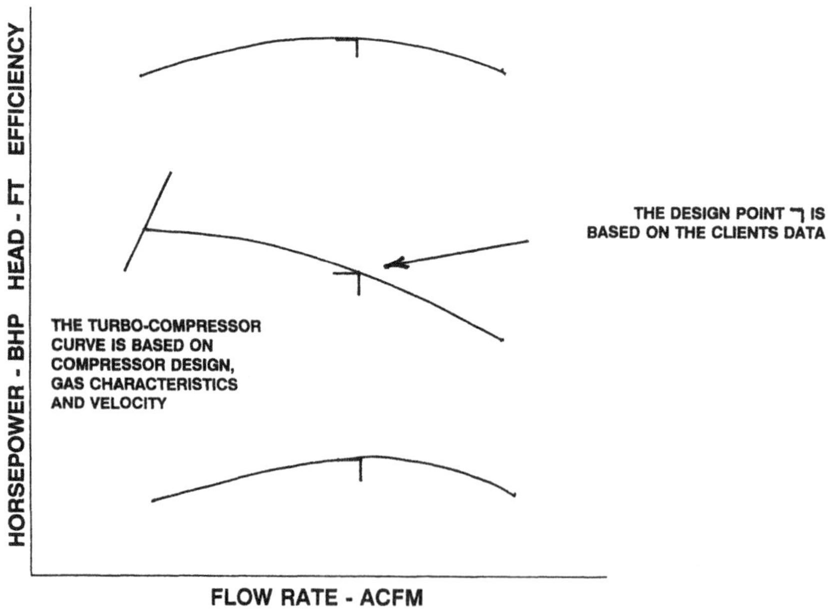

Figure 4.6 Turbo-compressor characteristics for a constant speed – the design point

operating point. Figure 4.6 shows that the compressor curve selected is based only on a single operating point specified by the client. Industry specifications only require that one operating point be guaranteed. Therefore the compressor curve shape is not guaranteed.

The operating point

The operating point is defined as the equilibrium condition that exists between the head produced by the equipment and the head required by the process system. (See Figure 4.7).

The operating point

Is determined by the intersection of the turbo-compressor and system resistance curves – both can change and do!

Figure 4.7 The operating point

If the process conditions that actually exist are properly specified and if the equipment is properly selected the operating point will occur at the best efficiency point. Figure 4.8 shows a case where the head required by the process is greater than the head produced by a centrifugal compressor at the rated point.

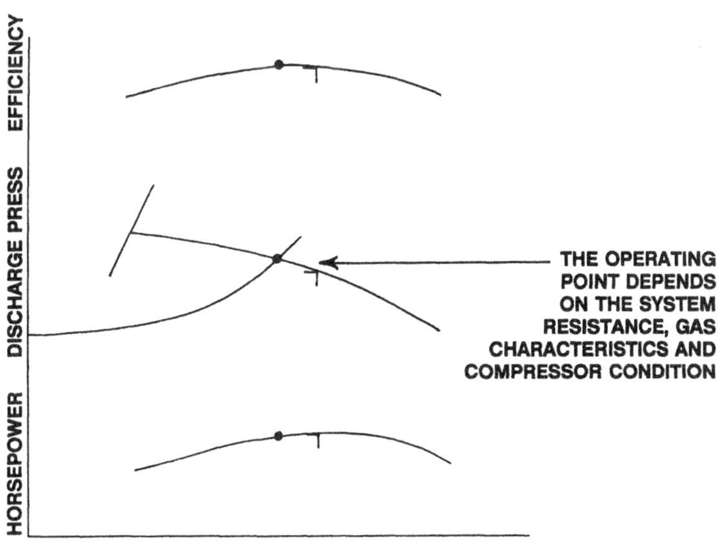

Figure 4.8 The operating point

The result is that the equilibrium point falls to the left of the best efficiency point resulting in a lower efficiency, lower flow rate and lower horsepower.

A positive displacement compressor in the process system

Referring back to Figure 4.2 of this chapter, draw any of the three typical system resistance curves on the pressure ratio vs flow graph. Notice that regardless of the change of resistance, the flow rate of a positive displacement compressor does not change. In addition, the flow rate will remain constant regardless of the density of the gas (molecular weight, inlet temperature or inlet pressure). Refer to Figure 4.9 which presents the characteristics of a positive displacement compressor.

Compressor characteristics

Positive displacement

- Constant volume delivery
- Variable head capacity
- Not self limiting
- Not flow sensitive to system resistance
- Not flow sensitive to gas density

Figure 4.9 Compressor characteristics

Based on the results drawn in Figure 4.2, a positive displacement compressor is not flow sensitive to either system resistance or gas density.

A dynamic compressor in the process system

Again, refer to Figure 4.2 and observe how the system resistance curves previously drawn affect a dynamic compressor's operating point. A dynamic compressor's flow rate is affected by system resistance change. In addition, a change in the gas density will also influence the system resistance (head required curve) and therefore change the flow rate.

Refer to Figure 4.10 which presents the characteristics of a dynamic compressor.

Compressor characteristics

Dynamic

- Variable volume delivery
- Fixed head capacity (for a certain flow)
- Self limiting
- Is flow sensitive to system resistance
- Is flow sensitive to gas density

Figure 4.10 Compressor characteristics

Based on the results of Figure 4.2, a dynamic compressor is flow sensitive to both system resistance and gas density.

5 The concept of fluid head

- Introduction
- Definition
- Paths of compression
- The different types of gas head
- Dynamic compressor curve format

Introduction

Without a doubt one of the most confused principles of turbo-compressor design in my experience has been that of fluid head. I have found that understanding this concept can best be achieved by recognizing fluid head is the energy required to achieve specific process requirements.

Head is the energy in foot-pounds force required to compress and deliver one pound of a given fluid from one energy level to another. One of the confusing things about this concept is that the industry persists in defining head in feet. Head should be expressed in foot-pounds force per pound mass or British Thermal Units per pound. A British Thermal Unit per pound of fluid is equal to exactly 778 foot-pounds force per pound mass of that fluid.

Remember, when we deal with fluid head, a fluid can be either a liquid or a gas depending upon the conditions of the fluid at that time. Ethylene for instance, can be either a liquid or a gas depending on it's pressure and temperature. If it is a liquid, an ethylene pump will be used and the energy required to increase the pressure of the liquid from P_1 to P_2 will be defined as head in foot-pound force per pound mass. Conversely, if the conditions render it a vapor, an ethylene compressor would be used to achieve the same purpose. We will see in this section that the amount of energy required to compress a liquid or a gas the same amount will be significantly higher in favor of the gas because the gas is at a much lower density than that of a liquid. Understanding a Mollier Diagram is an important aid to understanding the concept of fluid head. Every fluid can have a Mollier Diagram drawing which expresses energy on the X axis and pressure on the Y axis for various temperatures. Increasing the pressure of a vapor will result in increased energy required.

Having defined the concept of head as that of energy, one must now investigate the different types of ideal (reversible) gas heads, namely isothermal, isentropic or adiabatic and polytropic. All of these types of fluid head simply describe the path that the gas takes in being compressed. It must be remembered that any type of head can be used to describe a reversible compressor path as long as the Vendor uses the appropriate head and efficiency in his data reduction calculations. In this section we will show the assumptions for various kinds of heads and the relative difference in their values. In addition, the definitions of each type will be stated.

Definition

The definition of head required by any fluid compression process is presented in Figure 5.1.

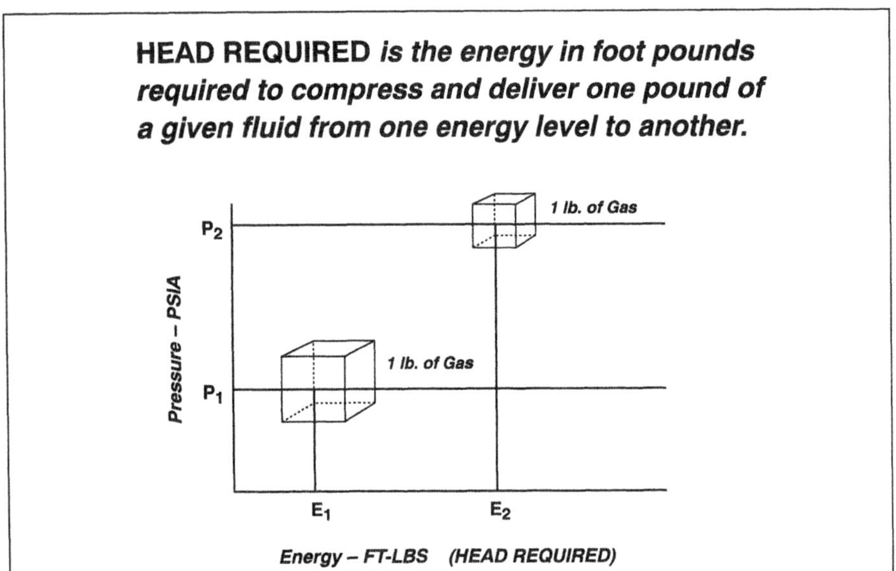

Figure 5.1 Head required definition

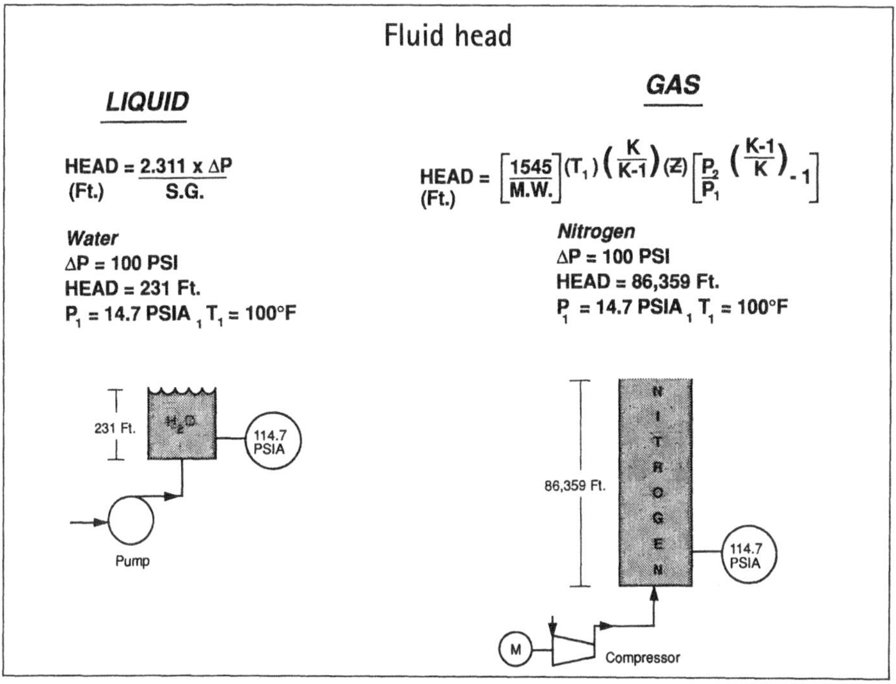

Figure 5.2 Fluid head definition

In any compression process, the amount of energy required to compress one pound of mass of a specific gas, at a given temperature from compressor suction flange (P_1) to the discharge flange (P_2) is defined as head required.

Figure 5.2 demonstrates how the density of a fluid significantly affects the amount of energy (head) required in a compression process.

Water with a density of 62.4 lbs/ft^3 requires only 231 ft-lb force per lb mass to compress the liquid 100 PSI. Note also, the equation for head required by a liquid is independent of temperature. On the other hand, nitrogen with a density of 0.07 lbs/ft^3 requires approximately 350 times the energy! This is because in both the case of water and nitrogen, the process requires that the fluid be compressed 100 PSI. However, since the mass of nitrogen is only 0.1% of the mass of water, a much greater amount of energy is required to compress the gas.

Figure 5.3 shows the head produced characteristics of positive displacement and dynamic compressors.

Note that regardless the amount of head required by the process, the flow rate of a positive displacement compressor is not affected. On the other hand, a dynamic compressor's flow rate is significantly affected by changes in the head required by the process. This is because the

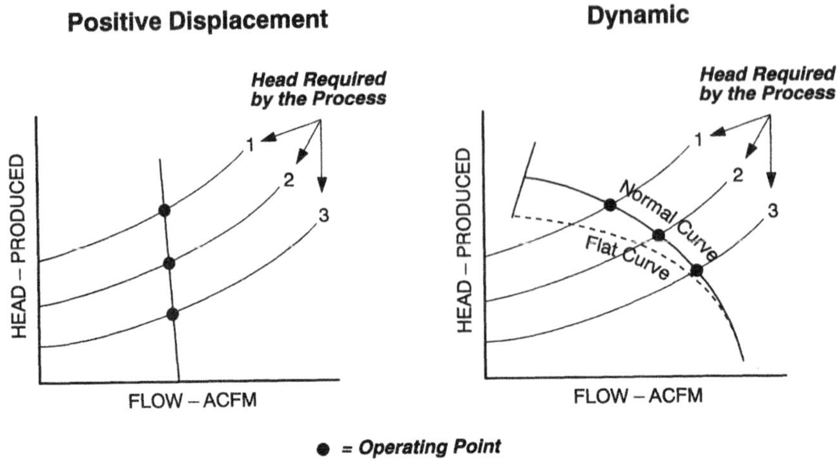

Figure 5.3 Compressor head produced characteristics

characteristic of any dynamic compressor is that it can only produce a greater amount of energy at a lower flow rate. The reason for this characteristic will be explained in a subsequent module. Therefore, any increase in the head required by a process will reduce the flow rate of a dynamic compressor. This is an extremely important fact because reduced flow rate in a dynamic compressor can lead to extreme, long term mechanical damage to the compressor unit. Also note in Figure 5.3 that the flatter the head produced curve a dynamic compressor possesses, the greater the effect of head required upon flow rate.

Head required

Thorough understanding of the concepts of head required by the process and head produced by the compressor is absolutely essential if dynamic compressor operation is to be understood.

It has been my experience that a lack of understanding exists in the area of dynamic compressor performance and often leads to a much greater emphasis upon the compressor's mechanical components (impellers, labyrinths, seals, bearings and shafts). In many cases, the root cause of dynamic compressor mechanical damage is that the head required by the process system exceeded the capability of the dynamic compressor.

Figure 5.4 presents the factors that determine the head (energy) required by any process.

Note that the head required by the process is inversely proportional to the gas density. If the gas density decreases, the head required by the process will increase. Gas density will decrease if gas temperature increases, inlet pressure decreases or molecular weight decreases. If the

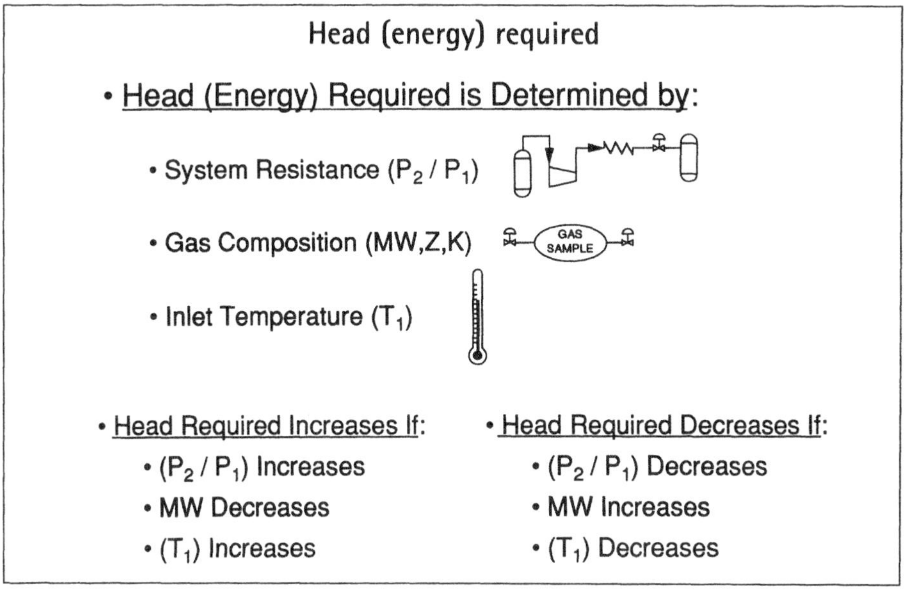

Figure 5.4 Head (energy) required

head required by the process increases, the flow rate of any dynamic compressor will decrease as shown in Figure 5.3. If the gas density increases, the head required by the process will decrease and dynamic compressor flow rate will increase.

Head produced

In Figure 5.5, the factors that determine the head produced by a dynamic compressor are presented.

Simply stated, for a given impeller vane shape, head produced by a dynamic compressor is a function of impeller diameter and impeller speed. Once the impeller is designed, it will produce only one value of head for a given shaft speed and flow rate.

The only factor that will cause a lower value of head to be produced than stated by the compressor performance curve is if the compressor has experienced mechanical damage or if it is fouled.

Figure 5.6 shows how the need for dynamic compressor inspection can be determined.

If for a given flow rate and shaft speed, the head produced falls below the value predicted by greater than 10%, the compressor should be inspected at the first opportunity. Having explained the concept of head required by the process, the method of calculating the head required by a process system needs to be discussed.

Head (energy) produced

- Head (Energy) Produced is Determined by:
    - Compressor Impeller Design

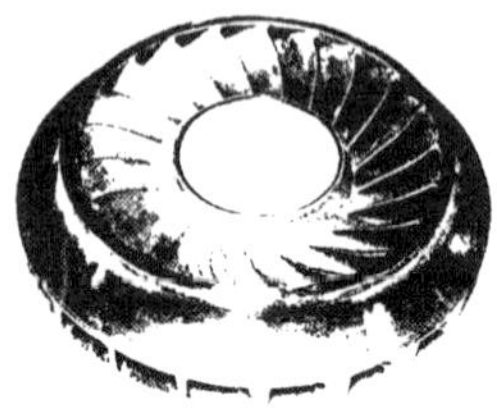

- Head Produced by the Impeller,
    - Increases with Tip Speed
        - Impeller Diameter
        - Compressor Speed (RPM)
    - Increases with Decreasing Flow

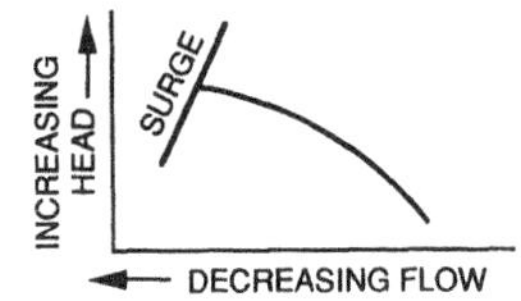

Figure 5.5 Head (energy) produced

The turbo-compressor curve

- The Turbo-Compressor Head (Energy) vs. Flow Curve is *Fixed by Compressor Design*

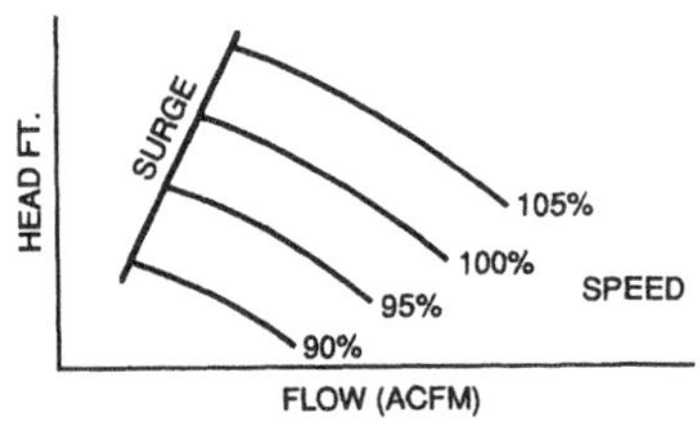

- For a Given Speed and Flow, If an Operating (Test) Point *Does Not Fall on the Curve*,* the Unit Requires Maintenance

* *Within ± 10% (Accounting for Instrument Calibration Error)*

Figure 5.6 The turbo-compressor curve

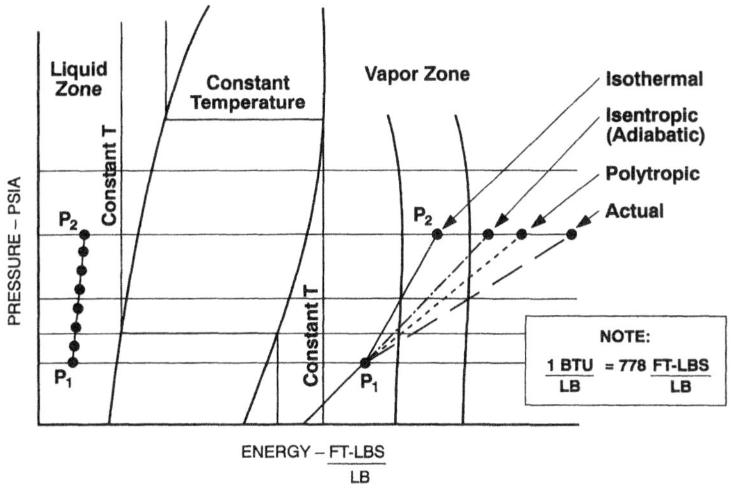

Figure 5.7 Ideal paths of compression

Paths of compression

Figure 5.7 presents a typical Mollier Diagram plotted pressure vs energy.

A Mollier Diagram can be drawn for any pure fluid or fluid mixture. Usually, Mollier Diagrams are prepared only for pure fluids since any change in fluid mixture will require a new Mollier Diagram to be prepared. The Mollier Diagram can be used to determine the head required by the process system for any liquid, saturated vapor or vapor compression process. Observe that for liquid compression, the amount of energy required to increase the pressure from P_1 to P_2 is very small. However, for compression of a vapor, the amount of energy required to compress from P_1 to P_2 is very large as previously explained. Refer back to Figure 5.3 of this module and study the equations that are used to determine the head required for a liquid and a vapor. In addition to many more parameters being required for calculation of head required for a vapor, certain ideal assumptions must be made.

The different types of gas head

A vapor can be ideally compressed by any one of the following reversible thermodynamic paths:

- Isothermal – constant temperature
- Isentropic (adiabatic) – no heat loss

- Polytropic – temperature not constant and heat lost

The actual path that any compressor follows in compressing a vapor from P_1 to P_2 is equal to the reversible path divided by the compressor's corresponding path efficiency. Therefore, we can write the following equation:

$$Actual\ Head = \frac{Head\ Isothermal}{Eff'y.\ Isothermal} = \frac{Head\ Isentroic}{Eff'y.\ Isentropic} = \frac{Head\ Polytropic}{Eff'y.\ Polytropic}$$

When evaluating compressor bids from different vendors quoting different types of reversible heads the above equation proves useful. If the head produced by one vendor divided by the corresponding efficiency is not equal to the corresponding values quoted by the competition . . . Better start asking why!

Figure 5.8 defines the different types of ideal heads and shows the relative difference between their values as compared to polytropic head.

The definition of polytropic head is confusing and difficult to understand if not investigated further.

The ideal gas head equations are described in Figure 5.9.

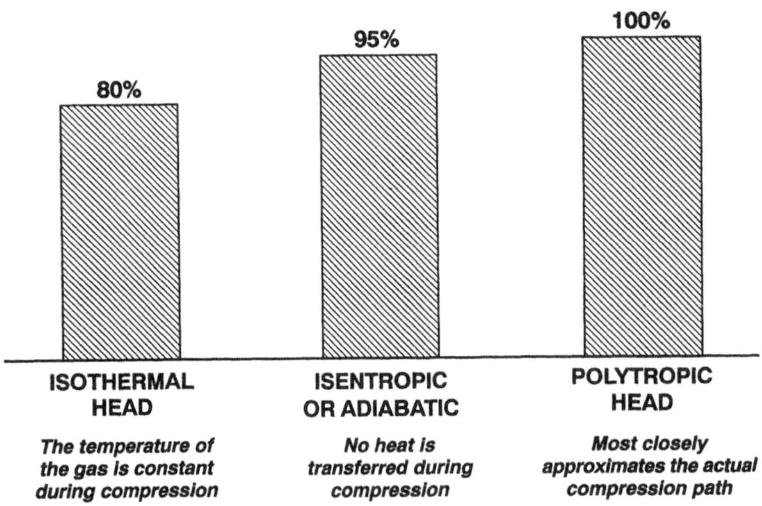

Figure 5.8 Ideal compression heads – relative values

Ideal gas head equations

Isothermal	Isentropic (Adiabatic)	Polytropic
$HD = (\frac{1546}{M.W})(T_1)(Z_{AVG})\left[L_N(\frac{P_2}{P_1})\right]$	$HD = (\frac{1545}{M.W})(T_1)(\frac{K}{K-1})(Z_{AVG})\left[(\frac{P_2}{P_1})^{\frac{(K-1)}{K}} - 1\right]$	$HD = (\frac{1545}{M.W})(T_1)(\frac{n}{n-1})(Z_{AVG})\left[(\frac{P_2}{p1})^{\frac{(n-1)}{n}} - 1\right]$

Where:

$\frac{1545}{M.W}$ =	Gas constant "R"	P_2 =	Discharge Pressure - PSIA
T_1 =	Inlet temp. - °R	P_1 =	Inlet pressure - PSIA
°R =	460 + °F	K =	Ratio of specific heats C_p/C_v
Z_{AVG} =	Average compressibility $\frac{(Z_1 + Z_2)}{2}$	n =	Polytropic exponent
		$\frac{(n-1)}{n}$ =	$\left(\frac{K-1}{K}\right)\left(\frac{1}{\eta_{POLY}}\right)$
Ln =	Log to base e	η_{POLY} =	Polytropic efficiency

Figure 5.9 Ideal gas head equations

Note that the only difference between isentropic and polytropic head is the values;

$$\frac{K-1}{K} \quad \text{and} \quad \frac{n-1}{n}$$

Also note that $\frac{n-1}{n} = \frac{\frac{K-1}{K}}{\eta \text{ poly}}$

Now if η poly = 100%, $\frac{n-1}{n} = \frac{K-1}{K}$ or

Polytropic Head = Isentropic Head

Therefore, I think of $\frac{n-1}{n}$ as a correction

Factor to $\frac{K-1}{K}$ that will most closely approximate the actual

compressor path for a given compressor (refer back to Figure 5.7).

Remember, polytropic head is an ideal reversible compression path. Today, most compressor vendors have adopted polytropic head as their standard for multistage compressors. The main reason is that polytropic head, since it contains an efficiency term, allows individual impeller (stage) heads to be added.

Dynamic compressor curves format

Finally, Figure 5.10 presents the different ways that compressor compression performance can be formatted.

Head vs flow is always preferred because the head produced by a dynamic compressor is not significantly affected by gas density. However, compression ratio or discharge pressure are! That is, compression ratio and discharge pressure curves are invalid if the inlet gas temperature, inlet pressure or molecular weight changes! This may seem confusing at first, but refer back to Figure 5.5 which shows how head is produced by a dynamic compressor. It is a function only of impeller diameter and shaft speed. Gas density influences the head required by the process (refer to Figure 5.4).

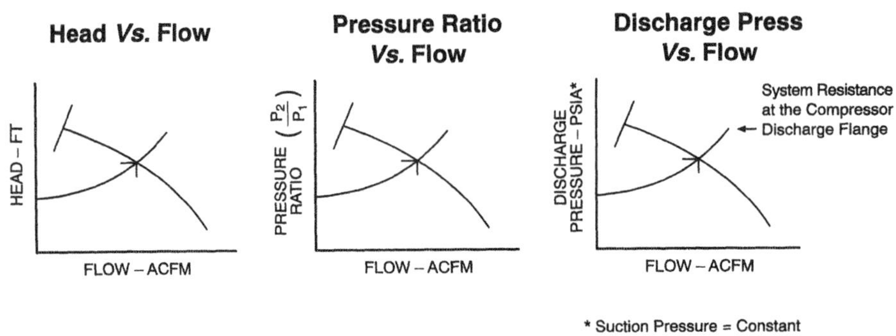

Figure 5.10 Dynamic compressor curves format

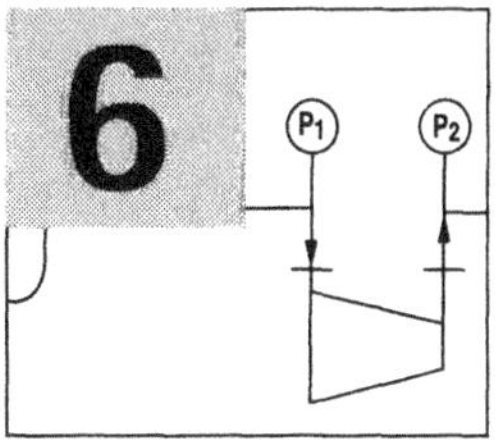

6 Performance relationships

- Introduction
- Satisfying the objective
- Gas characteristics
- Compression head
- Impeller types and specific speed
- Efficiency
- Horsepower
- The Fan Laws

Introduction

In this section we will cover the relationships that the COMPRESSOR VENDOR uses to determine the head produced, efficiency, horsepower required and overall design for a particular compressor application. As previously mentioned, the END USER's or the PURCHASER's objective is to deliver a specified amount of a given gas to the process. Therefore, the data that the compressor vendor obtains is required mass flow, inlet pressure, temperature conditions and gas composition. With this data a compressor manufacturer will calculate actual flow, the ideal energy required and the horsepower required to achieve that objective. The calculation for horsepower will require a specific compressor efficiency as well as compressor mechanical losses, that is bearing friction losses, seal losses and disc friction losses. Gas characteristics are defined in this chapter and useful relationships are presented to enable the reader to calculate various compressor requirements. Once the VENDOR obtains the data, the gas head can be calculated. Once the head and required flow are known, the impeller can be selected.

The principle of impeller design is chiefly based on that of specific speed. Specific speed is defined as the ratio of speed times the square root of the actual flow divided by head raised to the three quarters power. It can be shown that increasing values of specific speed will result in increasing impeller efficiencies. Therefore, having been given the required flow and energy (head) the only source of obtaining higher specific speed for the vendor is to increase the compressor speed. This fact is very significant, because while compressors have increased in efficiency over the years, the mechanical requirements have also increased significantly, I.E., higher bore impeller stresses, etc. resulting in potential reliability problems. Therefore, the design of the impeller is a very fine balance between the performance requirements and the mechanical constraints of the components used in the compressor design.

Efficiency is presented as a ratio of ideal energy to actual energy as depicted on a typical Mollier Diagram. In addition, the fan laws are presented showing how increased impeller energy can be obtained via speed change in a compressor application.

Satisfying the objective

The objective of the end user is to deliver a specified amount of a given gas. Refer to Figure 6.1 and note that his objective can best be stated by the relationship:

Gas Flow Produced = Gas Flow Delivered

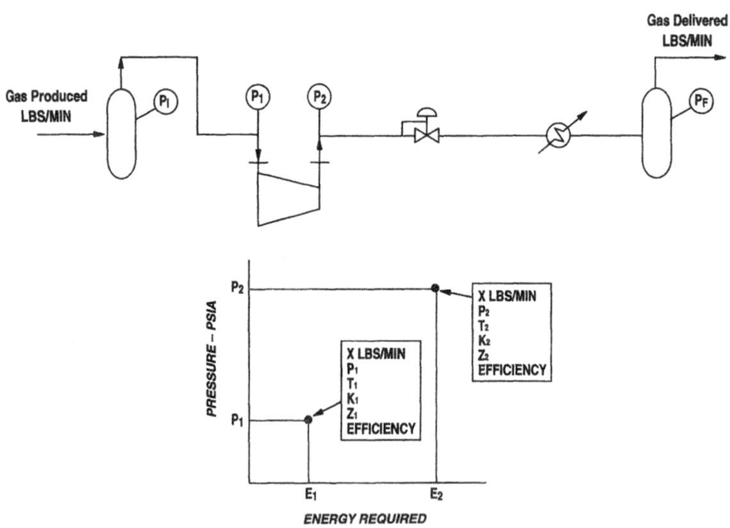

Figure 6.1 The objective: to deliver a specified amount of a given gas

This incidentally is the reason why most process control systems monitor pressure in the process system and install a controller to either modulate flow via a control valve (change the head required by the process) or vary the speed of the compressor (change the head produced by the compressor).

The vendor then, determines the head required by the process based on the parameters given by the contractor and end user on the equipment data sheet. It is very important to note that all possible sources should be used to confirm that the conditions stated on the data sheet are correct and realistic. This fact is especially true for dynamic compressors, since erroneous process conditions will impact the throughput of the compressor.

Gas characteristics

Figure 6.2 presents the relationships used to calculate the design parameters for the compressor. Note that the same relationships are used regardless of the type of compressor (Positive Displacement or Dynamic).

To achieve the client's objective the compressor vendor must calculate the actual flow to the compressor inlet, the actual energy and work required

Actual flow

$$\text{Volume flow rate } (FT^3/MIN) = \text{mass flow rate } \left(\frac{LB's}{Min}\right) + \text{density } \left(\frac{LB}{FT^3}\right)$$

gas density depends on P_1, T_1, Z, MW

Energy (ideal)

$$\text{Compression } HEAD_{POLYTROPIC} = \frac{FOOT\text{-}LB's}{LB\ MASS}$$

Energy (ideal) to compress and deliver one LB of gas from P_1 to P_2

Depends on P_1, T_1, Z_{avg}, K_{avg}, MW P_2, efficiency

Work

$$\text{Power (horsepower)} = \frac{\text{ideal energy } \left(\frac{FT\text{-}LB's}{LB}\right) \text{ mass flow } \left(\frac{LB}{Min}\right)}{33{,}000 \left(\frac{FT\text{-}LB's}{min\text{-}HP}\right) \times \text{efficiency } (\%)}$$

Figure 6.2

The gas characteristics used in the determination of design parameters are defined in Figure 6.3.

Gas characteristics

Compressibility (Z)	–	Accounts for the deviation from an ideal gas
Specific heat (C)	–	The amount of heat to raise one mass of gas one degree
C_P and C_V	–	Specific heat at constant pressure and volume respectively
Specific heat ratio (K)	–	C_P/C_V
MW	–	Molecular weight
Polytropic exponent (n)	–	Used in polytropic head calculations

$$\frac{n-1}{n} = \frac{k-1}{k} \times \frac{1}{\eta \text{ polytropic}}$$

Figure 6.3 Gas characteristics

Figure 6.4 shows useful relationships used in compressor calculations as well as the definitions for constants used.

Useful relationships

Actual flow – FT^3/minute

$$ACFM = \frac{\text{mass flow} \quad (LBS/MIN)}{\text{Density} \quad (LBS/FT_3)}$$

$$\text{Density } (LBS/FT^3) = \frac{(P)(144)}{ZRT}$$

$$ACFM = SCFM \times \frac{(14.7)}{P} \frac{(T)}{520}$$

Energy (Ideal) – FT-LB/LB Mass

Use head equation,
Polytropic is usually used

Efficiency – %

Derived from impeller test results – does not include mechanical losses

Work – horsepower

Brake horsepower = gas horsepower + mech. losses

$$\text{Gas horsepower} = \frac{(HD)(\text{Mass flow})}{(C)\ (\text{eff'y})}$$

where:

$$C = 33{,}000 = \frac{FT\text{-}LBS}{Min\text{-}H.P.}$$

$$HD = \frac{HEAD\ FT\text{-}LBS}{LB}$$

$$\text{Mass flow} = \frac{LBS}{Minute}$$

Eff'y = corresponding efficiency (polytropic, isentropic, etc)

P = pressure – psia

T = temperature – R*

*R = °F + 460

Z = compressibility

R = 1545/mol. wgt

SCFM = standard FT^3/min referenced to 60°F and 14.7 psia

Figure 6.4 Useful relationships

Compression head

The ideal gas head equations are again defined in Figure 6.5. As previously stated, polytropic head is the usual choice among compressor vendors.

Ideal gas head equations

Isothermal

$$HD = \left(\frac{1546}{M.W}\right)(T_1)(Z_{AVG})\left[L_N\left(\frac{P_2}{P_1}\right)\right]$$

Isentropic (Adiabatic)

$$HD = \left(\frac{1545}{M.W}\right)(T_1)\left(\frac{K}{K-1}\right)(Z_{AVG})\left[\left(\frac{P_2}{P_1}\right)^{\frac{(K-1)}{K}}-1\right]$$

Polytropic

$$HD = \left(\frac{1545}{M.W}\right)(T_1)\left(\frac{n}{n-1}\right)(Z_{AVG})\left[\left(\frac{P_2}{p1}\right)^{\frac{(n-1)}{n}}-1\right]$$

Where:

$\frac{1545}{M.W}$	=	Gas constant 'R'	P_2	=	Discharge Pressure – PSIA
T_1	=	Inlet temp. – °R	P_1	=	Inlet pressure – PSIA
°R	=	460 + °F	K	=	Ratio of specific heats C_p/C_v
Z_{AVG}	=	Average compressibility $\frac{(Z_1 + Z_2)}{2}$	n	=	Polytropic exponent
			$\frac{(n-1)}{n}$	=	$\left(\frac{K-1}{K}\right)\left(\frac{1}{\eta_{POLY}}\right)$
Ln	=	Log to base e	η_{POLY}	=	Polytropic efficiency

Figure 6.5 Ideal gas head equations

Impeller types and specific speed

Various types of radial (centrifugal) impellers are shown in Figures 6.6 & 6.7.

Open impellers

Open impellers are shown in Figure 6.6.

The advantages of open impellers is their ability to operate at higher tip speeds and thus produce greater head than closed impellers. Open impellers can produce 15,000 – 25,000 ft-lbs/LB of head per stage. This is because a side plate is not attached to the inlet side of the vanes which results in significantly lower blade stresses. The disadvantages of open impellers are their lower efficiency due to increased shroud (front

Figure 6.6 Compressor impellers

side) leakage and increased number of blade natural frequencies resulting from the cantilevered attachment of the blades to the hub. Most end users restrict the use of open impellers to plant and instrument air applications since the high speeds and intercooling offset the efficiency penalties caused by shroud leakage. Older design multistage centrifugal compressors frequently used open impellers in the first stages since the high flows caused unacceptable side plate stresses in closed impeller design. Modern calculation (finite element) methods and manufacturing methods (attachment techniques – machine welding, brazing, etc.) today make possible the use of enclosed first stage impellers for all multistage compressor applications. Finally, radial bladed impellers (whether open or enclosed) produce an extremely flat (almost horizontal) head curve. This characteristic renders these impellers unstable in process systems that do not contain much system resistance. Therefore, radial impellers are to be avoided in process systems that do not contain much system resistance (plant and instrument air compressors, charge gas compressors and refrigeration applications with side loads.

Enclosed impellers

Enclosed impellers are shown in Figure 6.7.

Note that the first stage impeller in any multistage configuration is always the widest. That is, it has the largest flow passage. As a result, the first stage impeller will usually be the highest stressed impeller. The exception is a refrigeration compressors with side loads (economizers).

Dynamic compressor vendors use specific speed to select impellers based on the data given by the contractors and end user. The vendor is given the total head required by the process and the inlet volume flow. As previously discussed, at the stated inlet flow (rated flow) the head

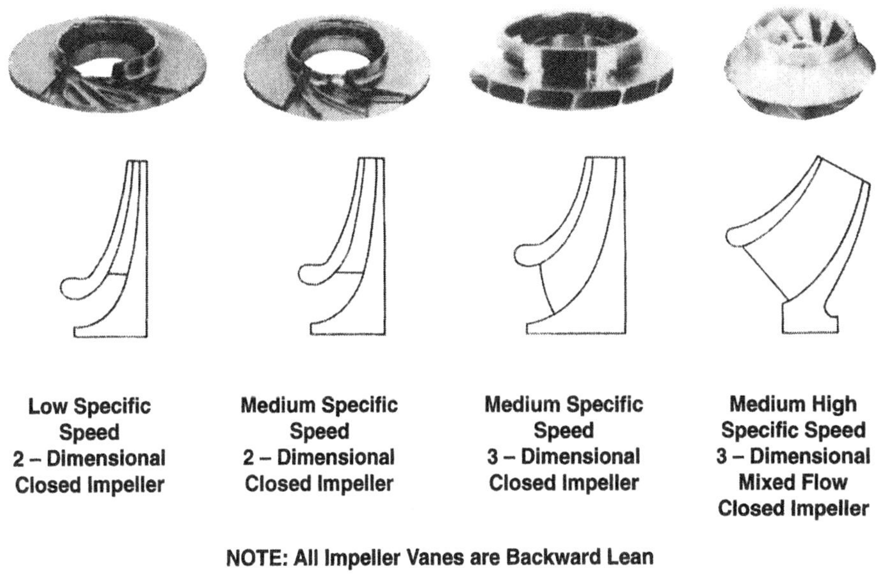

Figure 6.7 Enclosed impellers (Courtesy of IMO Industries, Inc.)

required by the process is in equilibrium with the head produced by compressor. Vendor calculation methods then determine how many compressor impellers are required based on mechanical limitations (stresses) and performance requirements (quoted overall efficiency). Once the head required per stage is determined, the compressor speed is optimized for highest possible overall efficiency using the concept of specific speed as shown in Figure 6.8.

It is a proven fact that the larger the specific speed, the higher the attainable efficiency. As shown, specific speed is a direct function of shaft speed and volume flow and an inverse function of produced head. Since the vendor at this point in the design knows the volume flow and head produced for each impeller, increasing the shaft speed will increase the specific speed and the compressor efficiency.

However, the reader is cautioned that all mechanical design aspects (impeller stress, critical speeds, rotor stability, bearing and seal design) must be confirmed prior to acceptance of impeller selection. Often, too great an emphasis on performance (efficiency) results in decreased compressor reliability. One mechanical design problem can quickly offset any power savings realized by designing a compressor for a higher efficiency.

Referring back to Figure 6.8, calculation of specific speed for the first impeller by the contractor or end user will give an indication of the type of dynamic compressor blading to be used. One other comment,

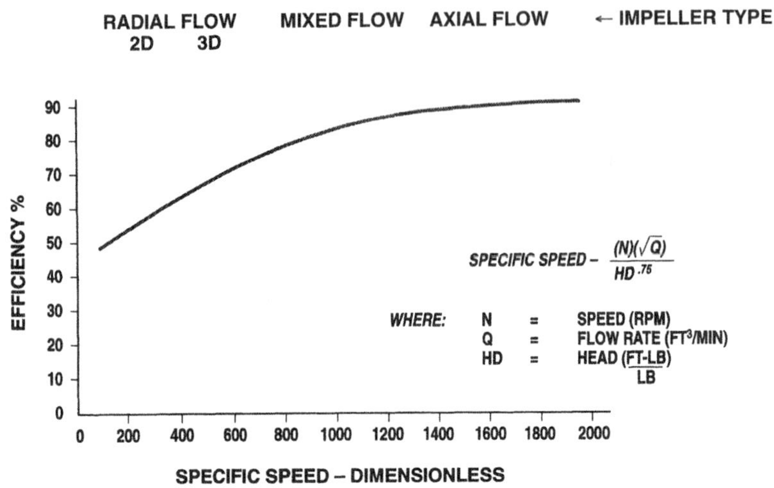

Figure 6.8 Impeller geometry vs. specific speed

Sundstrand Corporation successfully employs an integral high speed gear box design for low flow, high head applications or for low specific speed applications. The use of a speed increasing gear box (for speeds up to 34,000 RPM) enables the specific speed to be increased and therefore resulting in higher efficiency and less complexity than would be obtained with a multistage compressor design approach.

Efficiency

Compressor efficiency, regardless of the type of compressor, can best be understood by referring to a typical Mollier Diagram as depicted in Figure 6.9.

All produced heads shown on performance curves (isothermal, isentropic and polytropic) represent the ideal reversible head produced to compress a given gas from P_1 to P_2. This then is the theoretical compression path of the gas. That is, the energy required to compress a gas if the efficiency is 100%.

However due to friction, sudden expansion etc., the efficiency is less than 100%. Therefore the actual compression path requires more head (energy) to compress the gas from P_1 to P_2. The efficiency then is equal to:

$$\text{Efficiency} = \frac{\Delta E \text{ Ideal } (E_2 \text{ Ideal} - E_1)}{\Delta E \text{ Actual } (E_2 \text{ Actual} - E_1)}$$

$$\textbf{Efficiency} = \frac{\Delta E \text{ Ideal } (E_2 \text{ Ideal} - E_1)}{\Delta E \text{ Actual } (E_2 \text{ Actual} - E_1)}$$

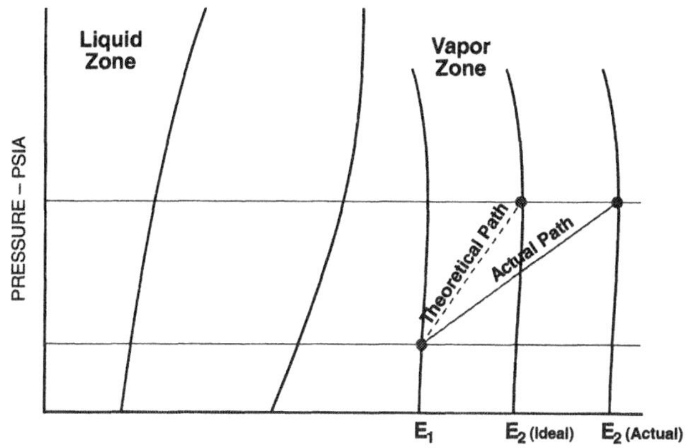

Figure 6.9 Efficiency

Note that ( ) is used to represent any ideal reversible path (isothermal, isentropic, polytropic).

Horsepower

Horsepower is defined as the total actual energy (work) required to compress a given gas from P_1 to P_2 when compressing a given mass flow:

$$\text{G.H.P.} = \frac{\text{Head}\,(\,)\,\frac{\text{ft-lbf}}{\text{lbm}}\,(\text{Mass Flow - lb/min})}{(33{,}000)\,(\text{Efficiency})(\,)}$$

Note: ( ) must be for the same ideal reversible compression path.

The brake horsepower is the sum of the gas horsepower and the mechanical losses of the compressor.

B.H.P. = G.H.P. + Mechanical losses

The mechanical losses are the total of bearing, seal and windage (disc friction) losses and are provided by the compressor vendor. For estimating purposes, a conservative value of mechanical losses for one centrifugal or axial compressor case would be 150 H.P.

$$\frac{Q_F}{Q_I} = \frac{N_F}{N_I}$$

$$\frac{HD_F}{HD_I} = \left(\frac{N_F}{N_I}\right)^2$$

$$\frac{BHP_F}{BHP_I} = \left(\frac{N_F}{N_I}\right)^3$$

Where:

Q = Flow Rate (FT^3/MIN)

N = Speed (RPM)

HD = Gas Head (FT-LB/LB)

BHP = Horsepower

I = Initial

F = Final

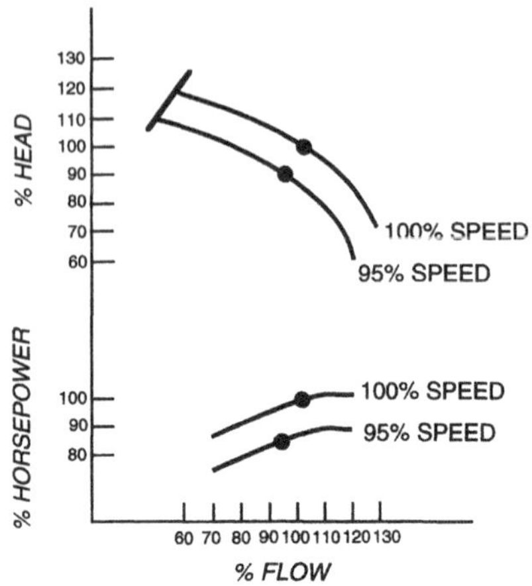

Figure 6.10 The Fan Laws

The Fan Laws

These familiar relationships, sometimes called the affinity laws for pumps were originally derived for a single stage fan which is a low pressure compressor. The Fan Laws are presented in Figure 6.10.

As shown, if speed is changed, the flow, head and horsepower vary by the first, second and third power of speed ratio respectively.

The reader must be cautioned however that the Fan Laws are only an approximation to be used as an estimating tool. Their accuracy significantly decreases with increasing gas molecular weight and increase in the number of compression stages.

What determines the turbo-compressor curve shape

- Introduction
- The compressor stage
- Impeller with side plate removed
- Impeller discharge velocities
- Blading types

Introduction

In this chapter we will define how the turbo-compressor curve is generated, specifically we will see how all turbo machines produce head, whether they process a compressible or incompressible fluid, by imparting a velocity to the fluid.

The compressor stage

To begin our discussion, let us observe a typical compressor stage shown in Figure 7.1.

A compressor stage is defined as one (1) impeller, the stationary inlet and discharge passages known as the inlet guide vanes and the diffuser respectively and the seals, namely the eye labyrinth seal and the shaft labyrinth seal. Each compressor stage at a given flow and impeller speed will produce a certain amount of head (energy) and have a specific stage efficiency. It can be observed that any dynamic curve (turbo-compressor or pump) has the characteristic of producing increased

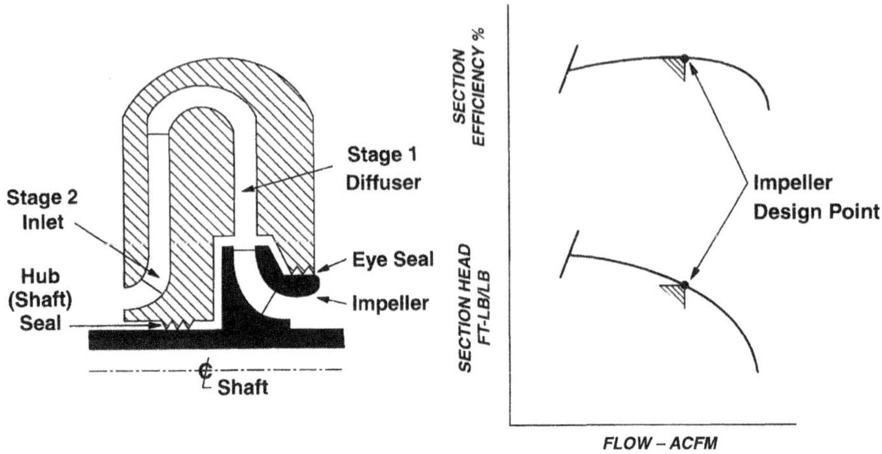

Figure 7.1 The compressor stage and characteristic curve

energy only at a lower fluid flow assuming the inlet speed and the inlet gas angle are constant.

Before we continue, a few important facts and relationships need to be presented. These relationships are: the definition of a vector, tip speed, flow as a function of velocity and flow related to conditions and the concept of actual flow. In addition, there is the important concept of an equivalent orifice. Refer to Figure 7.2.

Given any impeller configuration, specific areas can be reduced to equivalent orifices. The eye or inlet area, the discharge area between

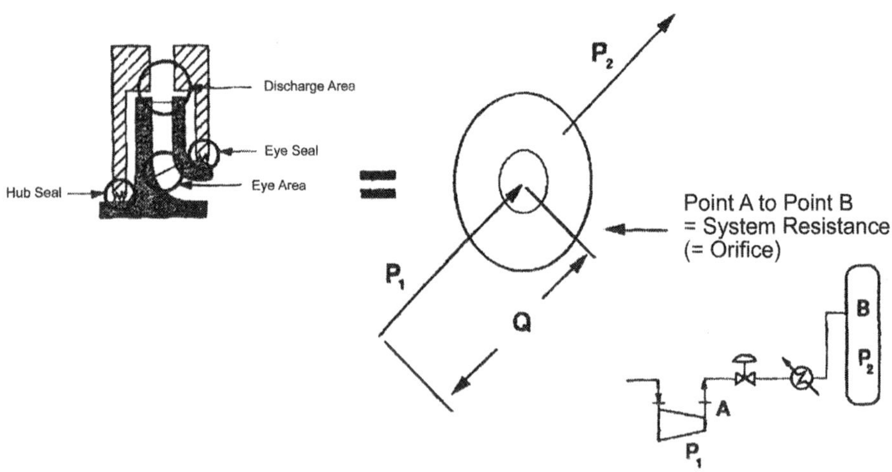

Figure 7.2 Reduce it to an equivalent orifice

any two vanes, the eye seal and the hub or shaft seal. This concept makes it much easier to understand that for a give area gas flow will change directly proportional to the differential pressure and the compressor stage. It can be seen in Figure 7.2 that there is an optimum design velocity for the inlet of the impeller and the discharge of the impeller. These velocities are controlled by selection of a proper inlet eye area and discharge area based on the impeller flow requirement. Again, the concept of an equivalent orifice is helpful to understand that the gas velocity is dependant on the geometry of the specific impeller. Also, the process system can be reduced to a simple orifice. In any process the suction side of the process and the discharge side of the process can be conceived as an orifice placed at the inlet and discharge of the compressor flanges for a given flow condition.

Figure 7.3 presents the definitions of facts and relationships necessary for the discussion that follows. Please note that the relationships presented are in British Units, metric units are not presented in this section, but can be easily derived referring to appropriate conversion tables.

Impeller with side plate removed

Facts and relationships

- A vector describes magnitude and direction →
- Tip speed 'U' = $\frac{(D)(N)}{229}$
- Flow related to velocity 'Q' = (A)(V)(60)
- Flow related to conditions (compressible flow) $Q_F - Q_I \times \frac{P_I}{P_f} \times \frac{T_f}{T_I} \times \frac{Z_f}{Z_I}$

Where					
U	=	Tip velocity (ft/sec)	f	=	Final condition
D	=	Diameter (in^2)	I	=	Initial condition
N	=	Speed (rpm)	P	=	Pressure (psia)
Q	=	Flow rate ft^3/min)	T	=	Temperature (°R)
A	=	Area (ft^2)	°R	=	°F + 460
V	=	Velocity (ft/sec)	Z	=	Compressibility

Figure 7.3 Facts and relationships

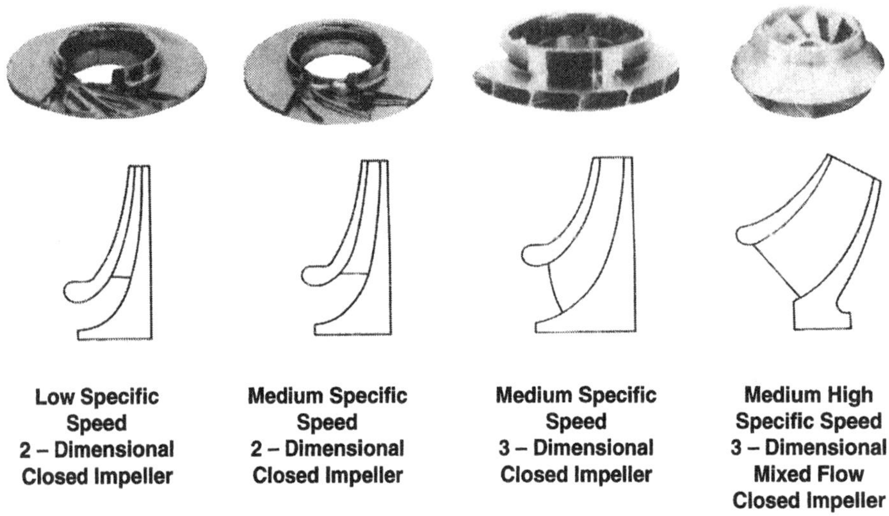

Figure 7.4 Typical impellers (Courtesy of IMO Industries, Inc.)

To begin our discussion, assume that we are operating at the impeller design point (as shown in Figure 7.1) and that we have removed the side plate of the impeller and are examining the flow between any two vanes. Typical impellers are shown in Figure 7.4 and the schematic of any impeller for our purposes showing the upper half of the impeller with the side plate removed is shown in Figure 7.5.

In Figure 7.5 we can see that only two velocities need to be considered to properly describe the generation of head (energy). At the tips of the vanes there are two velocities that are present. The blade tip velocity,

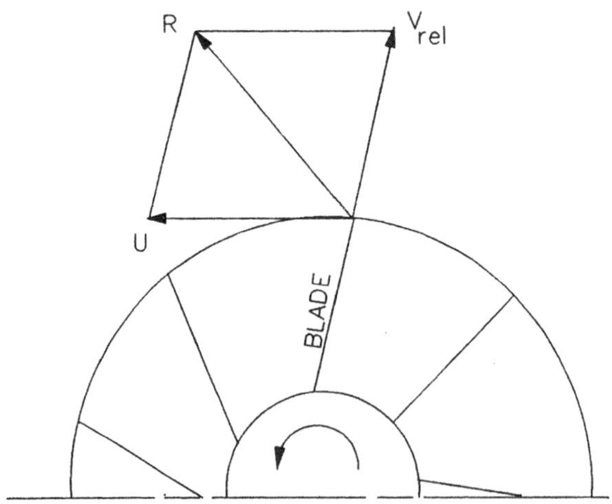

Figure 7.5 Impeller with side plate removed

identified as U and the velocity relative to the blade identified as V_{REL}. The blade tip velocity is the function of the diameter of the blade and the blade rotational speed. The velocity relative to the blade (V_{REL}) is a function of the area between the blades, the flow rate at that location and the angle of the blade at the discharge of the impeller. Summing these two velocities, the resultant or absolute velocity defines the magnitude and the direction of the gas as it exits the blade. For this discussion, we assume that the velocity relative to the blade exactly follows the blade angle, that is the slip is equal is zero. This assumption can be used since it will not impact the final conclusion of our discussion.

Impeller discharge velocities

If we now resolve the absolute velocity noted in Figure 7.5 (R) into x and y components, the x axis projection of the component is the tangential velocity of the gas at the impeller discharge (refer to Figure 7.6). Eulers' energy equation states 'The energy created by any turbo machine is proportional to the product of the tip speed and the tangential velocity'.

Let us now assume that the head required by the process changes such that the flow V_{REL} through the impeller reduces. Referring to Figure 7.7 let us again examine the discharge velocity to see what happens at this reduced flow condition.

Assuming that the rotor speed is constant, it can be seen that the tip speed value does not change since tip speed is a function of impeller

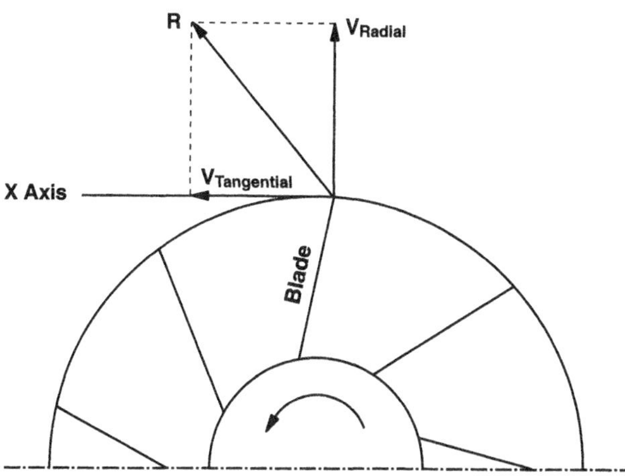

Figure 7.6 Impeller with side plate removed

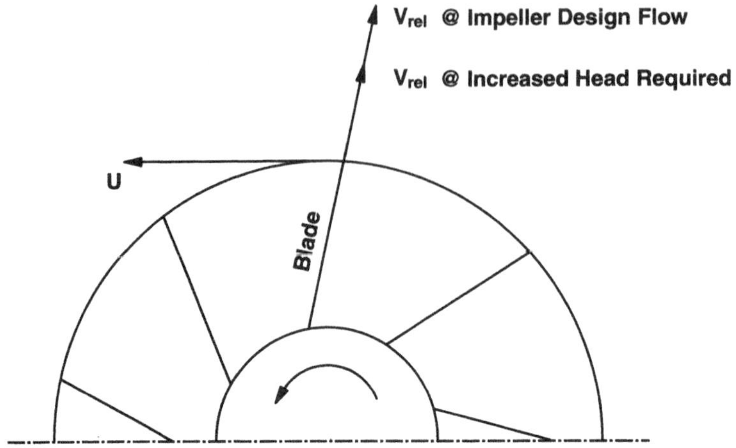

Figure 7.7 Impeller with side plate removed

diameter and shaft speed. However, the velocity relative to the blades (V_{REL}) will be reduced as a result of a lower volume flow passing through a fixed area, resulting in a low velocity relative to the blade at the discharge. If we again sum the velocity vectors to obtain the absolute velocity R (refer to Figure 7.8), we can see that the angle of the gas exiting the blade is significantly reduced and the X projection of the tangential velocity will be greater than the previous value (refer to Figure 7.9).

Since the head (energy) produced by the blade is proportional to the tip speed (unchanged) and the tangential velocity (increased) we can see that the reduction of flow through the blade has resulted in increased head or energy imparted to the fluid. Practically, this makes

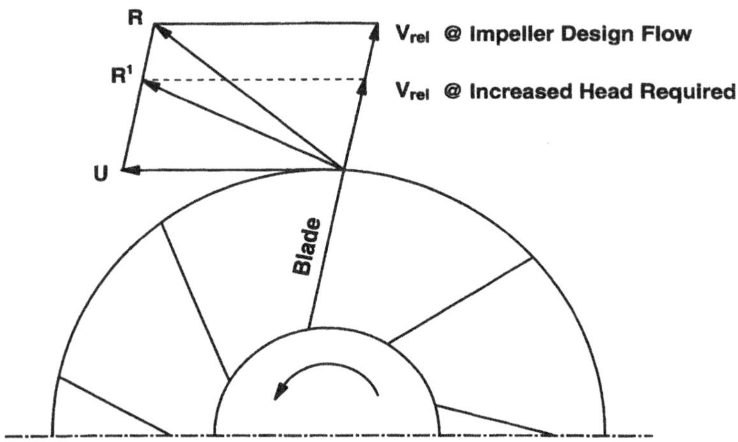

Figure 7.8 Impeller with side plate removed

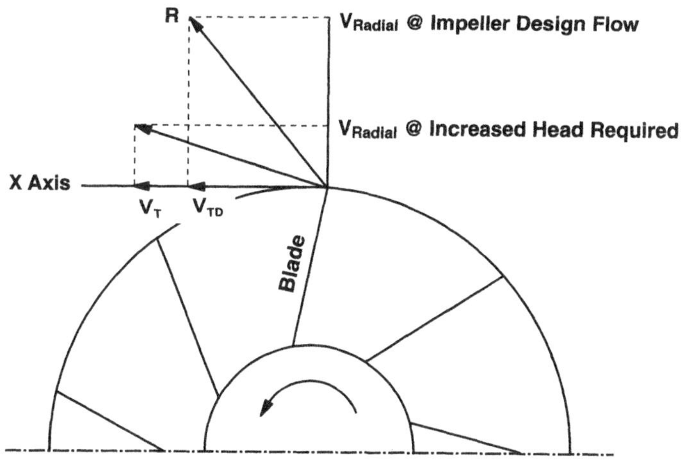

Figure 7.9 Impeller with side plate removed

sense, since the slower the gas proceeds through the vane, the more time it has to pick up energy imparted by the blades and as a result will increase the energy produced within the impeller. Therefore it can be seen for all dynamic blades and impellers which increase the energy of the fluid by the action of the vane on the fluid can increase fluid energy only at a lower flow rate, assuming the speed of the impeller and the inlet angle of the fluid to the blade remain unchanged.

Blading types

Backward lean

The previous discussion was focused on the characteristic of a backward leaning vane. Most turbo machinery vanes are backward leaning since they produce a greater head rise from impeller design point to the low flow operating point. The low flow limit of operation for centrifugal compressors is known as surge. Head rise is defined as the head produced by the impeller at the low flow operating point divided by the head produced by the impeller at the impeller design point. Today, the industry prefers backward leaning impellers with an external or exit blade angle of approximately 40 to 50 degrees. This blade angle will produce head rises in the range of 5–15% depending on the gas density.

Radial

Radial vanes are used in some older design open type first stage

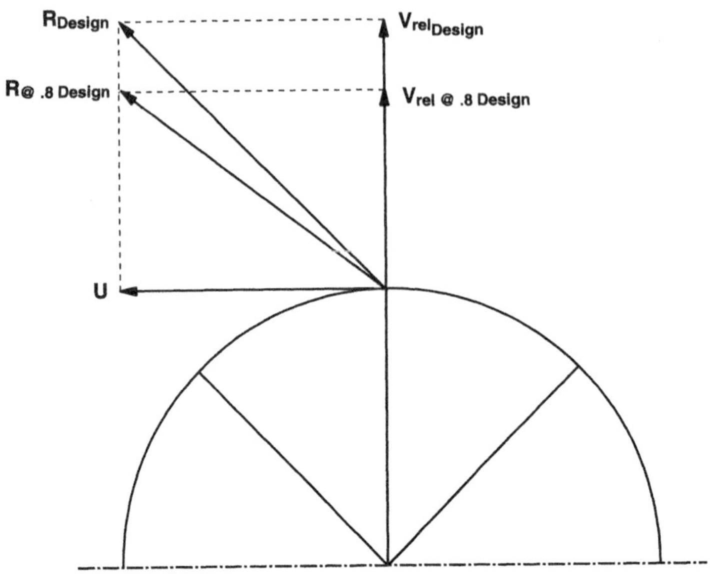

Figure 7.10 Radial blading

impellers, and in some modern impellers that operate at a very low flow. Let us now examine the effect of a radial blade on the performance curve. If we were to design an impeller with radial blades let us examine again what would happen when we changed flows from a rated point to a lower flow. At the rated point the blade tip speed and velocity relative to the blade will be as shown. Refer to Figure 7.10.

Notice that the velocity relative to the blade is completely radial assuming zero slip and consequently the absolute velocity is the sum of the two vectors. Again we project the tangential velocity on the x axis projection from the absolute velocity and note its value as shown in Figure 7.11.

At a lower flow, tip speed will remain constant (assuming constant shaft speed) and the relative velocity will decrease as in the case of the backward leaning blade. However, note that since the relative velocity follows the radial blade path, the magnitude of the tangential velocity remains constant regardless of the value of relative velocity. This is shown in Figure 7.12.

Since the energy generated by the blade is the product of tip speed (unchanged) and a tangential velocity (unchanged) the design head (energy) produced in a radial impeller will remain essentially constant. Therefore, the curve shape will be significantly flatter and will possess much less of a head rise than that of a non radial vane. In reality though, the effects of friction will in fact produce a curve shape that will increase from high flows to low flows but the effects will produce

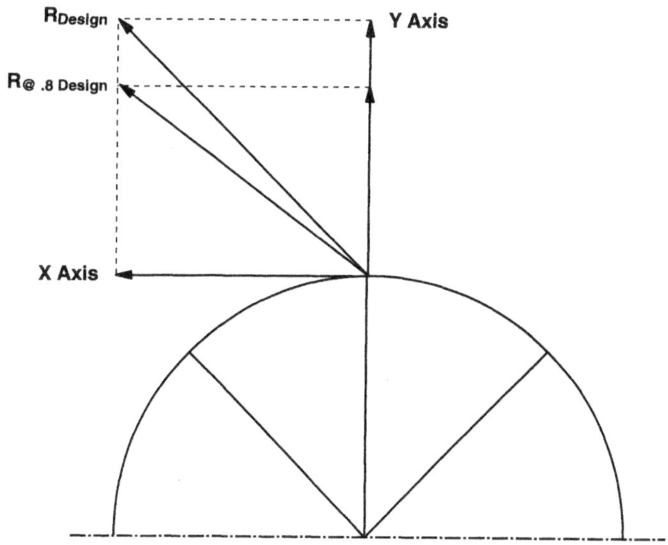

Figure 7.11 Impeller with side plate removed

much less of an energy increase. This value is typically approximately 3% head rise or less. This is an important fact to remember since the operating point of any dynamic machine will be the intersection of the head required and the machinery curve head produced. A characteristic curve with a low head rise will have greater sensitivity to process changes than a curve with higher head rise.

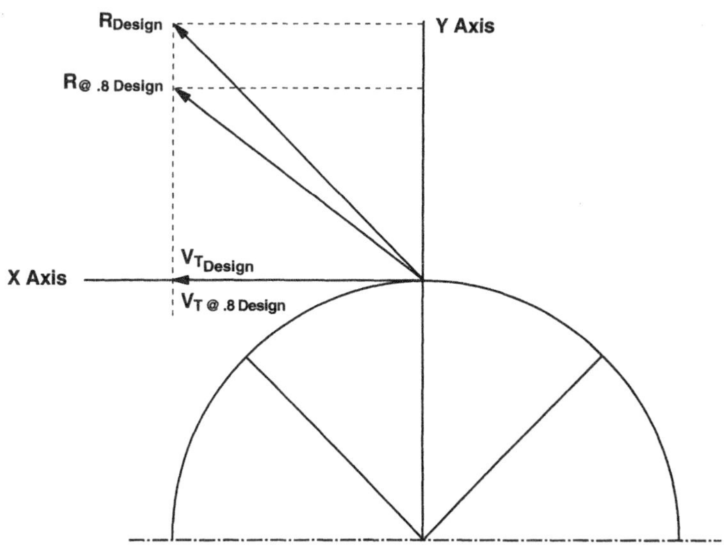

Figure 7.12 Impeller with side plate removed

In summary, it should be noted that the previous discussion can be equally applied to pump impellers since pumps also operate on a fluid (liquid). One very important thing to remember from this discussion however, is that regardless of the type of liquid used in pumps, velocity relative to the blade will never change since the fluid is incompressible. In the case of a turbo-compressor however this will not be true since the gas is compressible and the velocity relative to the blades at the discharge will change as a result of pressure and temperature of that gas at the exit. Therefore the statement that head (energy) produced by a compressor impeller will remain constant at a given speed is not totally true.

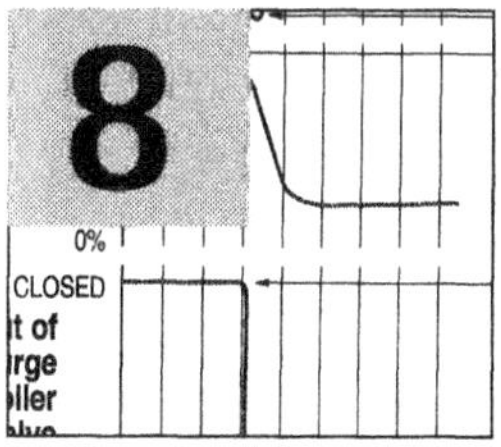

Surge (stall) and stonewall

- Introduction
- Surge facts
- The limits of the curve
- What causes surge
- What causes stonewall (choke)

Introduction

In this chapter, we will devote our attention to limits of the dynamic curve, namely surge and stonewall.

The first item that we will discuss will be low flow. This phenomena is known as surge. It should be pointed out immediately that surge is a system phenomena resulting from the inability of the turbo-compressor impeller to produce the amount of required energy that the process system requires.

Some facts concerning surge are that it is a high speed phenomena, flow reversals can occur in less than 150 milliseconds, pressure rapidly fluctuates, noise is generated, temperature increase can be very rapid and mechanical damage can occur.

The intensity of surge varies from application to application and is proportional to the density of the fluid. Higher pressure and higher molecular weight applications can result in greater mechanical damage. Low density applications (hydrogen) can cause surge damage that may not be detectable until the equipment is disassembled.

The cause of surge results from low velocity in the blade or impeller passages. It must be remembered that every dynamic blade or impeller is designed for one and only one operating point. Energy is produced by the action of the vanes and flow between blades or vanes. As the flow through the vanes is reduced, the velocity between the vane passages is also reduced.

As a result, flow changes from laminar (smooth) to turbulent. This causes flow discontinuities. There are many causes of surge. Referring to a cross section of an impeller and a stationary diffuser, we can see and remember from previous discussions that the gas angle of the absolute velocity of the gas exiting an impeller will reduce as the flow decreases. The gas exits the diffuser in a logarithmic spiral fashion. As the exit gas angle becomes less, the path the gas must follow becomes longer and eventually can turn back into the impeller thus causing a backflow. This is one example of a flow discontinuity that limits the amount of energy an impeller can produce. Another example of this phenomena could occur in the impeller or blade passage itself and result in backflow. Regardless of the initial cause, the inability of a rotating blade or impeller to produce the amount of energy required by the system will result in the gas acting on the blade or impeller rather than the blade or impeller acting on the gas. At this point, backflow occurs.

If we examine a typical compressor system and look at the discharge system from the compressor discharge flange to the discharge checkvalve, we can see that at the point of maximum blade or impeller energy production, the pipe and discharge pressure of the compressor can be thought of as an equivalent vessel. As long as the flow into this vessel is equal to the flow out, the pressure in that vessel remains constant. At the onset of surge, the compressor will no longer produce flow into the vessel and stored energy in this vessel now acts on the blade resulting in a flow reversal from the vessel back towards the suction of the compressor. Examining a typical compressor performance curve, we can see that as the head required increases, the operation point will move towards the surge or instability line. When the operating point intersects the surge line, a flow reversal will occur causing the operating point to enter the negative flow region of the performance curve. The equivalent vessel therefore will become evacuated during this reversal. Once the equivalent vessel is evacuated, the system pressure observed by the compressor rapidly reduces thus enabling the compressor to achieve a new operating point in the positive region of the performance curve at high flow. Once the vessel becomes filled, the operating point moves again up the compressor curve to surge and the cycle begins anew. Remember that these cycles can occur as rapidly as six (6) times *per second!*

Naturally, this action produces large amounts of energy which are absorbed by the mechanical components of the compressor and can result in a significant amount of temperature increase inside the compressor. It must be understood that the gas will move back and forth in the compressor upwards of six (6) times per second and not be cooled, therefore temperature increases can be rapid and thermal expansion can cause major damage during surge. This phenomena can occur most easily in axial compressors. Therefore, surge is to be avoided most in axial machines due to their ability to heat up faster than centrifugals and also as a result of the blade configuration which can be more readily excited than centrifugals.

The high flow limit of the dynamic curve is known as choke, or stonewall and is defined as the maximum flow a given impeller can handle. This value will be obtained when the ratio of relative inlet gas velocity to the sonic velocity of the process gas being handled is equal to one. Referring back to the concept of an equivalent orifice, the impeller represents an equivalent orifice. Simply stated, the high flow limit of the curve is attained when the impeller itself acts as a resistance and will not allow additional flow to be passed. We can see by referring to the relationship of the relative mach number, which is the ratio of the inlet relative velocity to the sonic velocity of that gas at suction conditions, that higher molecular weight applications will result in a higher mach number. Therefore, it can be stated that high molecular weight applications result in a shorter range of operation of the compressor curve. We define range of operation as the flow from surge to stonewall.

Typical ranges for surge and carryout are: Surge range can vary between 60% to 80% of design flow and carryout range can vary between 110% to 130% of design flow.

Surge facts

Surge can be a very dangerous phenomena if allowed to continue. Damage caused by surge in any type of dynamic compressor can render that compressor inoperable for periods of time in excess of two (2) months. Considering that most dynamic compressors are not spared, this can cause a significant loss of revenue. Typical daily product revenues for a world class process unit can exceed US$1,000,000 per day! Figure 8.1 shows some of the most common effects of surge.

Surge facts

- Surge is a high speed phenomenon. Flow reversals can occur in less than 150 milliseconds.
- Reversal rate is 30 to 120 cycles/sec.
- Pressure rapidly fluctuates.
- Noise generated.
- Temperature increase (can be rapid).
- Mechanical damage can occur.
- Unit may trip.
- Intensity varies with the application.

Figure 8.1 Surge facts

Of all the effects listed, by far the most damaging is rapid temperature increase for this can cause internal rubs of the compressor at operating speed resulting in impeller breakage, diaphragm breakage, extreme labyrinth seal wear and possible case breakage.

Figure 8.2 shows the results of an actual surge test performed on a 2500 HP solar compressor.

It is the results obtained by a strip chart recorder during a surge cycle. Δp_o is the flow change and Δp_c is the pressure ratio change. Notice how

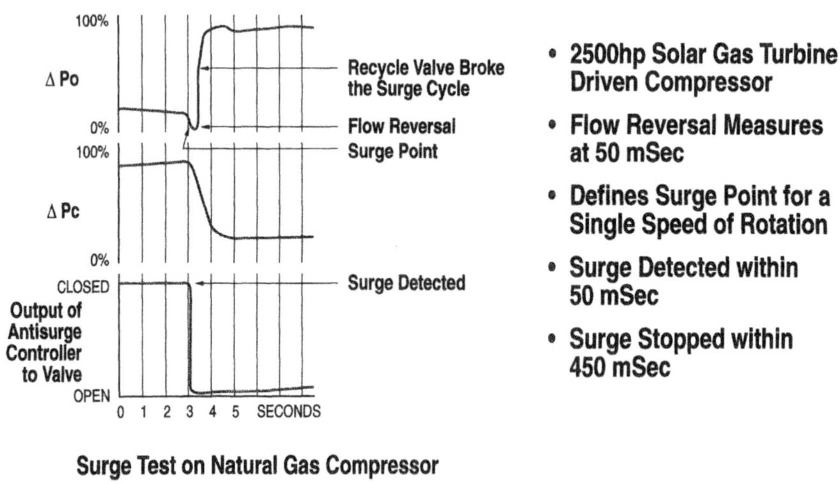

Figure 8.2 A surge test (Courtesy of Compressor Controls Corp.)

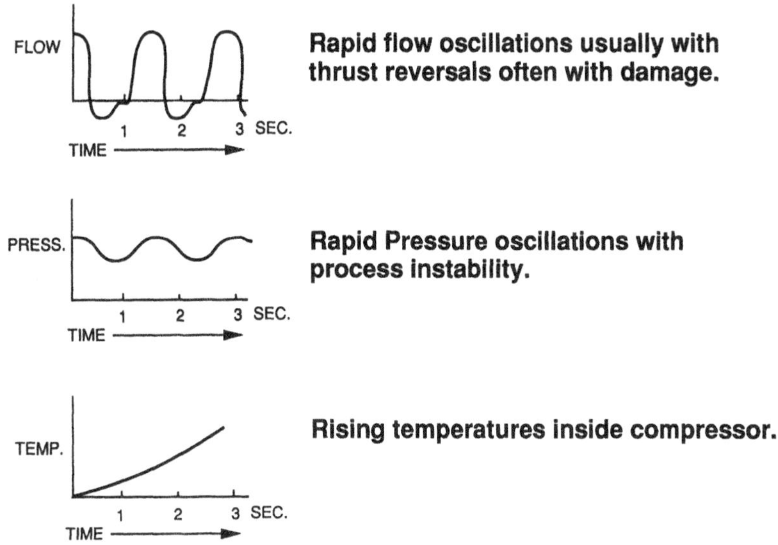

Figure 8.3 The surge phenomena

prior to the surge event that the pressure ratio required (Δp_c) increases while the compressor flow rate (ΔP_o) is decreasing. Observe that once a surge cycle is detected, the output of the controller works to open the anti-surge valve immediately. This fact is important to remember when trying to justify a field surge test to plant operations. A compressor will never be damaged by a surge test if conducted properly. As shown in Figure 8.2, strip chart recorders are used and the surge system is in automatic but the surge control line has been moved to the left of the actual surge location. The surge control line defines the set point of the surge controller and hence the point at which the surge control valve will open.

Figure 8.3 presents the characteristics of flow, pressure and temperature during surge. The degree of parameter change is directly proportional to the density of the gas. The higher the gas density, the greater the effect of the surge.

The limits of the curve

The limits of any type of dynamic compressor curve are the low flow limit (surge) and the high flow limit (choke flow). Surge, as we shall see, is caused by low flow turbulence while choke flow is caused by high velocity friction. Figure 8.4 shows a typical compressor curve and the side view of a compressor stage (top half view only).

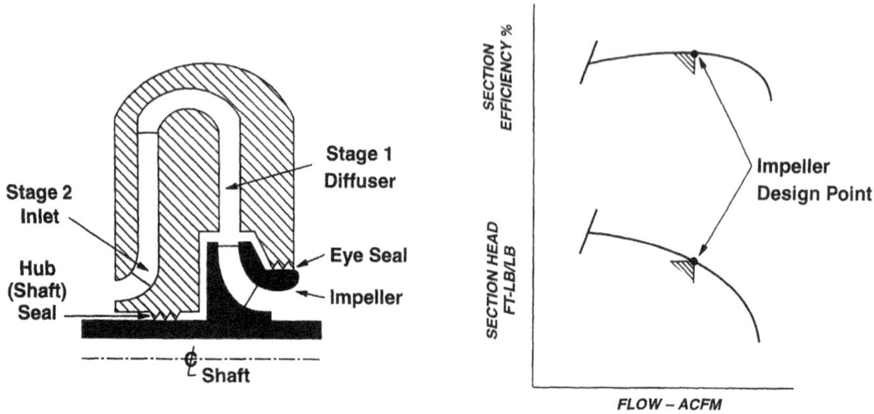

Figure 8.4 The compressor stage and characteristic curve

It can be seen that the limits of any dynamic compressor curve are a consequence of gas velocity in the compressor stage.

What causes surge

Surge is a process system phenomena that is the result of flow separation, caused by low gas velocity, anywhere in a compressor stage (inlet guide vane, impeller suction, impeller mid section, impeller discharge or diffuser). Centrifugal pumps experience the same flow separation phenomena at low flows which sometimes cause the liquid to vaporize resulting in recirculation which can cause cavitation.

As the process system requires more head in any type of dynamic compressor, the flow is reduced to a point that causes flow separation. This event is commonly known as stall. An example of diffuser stall is shown in Figure 8.5.

This diagram depicts the view of a simple impeller with the side plate removed and a vaneless diffuser. As the gas leaves the impeller via vector 'R', the resultant of the gas velocity through the impeller and the impeller tip speed, it passes through the diffuser in a path that approximates a logarithmic spiral. Since the relative gas velocity is the 'y' component of 'R' and will decrease with decreasing flow rate, the gas angle ∝ off the blades will decrease. As a result, the path of the gas off the blade will become longer with decreasing flow rate. At surge flow, the velocity off the wheel will be so low that the path of the gas will not leave the diffuser. This will cause a reduction in the head produced by the impeller. At this moment, the compressor is said to be

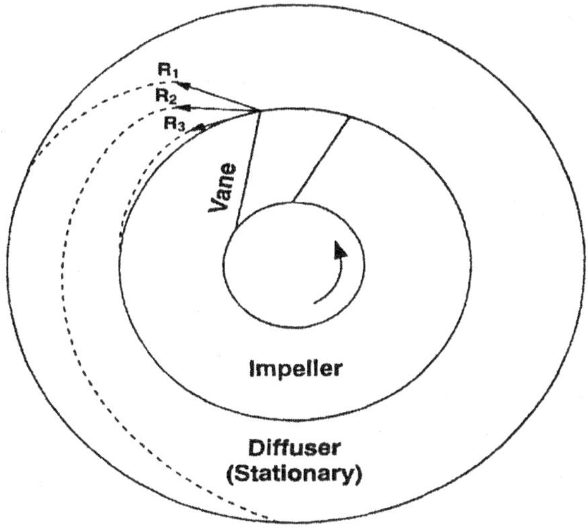

Figure 8.5 Diffuser stall

in a state of flow separation, flow instability or stall. Stall can be initiated by flow separation at any point within a compressor stage (inlet, mid-section, discharge, etc.). Regardless of the location of the flow separation within the compressor stage, a reduction in head produced by the impeller will occur. The dynamic compressor performance curve actually decreases to the left of the surge line since the flow separation (stall) increases losses in the stage and reduces head produced. Please refer to Figure 8.6.

Draw a decreasing head curve to the left of the surge line. On this curve, place an operating point. This point would be the initial point in the surge cycle. A closeup of the process system in Figure 8.6 is shown in Figure 8.7.

Once flow separation occurs, the head produced by any dynamic compressor decreases and the process gas present from the compressor discharge flange to the check valve flows backwards through the compressor. This backflow causes the volume shown in Figure 8.7 to be evacuated resulting in a low discharge pressure. Since the head (energy) required by the process system is a function of discharge pressure, the head required by the process will decrease allowing the compressor to operate in the high flow region of the performance curve. The reversal of dynamic flow caused by flow separation (stall) in the compressor stage and the recovery of flow resulting from reduced discharge pressure is defined as a surge cycle. This surge cycle will continue until either the head produced by the compressor is increased or the head

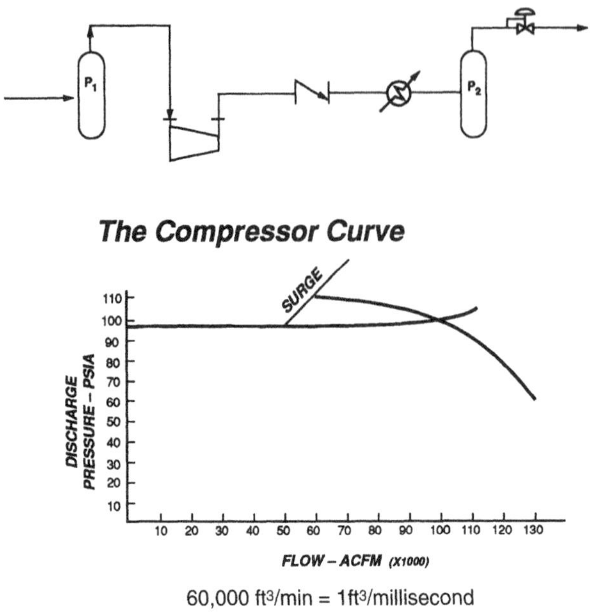

Figure 8.6 Surge – The effect of the system

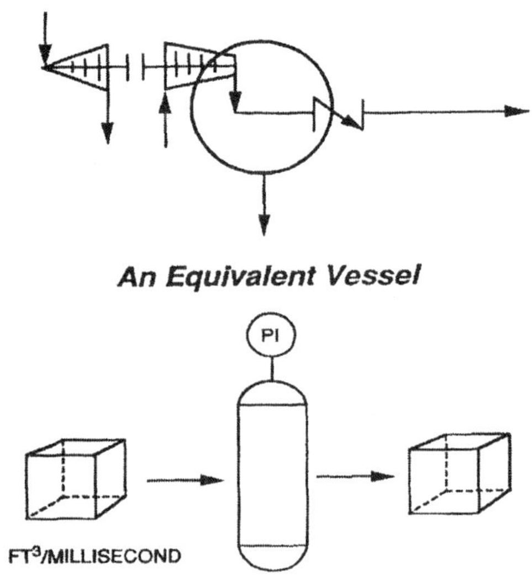

Figure 8.7 Flow in and out of a vessel

required by the process is reduced. The quickest way to eliminate surge is to rapidly reduce the discharge pressure by opening a blow-off or recycle valve in the discharge process system.

Please refer back to the 'Surge Facts' section of this chapter to review the damaging effects of surge.

What causes stonewall (choke)

The definition of stonewall or choke flow is presented in Figure 8.8.

Stonewall

Stonewall or choke flow is defined as the maximum flow a given impeller can handle.

This value will be attained when the ratio of the relative inlet gas velocity to the sonic velocity of the process gas being handled is equal to 1.0.

Figure 8.8 Stonewall

The name 'Stonewall' comes from the fact that the compressor curve suddenly drops off and appears to have come up against a 'Stonewall' (refer to Figure 8.4). The cause of this phenomena is excessive relative gas velocity through the impeller. As the head (energy) required by the process system is reduced, the volume flow through the impeller will increase.

Figure 8.9 shows that any impeller stage is essentially an equivalent orifice with constant flow areas or orifices.

Since volume flow is proportional to areas and gas velocity, an increase in volume flow is a direct increase in gas velocity.

The limit of compressor high volume flow is controlled by the relative mach number which is defined in Figure 8.10.

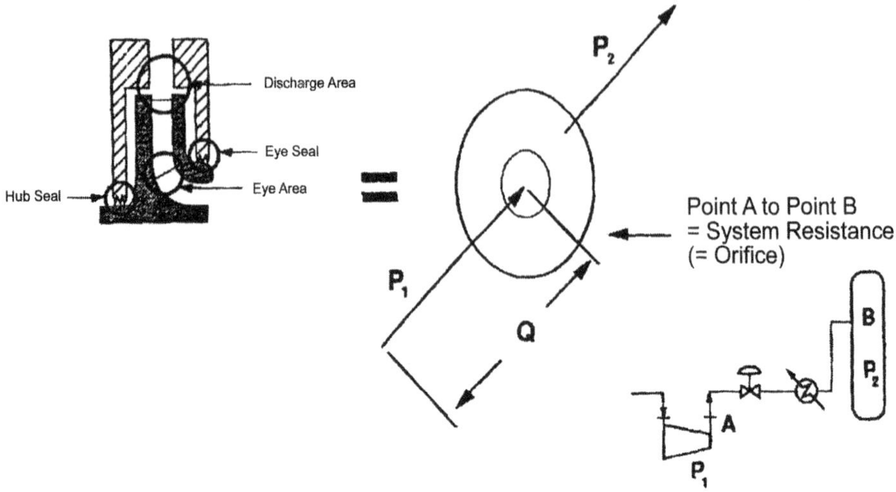

Figure 8.9 Reduce it to an equivalent orifice

Relative Mach Number

$$\text{Relative Mach Number} = \frac{V_{REL\ INLET}}{\sqrt{KgRT_1}}$$

Where::

V_{REL}=	=	Inlet velocity relative to the blade
K	=	Ratio of specific heats (CP/CV)
R	=	1545/Molecular Weight
T_1	=	Inlet temperature (°R)
°R	=	460 + °F
g	=	32 ft/sec^2

Figure 8.10 Relative mach number

When the relative mach number = 1, the maximum possible flow by any dynamic compressor is attained. Relative mach number is the ratio of gas impeller inlet relative velocity to the sonic velocity of the process gas. Figure 8.10 shows that the sonic velocity is directly proportional to K, g and T and inversely proportional to gas molecular weight. Therefore gases with high molecular weight will reach stonewall or choke flow sooner than gases with low molecular weights. This is why the flow range of compressors processing high molecular weight gases is always less than the flow range for compressor processing low molecular weight gases.

Unlike surge, stonewall is not a destructive phenomena. Since horsepower rapidly decreases in stonewall, most dynamic compressor mechanical tests are run in stonewall to reduce shop load horsepower. Also, dynamic compressors can only operate in stonewall if the head (energy) required by the process system is low enough to allow the compressor to operate in this high velocity region of the performance curve. For economic reasons, this is rarely the case since most engineering contractors optimize process pipe design to minimize pipe diameter and therefore increase the head required at high flow rates.

The only case of dynamic compressor damage caused by choke flow that I have experienced involved the axial compressor of a gas turbine. The design of the combustion chambers and turbine section allowed the axial compressor to operate in stonewall under certain operating conditions. The high velocity caused shock waves which excited a natural frequency of a compressor blade row causing blade breakage.

When sizing surge control valves for axial compressors, confirm that the maximum, full open surge valve flow will not excite any axial compressor blade natural frequencies. This is not usually a concern with centrifugal compressor impellers since the impellers are usually of the 'closed type' and therefore rigidly support the blades at both the hub and side plate. In the case of open compressor impellers (blades supported by only the hub), impeller natural frequency excitation should be checked. Open impellers are used in older designs for the first and second stages of large multistage compressors and for centrifugal plant and instrument air compressors.

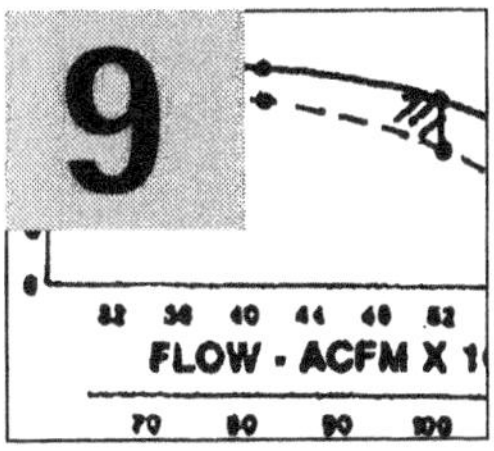

9 Turbo-compressor individual stage and overall performance

- Introduction
- The stage curve
- The overall curve
- Determining section performance

Introduction

Having previously discussed performance characteristics of a single compressor stage, we will now examine the effects that multistage compressor configurations have on the overall compressor performance curve.

The stage curve is defined as the curve representing the performance of one stage which consists of the impeller, the stage seals, the diffuser, the cross over and return passage. Each compressor blade row or impeller stage has a specific performance curve that is plotted as actual volume flow vs head (energy) and actual volume flow vs efficiency.

The amount of impeller stage head is limited by both mechanical and aerodynamic factors. Mechanical factors include impeller stresses and blade natural frequencies. Aerodynamic factors are tip speed, optimum efficiency, mach number and flow range considerations.

A compressor section is defined as the number of stages between turbo-compressor casing nozzles. A section can contain one or more stages. Its performance is defined by a section curve. The number of stages per section is limited by discharge temperature, process gas characteristics, casing and configuration, and rotor stiffness considerations.

We will examine stage and section performance for an ideal designed compressor case, a case with fouled impellers, and a case with varying molecular weight.

Any multistage compressor is designed on the basis that succeeding impellers will compress a lower volume flow resulting from the compressibility of the specific gas handled. Remembering that each application is designed for only one operating point, the compressor designer matches each successive impeller as closely as possible to achieve operation at the individual impeller design point such that each individual impeller will operate at its best efficiency point. It is important to understand that in many applications, operating at the compressor section rated point does not mean operating at each individual compressor impeller best efficiency (design point). Many older designed multistage compressors were built using specific impeller designs. For any given application, the succeeding impeller may not be at its optimum efficiency point (design point) but may be significantly far from that operating point. A good rule of thumb in selecting a multistage compressor is to assure that when operating at the compressor section rated point each impeller operates at approximately + or – 10% of its design point. Today some compressor manufacturers are using computer (cad cam) design to manufacture each impeller individually so that each impeller operates at its design point so there is a minimum of mismatch between impellers. Therefore, it can be seen that the overall performance section curve is the composite of the operation of each individual impeller performance curve.

If the performance of one or more of the impeller stages in a compressor section were to deteriorate as a result of fouling or increased labyrinth seal clearances, the overall performance curve will be affected. The amount of this effect will depend on the performance deterioration of the individual impeller stages. We have shown a case where the first and second impellers of a three stage compressor section become fouled. The resulting section operating curve can deteriorate significantly in the terms of head (energy), flow range and efficiency.

Another case to consider is the change in gas molecular weight. Remembering that once an impeller is designed, its energy production at a given flow point is essentially fixed, we can see that pressure rise will change as gas composition changes. It must be remembered that each gas composition requires a different amount of head (energy) to increase its pressure level to a given amount. An impeller or blade is designed for only one gas composition and therefore only a specific amount of energy is designed in a blade or impeller for a given application at a specific flow rate. In a multistage compressor, the effect of gas composition change can have a significant effect on the overall compressor curve. If for instance, a gas of higher molecular weight were to be handled, the pressure produced in the first stage of a multistage compressor section would increase. Since the gas is compressible, this would result in a reduced volume to the second stage. This impeller was initially designed for a higher volume but now will handle a lower volume. Also, a dynamic blade or impeller will produce higher energy at a reduced flow, therefore compounding the effect of the molecular weight increase with increased blade energy capability resulting in a further reduction in flow rate to the succeeding stage. This fact will result in a shifting of the section curve from the design point towards the surge point and could result in surge of a compressor with no significant change in system resistance.

The stage curve

Most of the dynamic compressors used today are multistage because the head required by most processes are in excess of the head produced by one stage of a dynamic compressor. Typical values for one stage of dynamic compression are:

- Centrifugal closed impeller 10,000 $\frac{\text{ft–lb}_f}{\text{lb}_M}$
- Centrifugal open impeller 15,000–25,000 $\frac{\text{ft–lb}_f}{\text{lb}_M}$
- Axial blade row 3,000 $\frac{\text{ft–lb}_f}{\text{lb}_M}$

Figure 9.1 shows the components associated with one stage of a centrifugal closed impeller.

For each impeller or blade stage, there exists a specific performance curve. In order to achieve maximum efficiency and flow range in a multistage dynamic compressor, each stage should be operating at its maximum efficiency point when the compressor is at its rated (guaranteed) point. This is not true for many older designs where only a limited number of compressor impeller or blade designs were available to select from. The amount of head produced by one stage is limited by certain mechanical and aerodynamic factors. These factors are presented in Figure 9.2.

A STAGE Consists of an Impeller, Stage Seals, Diffuser (Crossover and Return Passage – If in a Multistage Configuration)

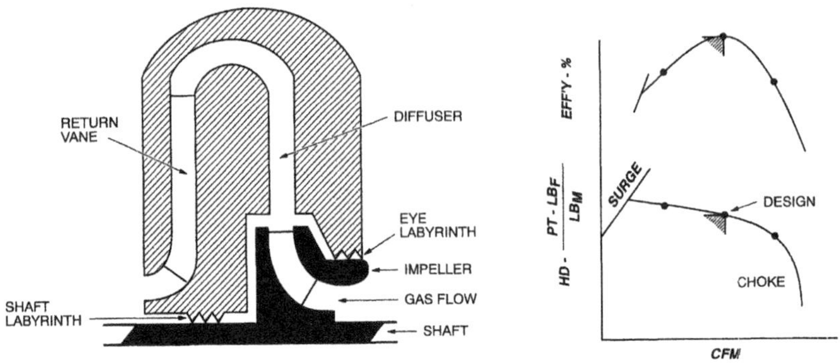

Figure 9.1 A stage consists of an impeller, stage seals, diffuser (crossover and return passage – if in a multistage configuration)

Compressor impeller stage head is limited by:

Mechanical factors

- Impeller stresses
- Blade natural frequencies

Aerodynamic factors

- Tip speed
- Optimum efficiency
- Mach number
- Flow range considerations

Figure 9.2 Factors limiting compressor impeller stage head

As an example, open impellers can produce as much as 250% greater head than a closed impeller because the impeller stresses at higher speeds are significantly reduced by omitting the impeller cover or side plate.

The overall curve

Frequently, there is much confusion concerning the terms compressor stage and compressor section. Process engineers and operators usually use the term stage to describe what properly is termed a compressor

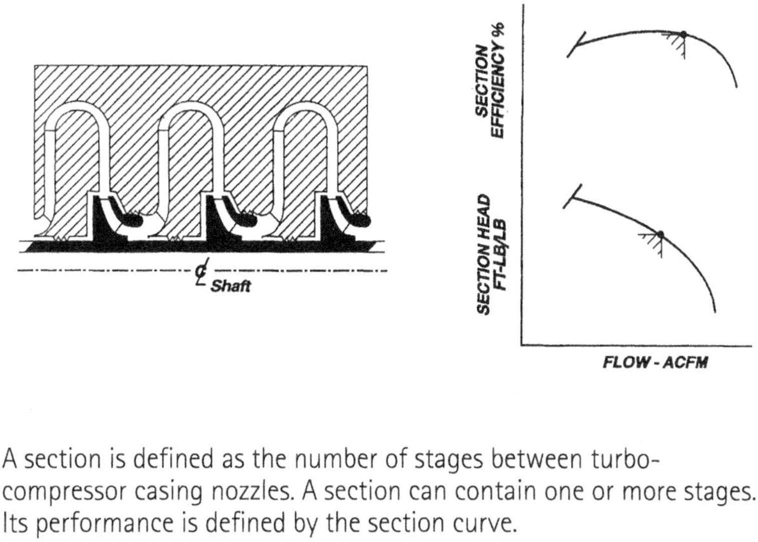

Figure 9.3 Compressor section-definition/performance curve

section. This is probably because process flow diagrams only show a stage and not the individual impellers in the usual block diagram format. Figure 9.3 defines a compressor section and shows a typical section performance curve.

In order to properly define compressor performance, a performance curve is required for each section. Each sectional performance curve is developed from the individual impeller stage curves. An important fact to remember concerning compressors is that the design of each succeeding stage is based on the predicted preceding stage performance. If the preceding stage performance is not as predicted, the next stage will be affected. The greater the number of stages, the greater this effect will be. It is commonly called mis-matching. Any change in gas density (pressure, molecular weight, temperature, compressibility) will affect the flow into the succeeding stage and thus effect the performance. The most effective way to minimize the effect of mis-matching is to require that each compressor stage operate as close as possible to its design or best efficiency point.

The number of stages per section for dynamic compressors are also limited by performance and mechanical factors (refer to Figure 9.4).

The number of stages per section – limiting factors

- Discharge temperature
- Process gas characteristics
- Casing configuration
- Rotor stiffness

Figure 9.4 The number of stages per section – limiting factors

Some processes can cause accumulation of solid materials (fouling) within the compressor impellers or stationary passages. Such a phenomena is usually temperature related and can influence the number of intercooled sections for a given application. Also, a large number of stages on a single rotor can reduce the rotor stiffness and thus reduce the natural frequency (critical speed) of the compressor.

Determining section performance

As previously discussed, the performance curve for any section is derived from the individual stage curves of each impeller in that section. Figure 9.5 shows an example of a three stage nitrogen compressor section.

Note how the inlet volume flow to each successive section is reduced and how it depends on the impeller head produced by the proceeding stage. Figure 9.6 shows the effect of fouling the first and second stage impellers in the same compressor.

Note how this affects the operating points on the third stage: • = original operating point, ○ = new operating point in fouled condition. The same affect would occur if the interstage labyrinth clearance increased from erosion or vibration on the first and second stages. Any reduction of head produced in a preceding stage will increase the volume flow in a succeeding stage and reduce the overall head produced by a compressor section.

In Figure 9.7, the effect of changing the molecular weight in the same compressor is shown.

Note that for molecular weight changes of less than 20%, the head produced by a dynamic compressor stage does not significantly change. The greater the density of a gas, the greater the discharge pressure and

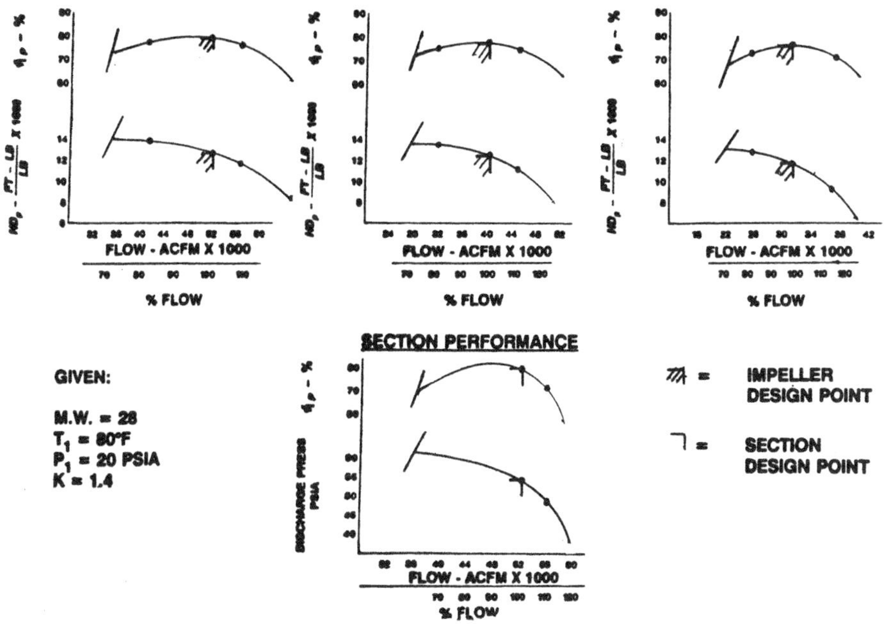

Figure 9.5 Stage and section performance

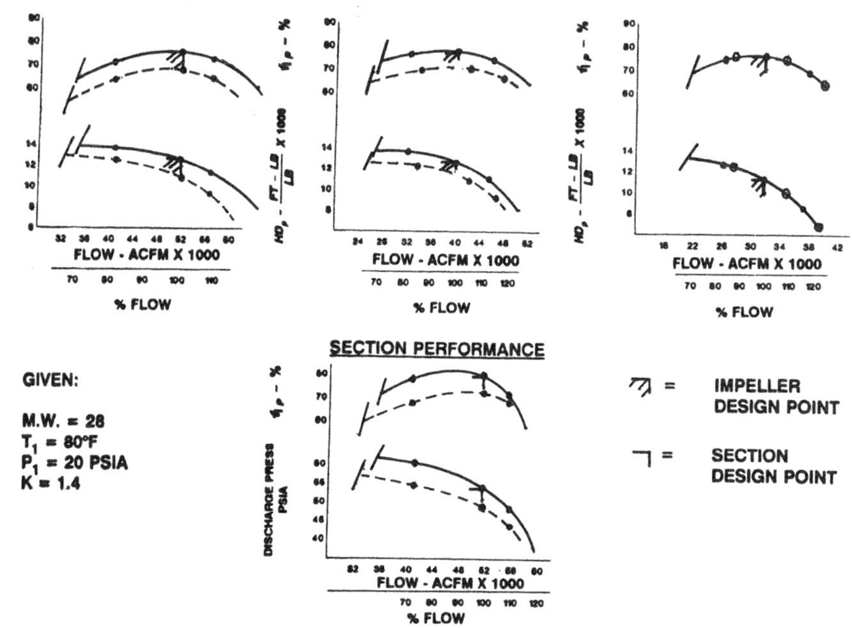

Figure 9.6 Stage and section performance (1st and 2nd impellers fouled)

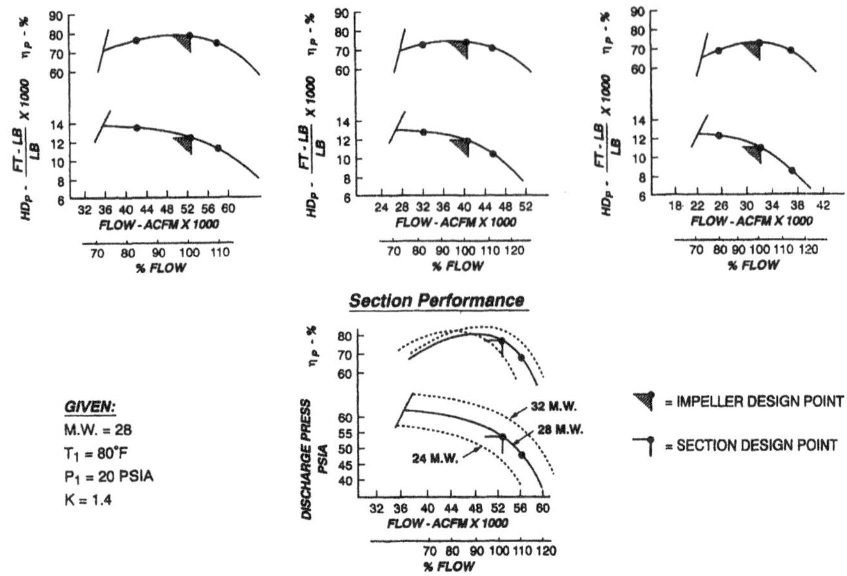

Figure 9.7 Stage and section performance (M.W. varies from 24–32)

the closer the surge point to the design point. This is because once any dynamic compressor stage is designed, the head (energy) produced for a given flow and speed is fixed. The greater the density of the gas (proportional to molecular weight), the higher the pressure produced and the lower the volume flow. Flow is inversely proportional to pressure (Boyles Law). Since surge is caused by low flow, a dynamic compressor handling a denser gas will surge sooner.

In the case of reduced gas density, the opposite effect will occur. The discharge pressure will be reduced and the surge point will move to the left, farther from the design point.

This concludes the chapter concerning individual stage and overall performance. One final comment. Individual stage performance curves are vendor propriety information and are not reproduced. The only opportunity that an end user has to review these curves are:

- Prior to an order
- During a design audit

Readers are encouraged to review this information during the bid phase of a contract to determine if all individual impeller operating points when operating at the guaranteed point (rated) are as close as possible to the individual impeller best efficiency points.

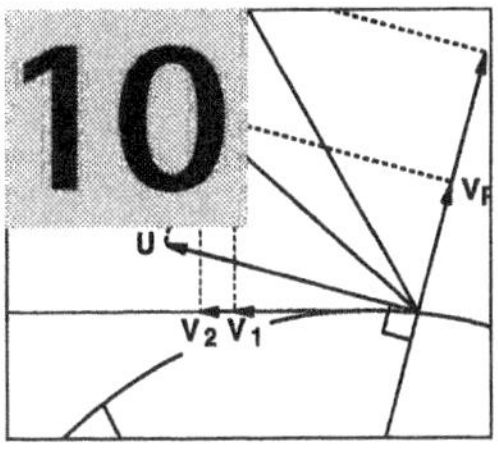

The effect of a gas density change

- Introduction
- The factors involved
- The effect on turbo-compressor pressure ratio
- The effect on the compressor head
- The effect on system resistance
- The effect on turbo-compressor flow rate
- The effect on power

Introduction

In this chapter we will discuss the effect of a gas density change on dynamic compressor performance.

The factors that define a given fluid are presented and we can readily observe that there are a significantly larger number of factors necessary to define a vapor since it is a compressible fluid. Referring again to the fluid head equations for a liquid and a vapor, one can see that the pressure ratio developed for a gas is dependent on temperature, molecular weight, the specific heat ratio of the gases and compressibility. The effect of a gas composition or temperature change on turbo-compressor compression ratio is shown and depicted in tabular form. By solving the head equation for pressure ratio, it can readily be seen that pressure ratio will increase with increasing molecular weight and decrease with decreasing molecular weight if the inlet gas temperature is held constant. Conversely, if the molecular weight is held constant, the pressure ratio will increase with decreasing gas inlet temperature and decrease with increasing gas inlet temperature. The effect of a gas composition change on head is assumed to be such that once an impeller is designed, it will remain constant. This is not entirely true for an impeller working on a vapor for the following reasons:

1. Head is generated by impeller tip speed and exit gas velocity relative to the blade.
2. Gas composition and inlet gas temperature changes affect the compression ratio.
3. Volume flow rate varies with pressure, temperature and compressibility.
4. Since the impeller exit area is fixed, a change in volume at the exit will produce a change in absolute gas velocity and gas tangential velocity. Therefore, head does change slightly.

NOTE: For molecular weight changes on the order of 20%, it is common to assume head is constant for a given flow and speed.

The turbo-compressor impeller produced head changes slightly and also the curve shape will change as a result of the changing velocity relative to the blade. A summary table is presented which shows the change of head and the limits of the compressor curve (surge point and choke point) for condition changes of molecular weight and temperature. Also shown is the change of velocity relative to the blade at the exit. One can see that an increasing molecular weight will result in increasing head and a smaller flow range while a decreasing molecular weight will decrease head and produce a greater flow range. On the other hand, increasing temperature will decrease head and increase the flow range. Whereas decreasing temperature will increase head and reduce the flow range. This fact can be explained when we observe the discharge velocity triangle of any compressor stage. Increasing fluid density will result in higher compression through the impeller or blade row. This will result in a reduced exit volume for a given volume of inlet flow. A reduced exit volume flow acting on the same vane area will result in a reduced velocity relative to the blade. A reduced velocity relative to the blade will produce a higher tangential velocity and a correspondingly higher produced head. This same exercise can be presented for a reduction of molecular weight and for temperature changes as well.

It should be mentioned that the effect of molecular weight does have a slight effect on system resistance and this is covered in this section. Also, gas composition will produce a change of the actual volume flow rate since different compression ratios will be produced and as a result, an effect on power will be experienced since both head, mass flow and efficiency will change. Finally, a multistage compressor example is shown to summarize the effect of molecular weight change on multistage compressors.

The factors involved

The parameters necessary to define a given fluid are presented in Figure 10.1. Note that only two parameters are necessary to define a fluid in the liquid state since it is incompressible. On the other hand, three times that number are required to define that fluid in its vapor state since the vapor is compressible.

What factors define a given fluid

Liquid (incompressible)	**Gas (compressible)**
Specific Gravity (S.G.)	Molecular Weight (M.W.)
Viscosity (υ)	Specific Heat Ratio (K)
	Compressibility (Z)
	Pressure (P-PSIA)
	Temperature (T - °R)
	Viscosity (υ)

Figure 10.1 What factors define a given fluid

Figure 10.2 shows the relationships used to determine the head (energy) required to increase the pressure of a fluid in it's liquid and vapor state. Note how much the density of the fluid influences the amount of energy required to meet a certain process requirement. When one considers that the additional amount of head produced as a centrifugal compressor's flow rate decreases from rated point to surge point is on the order of only 10%, it can be seen that a small change in gas density can result in a significant flow reduction and possibly compressor surge.

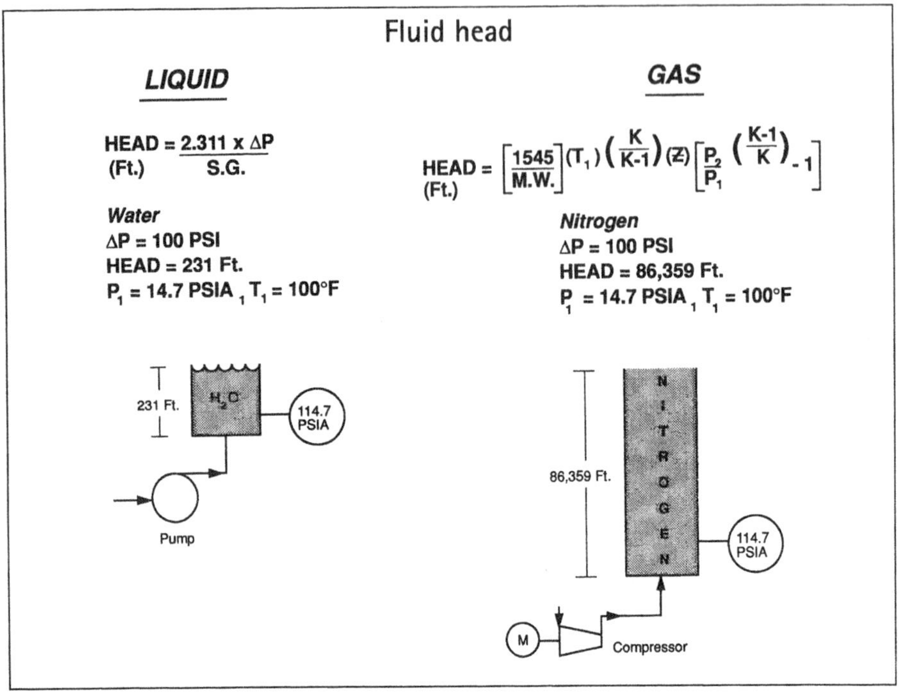

Figure 10.2 Fluid head

The effect on turbo-compressor pressure ratio

The pressure ratio produced by a dynamic compressor is affected by gas density. Figure 10.3 shows that for a given compressor flow and speed the head produced by a dynamic compressor is essentially constant. Therefore, any change in MW, T, K or Z will change the pressure ratio produced. This information is presented in tabular form for changes in molecular weight and inlet gas temperature.

The effect on the compressor head

It is commonly thought that dynamic compressor head produced is always constant for a given flow rate and speed. Figure 10.4 presents this fact for the same compressor operating on different gases (O_2 and N_2).

This statement is not true for a fluid in the vapor state since head in a dynamic compressor is produced by blade velocity and gas velocity. Gas velocity will change will change with gas density since a gas is compressible. These facts are presented in Figure 10.5.

THE EFFECT OF A GAS COMPOSITION AND TEMPERATURE CHANGE ON THE TURBO-COMPRESSOR PRESSURE RATIO

$HEAD_{ISENTROPIC}$ IS RELATED TO PRESSURE RATIO BY:

$$HD_{ISEN} = \left(\frac{1545}{M.W}\right)(T_1)\left(\frac{K}{K-1}\right)(Z)\left[\left(\frac{P_2}{P_1}\right)^{\frac{K-1}{K}}-1\right]$$

ASSUMING HD_{ISEN} IS CONSTANT FOR A GIVEN FLOW,

$$\frac{P_2}{P_1} = \left(1+\frac{(HD_{ISEN})\,(M.W)}{(1545)\,(T_1)\left(\frac{K}{K-1}\right)(Z)}\right)^{\frac{K}{K-1}}$$

THEREFORE THE FOLLOWING TABLE CAN BE DEVELOPED:

EFFECT OF GAS AND T CHANGES ON PRESS. RATIO		
MOLECULAR WGT.	INLET TEMP.	PRESSURE RATIO
INCREASES	CONSTANT	INCREASES
DECREASES	CONSTANT	DECREASES
CONSTANT	INCREASES	DECREASES
CONSTANT	DECREASES	INCREASES

Figure 10.3 The effect of a gas composition and temperature change on the turbo-compressor pressure ratio

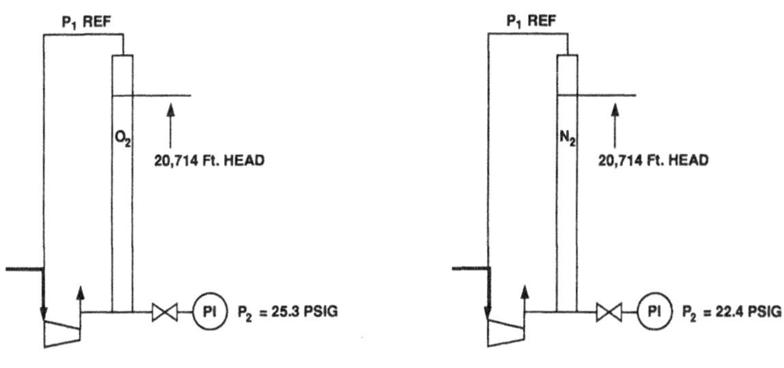

Oxygen Gas

Design Case
Inlet Pressure = 5.3 PSIG
Inlet Temperature = 100°F
Molecular Weight = 32
K = 1.401

Nitrogen Gas

Inlet Pressure = 5.3 PSIG
Inlet Temperature = 100°F
Molecular Weight = 28
K = 1.402

Figure 10.4 The effect of gas composition change on HEAD

The effect of a gas composition and temperature change on turbo-compressor head

The assumption that compressor head remains constant for a given flow with gas composition and temperature changes is not true because:

- Head is generated by impeller tip speed and exit velocity relative to the blade
- Gas composition and temperature changes affect the compression ratio
- Volume flow rate changes with pressure, temperature and compressibility
- Since the impeller exit area is fixed, a change in exit volume rate will produce a change in velocity

Note: For changes on the order of 20%, it is common practice to assume head is constant for a given flow and speed

Figure 10.5 The effect of a gas composition and temperature change on turbo-compressor head

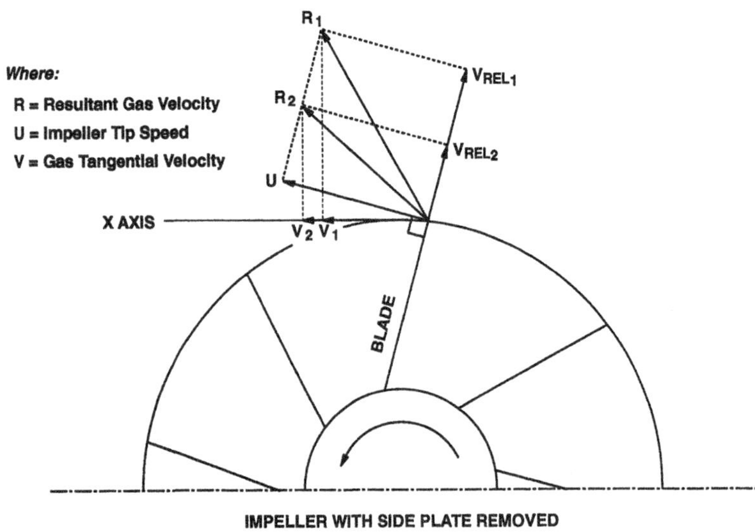

Figure 10.6 Head produced ∝ $(U)(V_T)$

Please refer to Figure 10.6 which shows the relationship between gas velocity (V_{rel}) blade tip speed (U) and tangential gas velocity in a centrifugal compressor.

CONDITION	VEL_{REL} EXIT	HEAD	SURGE POINT	CHOKE POINT
M.W. INCREASES	DECREASES	INCREASES	+ FLOW	- FLOW
M.W. DECREASES	INCREASES	DECREASES	- FLOW	+ FLOW
T_1 INCREASES	INCREASES	DECREASES	- FLOW	+ FLOW
T_1 DECREASES	DECREASES	INCREASES	+ FLOW	- FLOW

ABOVE CHANGES ASSUME A CONSTANT INLET ACTUAL VOLUME FLOW RATE

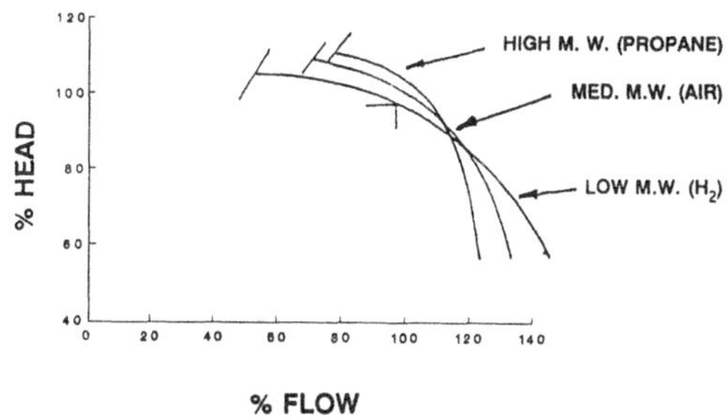

Figure 10.7 Turbo-compressor impeller head change and curve shape summary

Since the head produced by any dynamic impeller is proportional to blade tip speed and gas tangential velocity, reduced gas velocity through the impeller (V_{rel}) will increase the head produced as shown in Figure 10.6. This is the result of increased gas tangential velocity for a given impeller diameter and speed. As shown in Figure 10.5, gas velocity (V_{rel}) will vary with gas density.

Figure 10.7 presents the effect of gas density changes on impeller produced head, surge point and choke point. It can be seen that curve shape is influenced by gas density changes. Therefore, a low density gas will always have a greater flow range than a high density gas.

The effect on system resistance

Figure 10.8 presents the effect of gas density change on the system resistance curve. A slight change in the friction drop in pipes, fittings and vessels results from a change in gas density.

The effect on turbo-compressor flow rate

The effect of gas density changes on actual mass and standard flow rates is shown in Figure 10.9. Note that gas density changes will change the

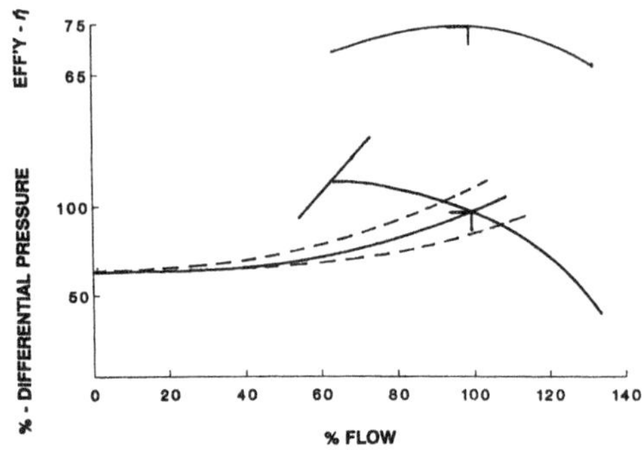

- FOR A GIVEN DIAMETER AND LENGTH OF PIPE, SYSTEM PRESSURE DROP IS A FUNCTION OF:

$$\sqrt{(DENSITY)X(PRESSURE)}$$

AND VELOCITY. THEREFORE, A GIVEN SYSTEM RESISTANCE CURVE WILL CHANGE WITH GAS COMPOSITION, PRESSURE, TEMPERATURE AND VELOCITY.

Figure 10.8 The effect of process changes on the system resistance curve

operating point of each compressor stage in a multistage compressor. Depending on the impeller selection, this change could have an adverse affect on the operation of a dynamic compressor causing surge and corresponding high vibration, temperature, flow changes, etc.)

The effect of gas composition on turbo-compressor flow rate

- The actual volume flow rate will vary as a result of the operating point change which is the intersection of the turbo-compressor curve (pressure vs. flow) and the system resistance
- The mass flow rate (lbs/min) will be the product of the new actual volume flow rate (ft^3/min) and the gas density (lb/ft^3) at the new gas conditions (M.W., P, T, Z)
- The standard volume flow rate (SCFM) will be the product of the new actual volume flow rate (ft^3/min) at its pressure temperature corrected for standard conditions (14.7 PSIA and 60°F)

Figure 10.9 The effect of gas composition on turbo-compressor flow rate

The effect on power

As shown in Figure 10.10, dynamic compressor required power increases directly with gas density up to the choke flow or stonewall region of the performance curve. In the choke flow region, the head produced by the compressor approaches zero since the gas velocity is equal to its sonic velocity.

The effect on power

$$\text{Power (BHP)} = \frac{\text{Head}\left(\frac{\text{Ft-Lb}}{\text{Lb}}\right) \times \text{Mass flow}\left(\frac{\text{Lb}}{\text{Min}}\right)}{33{,}000\left(\frac{\text{Ft-Lb}}{\text{Min-H.P.}}\right) \times \eta\ (\%)} + \text{Mech. losses (BHP)}$$

Figure 10.10 The effect on power

Figure 10.11 shows the affect on compressor section performance resulting from a change in the gas molecular weight. As previously discussed molecular weight changes can result in compressor stage mismatching which can cause significant mechanical damage to the compressor train.

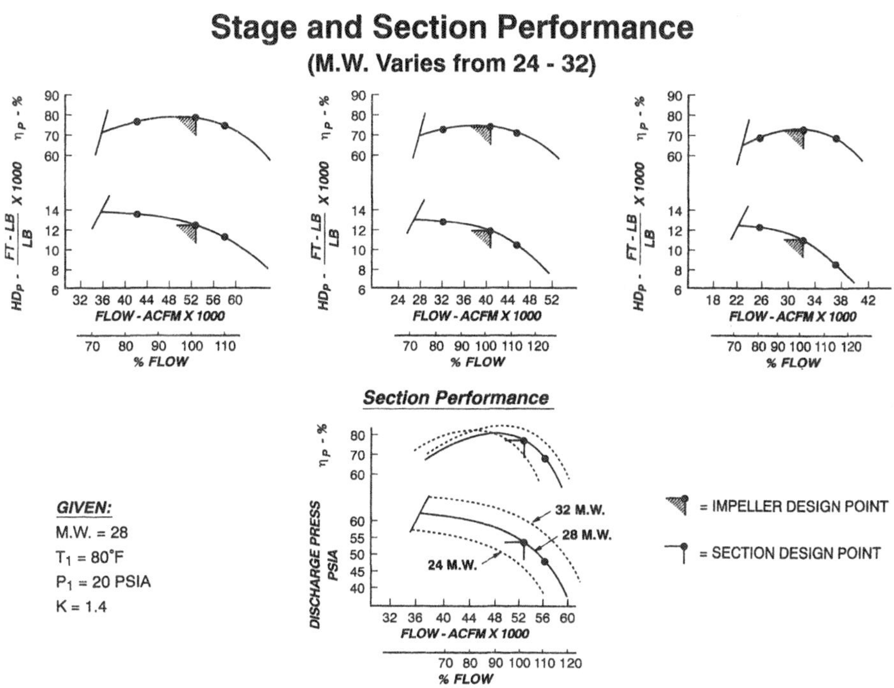

Figure 10.11 Stage and section performance

11 The effect of fouling

- Introduction
- The mechanism of fouling
- The effect of fouling on the operating point
- The causes of fouling
- Detecting fouling by condition monitoring
- Preventing and correcting fouling
- Corrective measures

Introduction

Thus far, we have investigated the cause of the compressor curve. That is, the method in which energy is produced in any blade row or impeller. Energy is produced by the change in two velocities, namely the difference between inlet and discharge velocity relative to the blade and the change between inlet blade tip speed and discharge blade tip speed. In this section we will explore what will happen to the energy production of a blade row or impeller if that blade row or impeller were to change in area. Therefore, we'll explain the mechanism of impeller fouling with examples of the effect of fouling at the operating point and discuss measures to prevent and correct fouling.

The mechanism of fouling

As mentioned earlier, one can reduce any blade row or impeller to a series of equivalent orifices. Flow is a function of area and velocity.

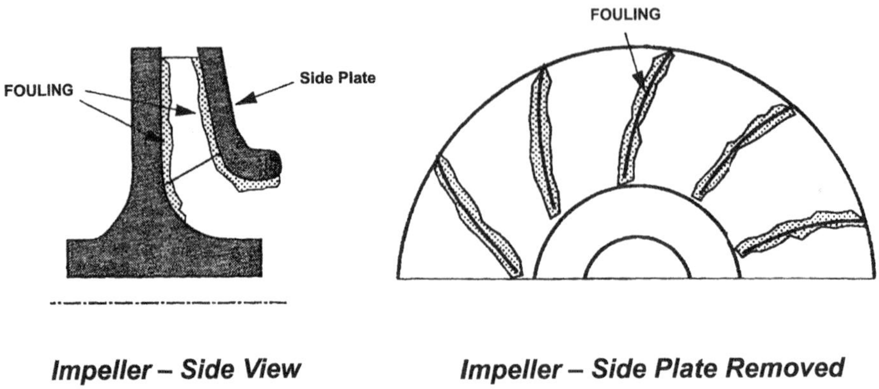

Figure 11.1 Fouling – the effect on the operating point. Left: impeller – side view. Right: impeller – side plate removed

Whenever any blade row or impeller is designed, the designer sets inlet and discharge blade areas such that optimum velocities relative to the blade will be achieved at teach location. By a series of tests and experience, designers have defined optimum relative velocity rather well. Therefore the resulting inlet and discharge areas will produce optimum velocities and corresponding optimum impeller efficiencies. If however, the areas were to change, and flow passages were to become rough and non-continuous, an impeller performance change would result. Figure 11.1 shows the effect of fouling on a closed centrifugal impeller.

Impeller fouling is defined as the accumulation of debris in the impeller or blade passage that reduces the flow area and roughens the surface finish. The distribution of the foulant on the impeller or blade row is non-uniform and usually changes with time. Flow patterns within the impeller or blade cause unequal distribution. In addition, the forces exerted on the foulant cause it to chip off with time as it becomes dry and brittle. This results in a change in rotor balance and a change in performance (head and efficiency).

The effect of fouling on the operating point

If we refer back to the previous example of a backward leaning centrifugal compressor impeller, the effect of fouling can be understood. Figure 11.2 shows the effect of fouling on the relative velocity.

Since the area of the flow passage is reduced when the impeller is fouled, V_{REL} will increase, the flow angle $\propto$ will increase and therefore result in an absolute velocity (increased R) as shown in Figure 11.3.

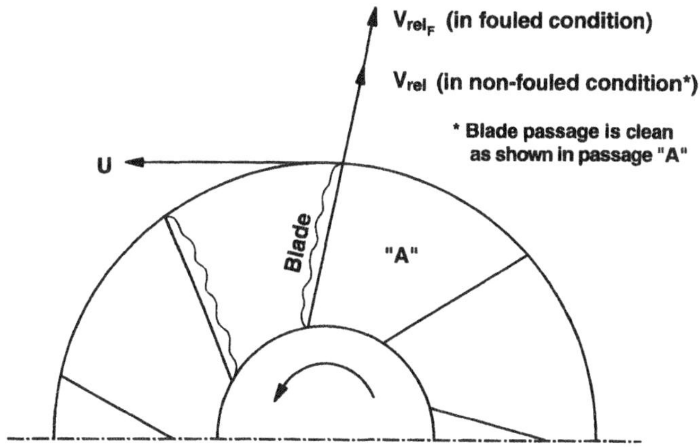

Figure 11.2 Impeller with side plate removed

The increase in ∝ and R due to fouling will reduce the tangential velocity of the gas as shown in Figure 11.4.

Since the head (energy) produced by the impeller is the product of the impeller tip speed "U", which does not change in the fouled condition, and the tangential velocity which is reduced, the head produced will be reduced in the fouled condition. In addition, the non-uniform distribution of the foulant will reduce the efficiency of the impeller stage.

Figure 11.5 shows the effect of fouling on the impeller stage curve.

Impeller fouling is the accumulation of material in the impeller passages that reduces flow area and roughens surface finish. It reduces impeller head capacity and efficiency.

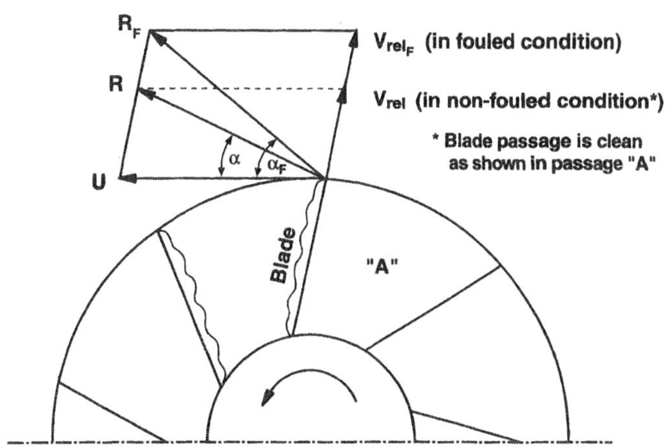

Figure 11.3 Impeller with side plate removed

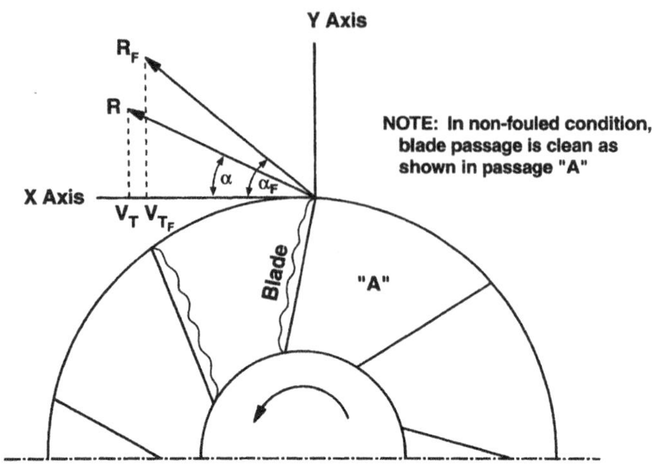

Figure 11.4 Impeller with side plate removed

Note that the surge margin actually increases slightly in the fouled condition. This is because the cause of surge is low gas velocity. Since the area of the flow passage is reduced, the gas velocity increases thus increasing the surge margin. The surge margin is defined as the flow at surge divided by the impeller design flow. However, the stage head produced by the impeller at any flow rate is reduced. Therefore, for the same process system head required, the impeller flow rate will be reduced thus forcing the operating point closer to the surge line.

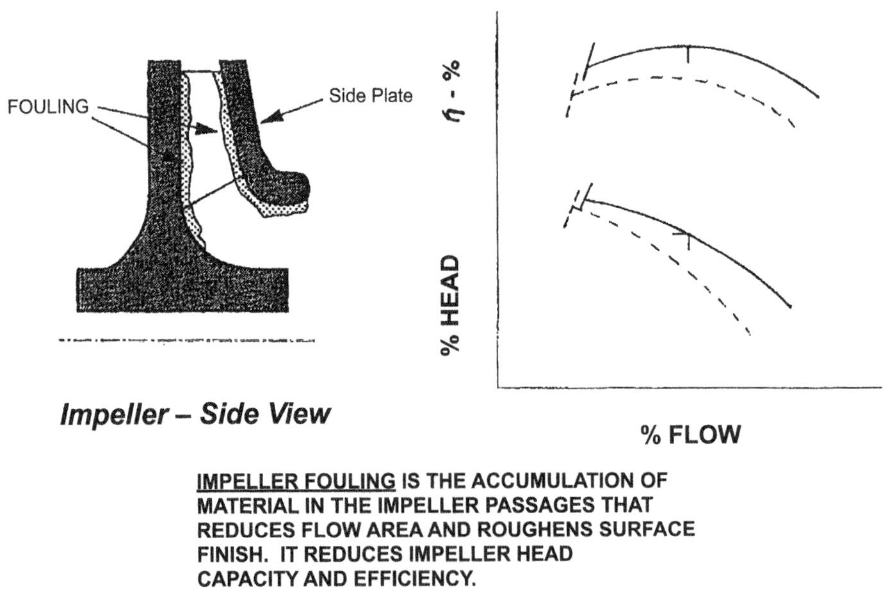

Figure 11.5 Impeller fouling

The causes of fouling

The causes of fouling are described in Figure 11.6.

The causes of fouling

- Air borne debris – air service
- Suction separator carry-over
- Process gas characteristics
  - Temperature sensitive
  - Pressure sensitive

Figure 11.6 The causes of fouling

Fouling is difficult to predict. In air services, it is a function of the environment which can change with time. Fouling of an air compressor or the air compressor section of a gas turbine can suddenly occur. Causes could be:

1. A new industrial plant in the area
2. A new furnace stack in the area
3. Change in prevailing wind direction
4. Land erosion caused by de-forestation

In process applications, fouling can also occur unannounced. One example that comes to mind is a reformer recycle compressor upgrade in which I participated. A new 30% larger recycle compressor was installed and immediately after start-up the entire compressor was fouled with ammonium chloride. The previous recycle compressor had not fouled in 10 years of operation! Figure 11.7 shows the removed rotor after only one week of operation.

Figure 11.7

Figure 11.8 shows a close-up of the fouled impeller looking in the discharge passage.

What was the root cause? After an exhaustive troubleshooting period, it was discovered that we forgot to consider the suction side of the process system in our upgrade of the reformer unit. It seems that a 30% increase in suction line gas velocity and an undersized, poorly designed suction drum demister pad resulted in an entrainment velocity sufficient in power to carry over ammonium chloride into the compressor. Redesign of the demister pad solved the problem.

In other processes, a slight change in operating conditions can cause polymers to be formed that will adhere firmly to the impeller passages. Polymerization can occur in Ethylene plant cracked gas compressors, wet gas compressors and linear low density polyethylene recycle compressors. In these services, the temperature of the process gas within the compressor must be kept below the critical temperature that will initiate the formation of polymers. This temperature varies with the process but generally should be below approximately 235°F to avoid problems. As we have learned in previous chapters of this book, process

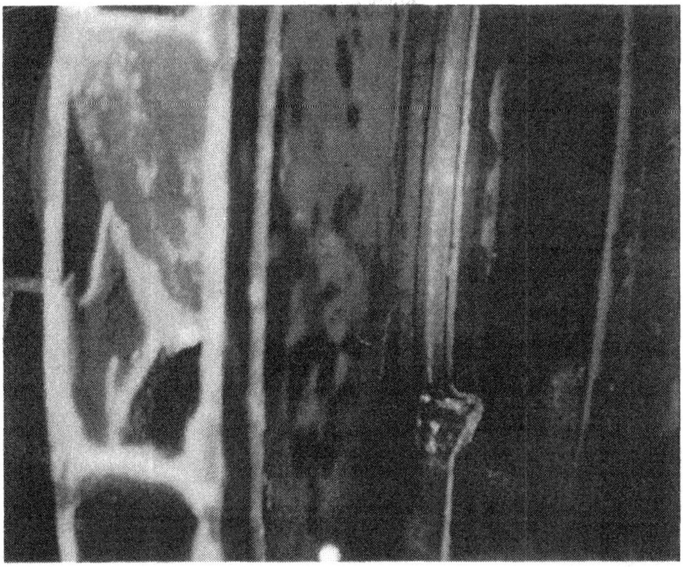

Figure 11.8 Close-up of the fouled impeller

gas temperature change within any compressor is a function of compressor efficiency. Therefore, compressor stage efficiency deterioration due to excessive stage seal clearances, impeller erosion, axial misalignment etc. can initiate fouling in these processes. Most centrifugal and all axial compressor have more than one stage. Fouling can occur in all stages, only the first stage or in any combination. The overall effect of fouling on compressor performance will be the accumulative effect of deterioration in each stage. Figure 11.9 shows an example of fouling in a three stage centrifugal compressor.

Therefore, periodic condition monitoring of compressor performance can provide advance warning to correct compressor internal deficiencies at the first opportunity.

Detecting fouling by condition monitoring

As previously discussed, the mechanism of fouling involves non-uniform accumulation and removal of debris from the impeller or blade row which results in a performance design change in the mass accumulation of the foulant on the impeller. This action will change the rotor balance and the phase angle (the heavy spot) of the unbalance. The suggested condition monitoring practice to detect and confirm fouling is presented in Figure 11.10.

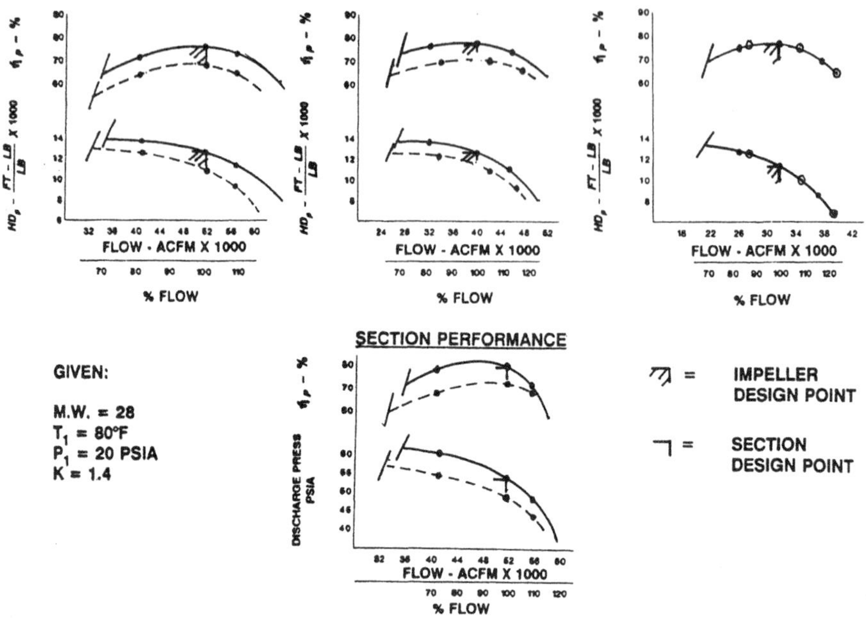

Figure 11.9 Stage and section performance (1st and 2nd impellers fouled)

Condition monitoring guidelines to determine fouling

- Trend radial vibration and phase angle - a random change in both parameters indicates fouling
- Trend compressor head and efficiency. Plot results on compressor test curve - an increasing deficiency margin confirms fouling

Figure 11.10 Condition monitoring guidelines to determine fouling

Preventing and correcting fouling

Thus far we have seen how fouling can occur and how it can cause surge. The detrimental effects of surge have been discussed in a previous module. How then can we prevent fouling if possible? And how can we correct fouling if and when it occurs? Can we correct fouling on-line? These are important questions that must be answered. Figure 11.11 presents solutions for preventing and correcting fouling.

Preventing and correcting fouling

- Preventive measures
  - Process control
  - Proper inlet air filter selection and maintenance
  - Condition monitoring (gas pressure, temperature and vibration)
  - On-line solvent injection
  - Impeller or blade coatings
- Corrective measures
  - Slow roll or stationary wash
  - On-line random washing

Figure 11.11 Preventing and correcting fouling

Preventing fouling

Listed below are proven methods for preventing fouling.

1. **Process control** – Accurate control of process conditions can prevent fouling in applications where polymers can be formed. Control of temperature is usually the most important. The following applications can be affected by excessive process temperature:

   - Ethylene cracked gas
   - Linear low density polyethylene
   - High density propylene
   - Fluid catalytic cracker off gas (wet gas)
   - Thermal catalytic cracker off gas (wet gas)
   - Coker gas

   The temperature below which fouling can be prevented varies with each process, compressor and application. Monitoring of process conditions is necessary to establish a threshold temperature in each case. In some cases, fouling cannot be prevented with the existing compressor. It may be necessary to modify the aerodynamic design and/or add additional cooling. Note that water injection is used quite regularly in ethylene unit cracked gas compressors to reduce gas temperature. However, an industry best practice is to keep the amount of injection water below 3% of the inlet mass flow.

2. **Proper inlet filter selection and maintenance** – In air compressor and gas turbine applications, filter selection is an important factor in preventing fouling. Site conditions must be accurately monitored

prior to installation. It is advisable to contact filter vendors to survey site conditions. Filters must be able to prevent debris and excessive moisture from entering the air compressor. In platform applications and installation close to water, special designs must be used to prevent excessive chloride (salt) build-up. It is the writer's experience that multi-stage air filtration must be considered. Fouled air compressor sections of gas turbines can reduce output power by 3–5%.

Another important consideration is filter maintainability. For crucial (unspared) services, filters with on-line cleaning capability should be considered. In recent years pulse type ("Huff and Puff") filters that use a reverse air pulse to clean filter cartridges when filter pressure drop becomes excessive have become popular.

Regardless of the type of air filtration used, or not used, remember environmental conditions can change as previously discussed in this module.

3. **Condition monitoring of compressor parameters** – Regardless of the service, condition monitoring of molecular weight, pressure and temperature will provide valuable information regarding process condition changes that can cause or increase fouling.

4. **On-line solvent injection** – Certain process applications have had success with on-line solvent injection. The objective of this measure is to *continuously* inject a small amount of solvent to reduce the friction coefficient of the blade and impeller surface and thus prevent the fouling from accumulating on the surface. The object of solvent injection is often misunderstood. The idea is to prevent foulant accumulation, not clean the impeller or blade on line. Non-continuous solvent injection will allow the impeller or blade surface to dry and thus promote fouling. Even if a unit injection is continuous, selection and location of injection spray nozzles is a critical factor in success. Insertion of spray nozzle(s) in the inlet pipe of a compressor section is usually not as effective as injection directly into each stage of the compressor. Knowledge of solvent vapor pressure and internal compressor temperatures is necessary to determine if section or stage solvent injection is required.

   Type of spray nozzle to be used can best be determined by discussions with process designers or other users having similar process applications.

   There is no question about it, solvent injection is expensive! Most solvents used are naphtha based. Many users have required flow meters at each injection point to determine the proper amount of injection. Typically, the total amount of solvent injection in one compressor case should be 1–2% of mass flow. In the writer's opinion, the total amount should never exceed 3% of weight flow. Excessive solvent injection will erode leading edge blade tips at the blade to cover junction on closed impellers and can cause impeller failure if not detected in time.

5. **Impeller or blade coatings** – In the last few years, some users have investigated and have used impeller and blade row coatings to reduce friction in lieu of continuous solvent injection. Based on the writer's experience this has met with mixed results. There has been success with coating impellers with Teflon based material in ethylene cracked gas service. It is recommended that the process designer and compressor vendor be contacted for additional information in this regard.

Corrective measures

If fouling cannot be immediately prevented, various off-line and on-line cleaning procedures are available. Off-line cleaning methods are definitely preferred to on-line cleaning methods since they do not expose the compressor to vibration excursions and have proven to be more effective if a compressor is severely fouled. Failure to immediately remove all foulant from the rotor can cause catastrophic failure since a large amount of unbalance can be instantaneously introduced in the rotor system. General guidelines for an off-line or "crank" washing procedure are presented in Figure 11.12.

Off-line cleaning guidelines

1. Determine proper solvent by solubility test of foulant material.
2. Confirm solvent does not harm compressor internals (aluminum or Teflon based labyrinth).
3. Fill and vent case to be sure it is completely filled with solvent. Note: use warm solvent (not to exceed 250°) if possible. Be sure to confirm solvent is acceptable for use at this temperature.
4. Use turning gear or small belt drive assembly to turn case at 40–60 rpm during wash.
5. Measure conductivity of solvent solution prior to start of wash.
6. Change wash fluid frequently and continue wash until conductivity of wash liquid is equal to conductivity of initial value measured in step 5 above.

Figure 11.12. Off-line cleaning guidelines

If an on-line wash is to be used, it should be employed periodically – usually at least once a week and begun when the rotor is in new (clean) condition. As described previously, the injected amount of wash fluid should not exceed 3% of mass flow. Process and compressor designers should be consulted regarding the exact procedure to be used. It is advisable that a radial vibration trip system be employed if on-line washing procedure is to be implemented. This action will prevent catastrophic compressor failure in the event of an incomplete on-line wash.

We have included some examples of fouling along with preventive measures and cleaning methods in Figure 11.13.

Examples of fouling

Service	Preventive action	Cleaning method
Air	Good inlet filtration	Water wash catalyst or nut shell
Cracked gas	Process gas temperature control/continuous injection	Solvent wash
FCC wet gas		
TCC wet gas		
Coker gas	Continuous injection	Solvent wash
Polyethylene	Control process conditions and possible impeller coating	Stationary impeller Heating (off-line)
Reformer recycle	Prevent carry-over	Steam/water wash

Figure 11.13 Examples of fouling

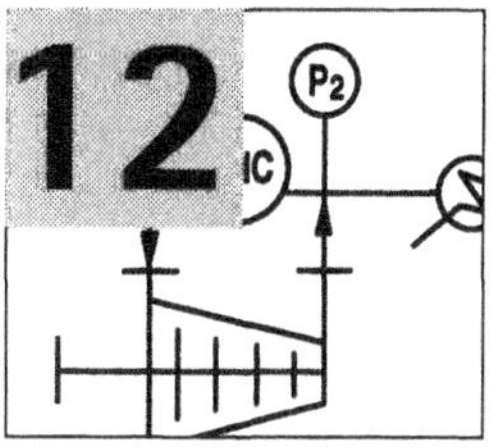

12 Turbo-compressor control overview

- Introduction
- The objectives
- Adjusting to system changes
- Adjusting the turbo-compressor performance curve
- Turbo-compressor protection

Introduction

Thus far we have discussed turbo-compressor operation in the uncontrolled state. That is, without any automatic process control or protection in the system. Having discussed those factors involved in turbo-compressor performance, design and operation, we will now overview both turbo-compressor control and protection.

As previously discussed, the objectives of the end user in any application are to purchase and operate a turbo-compressor with maximum reliability, product throughput and at a minimum operating cost (maximum efficiency).

To achieve the user objectives, it becomes necessary to adjust the turbo-compressor throughput to certain values, since we have seen that the throughput will vary as a function of process system requirements (head required). There are essentially two options available to the user to achieve this objective:

- Adjust the process system energy requirements (head required)
- Adjust the turbo-compressor energy output (head produced)

Process energy requirements can be changed by either adjusting the suction or the discharge system resistance. The most common type of resistance adjustment is known as discharge throttling. This process results from adjusting the pressure drop of a control valve in the discharge to change the compression ratio of the turbo-compressor. Observing any turbo-compressor curve we can see that for a specified gas composition and inlet temperature there exists one and only one pressure ratio available from the turbo-compressor for a given throughput. This assumes the turbo-compressor speed and the inlet guide vane position (if applicable) are held constant. If we position a throttling valve on the discharge of the compressor system shown we will see that the desired flow throughput can be obtained by throttling the valve. Obviously, energy is lost during this exercise via the throttled pressure drop across the valve.

Another alternative would be to position the throttling valve in the compressor suction process system. We will demonstrate that this method of increasing system resistance is less energy exhaustive and will result in lower operating costs than in the case of discharge throttling. An important fact to remember is that both cases increase the system resistance. In the case of suction throttling, the suction system resistance is increased but this results in a lower suction pressure at the compressor flanges. One fact that is often misunderstood is that the turbo-compressor curve describes the energy increase from the turbo-compressor **inlet to discharge flange** and pressure must be measured at the inlet and outlet of the turbo-compressor to directly assess its operation. Frequently, suction pressure is measured upstream of the suction throttle valve and the operators are not aware of the actual pressure ratio that the compressor experiences. In summary, suction throttling will control the operating point of the turbo-compressor but will expend less energy and reduce operating costs.

The other control alternative available is that of varying the produced energy of the turbo-compressor. It is much more effective to change the turbo-compressor produced energy or curve than to throttle to accommodate operation changes. Available methods of adjusting the turbo-compressor curve are:

- Curve change
- Adjustable guide vane changes

Speed change adjusts compressor energy (head) and throughput by following the fan laws as previously discussed. It can be accomplished with variable speed turbines, variable speed drives (hydraulic clutches) and variable speed motors. This latter type driver is becoming more popular as capital investment costs for variable frequency motors is reducing.

In the case of a constant speed motor driver, adjustable guide vanes can be employed to result in the same turbo-compressor operational flexibility. It is important to note that inlet guide vanes **are not throttle devices** but change the inlet angle of the gas to the impeller. This results in different produced energy values for a given compressor inlet flow. It should be noted that inlet guide vanes in series configuration multi-stage turbo-compressors can be employed only in the first stage. This is due to the fact of the limited accessibility for the necessary guide vane linkages. Integral gear turbo-compressors however, can be designed with inlet as well as discharge adjustable guide vanes in each stage due to the configuration of these compressors.

The turbo-compressor protection systems will be discussed. The basics of surge protection will be presented and the following facts will be reviewed. The overall concept of a surge control system is merely to reduce the system resistance and thus prevent low velocities in the impellers. The concept of the surge control system will be discussed and the important fact of coordinating all components will be covered. All instrumentation, controllers and associated valves must be properly sized to enable the system to work efficiently.

The objectives

The objectives of the end user of rotating equipment to maximize profits are shown in Figure 12.1.

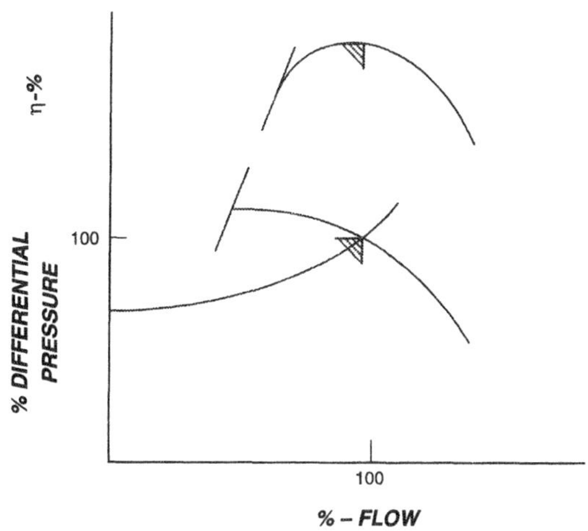

- Maximum reliability (on stream time)
- Maximum product throughput
- Minimum operating costs

Figure 12.1 The objectives

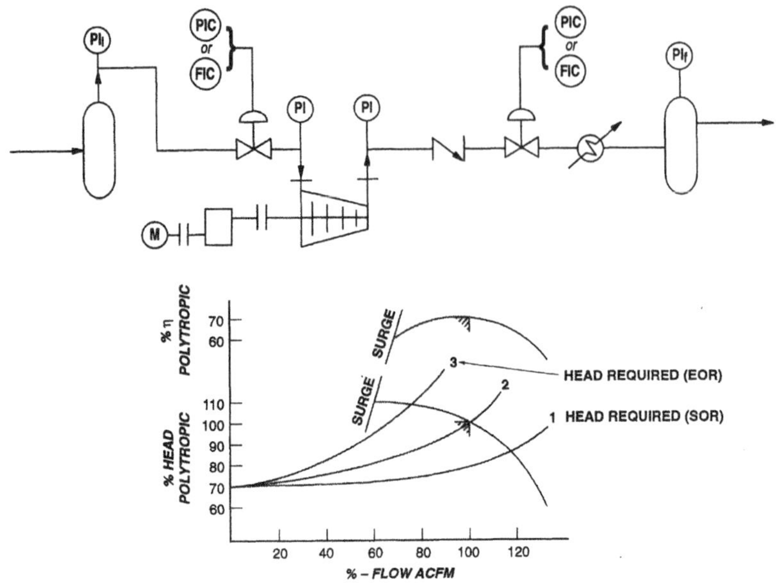

Figure 12.2 Controlling the process system

Automatic control and protection systems will achieve all of the stated objectives by controlling the equipment to operate at maximum throughput at maximum efficiency and protecting the equipment to prevent long term costly downtime.

Adjusting to system changes

A typical process system and performance curve for a constant speed (motor driven) centrifugal compressor are shown in Figure 12.2.

As previously discussed, the performance curve shows the head produced by the centrifugal compressor. In order to operate at the rated point, the head required by the process system must be in equilibrium with the head produced by the centrifugal compressor. This can be accomplished by positioning a control valve in the suction (suction throttling) or discharge process system (discharge throttling). The controlled process variable can be either pressure or flow. In either case, energy is lost since there will always be a pressure drop across the control valve.

Figure 12.3 shows a process system with suction throttling.

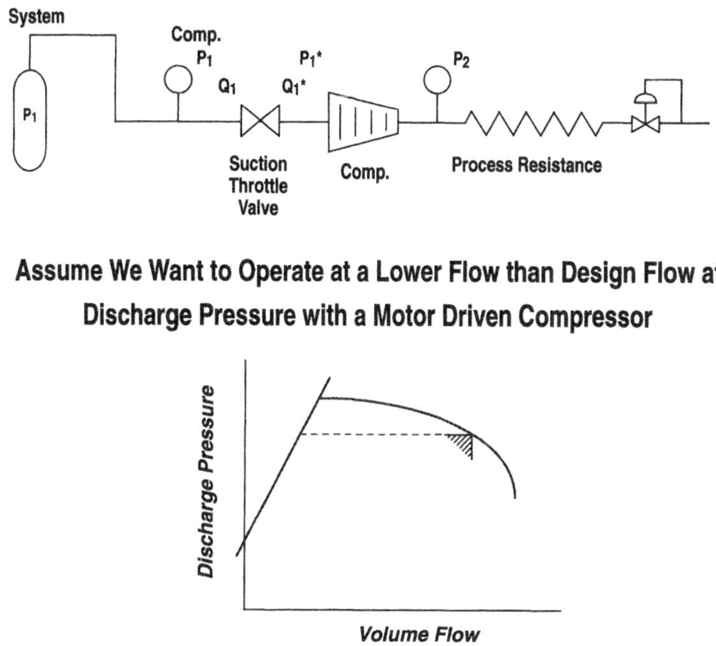

Figure 12.3 A process system with suction throttling

The procedure of suction throttling is frequently misunderstood because the suction pressure gauge is often upstream of the suction throttle valve as shown in Figure 12.3. This does not allow the actual compressor suction pressure to be observed. To operate at a lower flow than design flow as shown in Figure 12.3, the suction throttle valve will be slightly closed. This will reduce the pressure at the compressor suction which will increase the head required by the process since head required is defined as:

$$Hd_{POLYTROPIC} = \frac{1545}{MW} \times T_1 \times \frac{n}{n-1} \times Z \times \left[\left(\frac{P_2}{P_1} \right)^{\frac{n-1}{n}} - 1 \right]$$

In addition, there will be an increase in volume flow into the compressor. This will result in a lower value of head required and horsepower requires than for discharge throttling an equal amount. Usually, suction throttling requires 3–5% less horsepower than discharge throttling for the same reduction in flow rate. The greater the head rise of a performance curve, the greater the savings due to suction throttling over discharge throttling.

Adjusting the turbo-compressor performance curve

Using suction or discharge throttling as previously described will always result in process system energy losses. Figure 12.4 presents the facts concerning adjusting the turbo-compressor curve or head produced.

Adjusting the turbo-compressor curve

It is more efficient to change the turbo-compressor curve than to throttle to accommodate operation changes.

Available methods are:

- Speed change – utilizes fan law
  - Turbines
  - Variable speed drives
  - Variable frequency motors
- Adjustable guide vanes changes inlet gas angle

Figure 12.4 Adjusting the turbo-compressor curve

Since accurate control of the operating point can be attained without the use of throttle valves, this is the most efficient method of process control.

Figure 12.5 shows the same process system as seen in Figure 12.3 but with turbo-compressor speed control.

In this control scheme, there are no pressure control valves. As in Figure 12.3 the process can be flow controlled or pressure controlled in the suction or discharge process system. However, in Figure 12.5 the output of the controllers is run to the driver speed control system. In this manner, the exact amount of energy charge required to meet the process requirements will be applied thus optimizing the efficiency of the process. In this system the turbo-compressor produced energy is varied to meet changing process energy requirements. In Figure 12.3, the process energy requirements were varied by control valves to meet the required values of flow or pressure.

Many compressor applications today are considering the use of variable frequency electric motors (V.F.D.) for this reason. While the initial cost of a V.F.D. can be 2×–3× the cost of a fixed speed motor, the long term power savings can usually be justified.

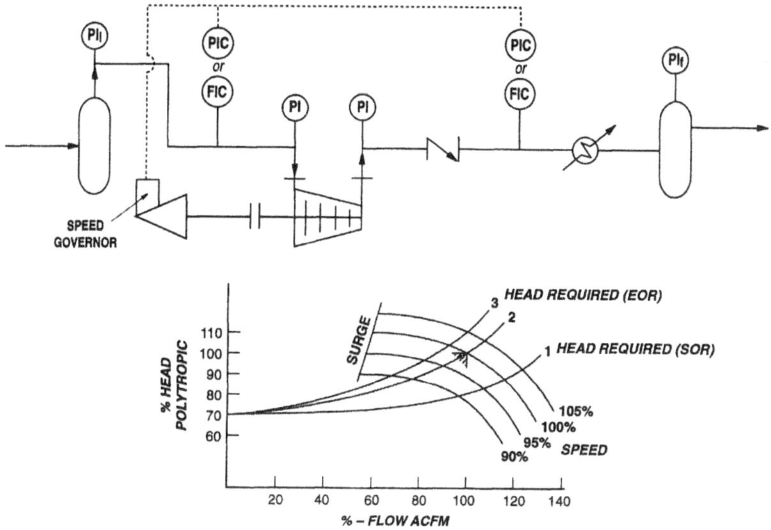

Figure 12.5 Controlling the turbo-compressor curve

The change of flow rate, produced energy (head) and required horsepower with speed are shown by the fan laws in Figure 12.6.

It is important to understand that these relationships are approximate for compressors and become less accurate with increasing number of stages and gas density. The exact values of produced head and required horsepower should always be taken from the compressor curve

$$\frac{Q_F}{Q_I} = \frac{N_F}{N_I}$$

$$\frac{HD_F}{HD_I} = \left(\frac{N_F}{N_I}\right)^2$$

$$\frac{BHP_F}{BHP_I} = \left(\frac{N_F}{N_I}\right)^3$$

Where:

Q = Flow Rate (FT^3/MIN)
N = Speed (RPM)
HD = Gas Head (FT-LB/LB)
BHP = Horsepower

I = Initial
F = Final

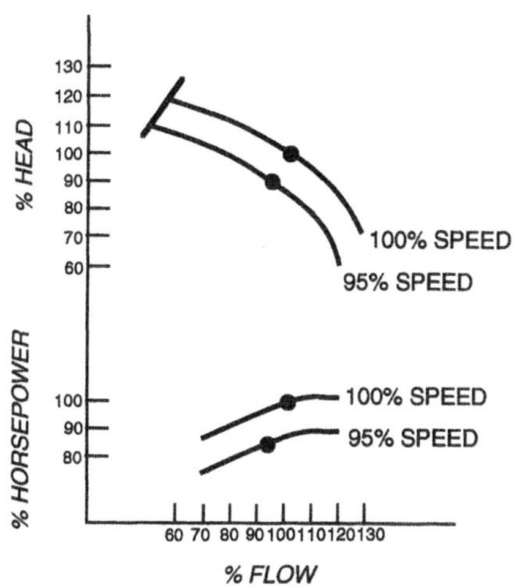

Figure 12.6 The fan laws

provided by the vendor. When designing control systems with variable speed drives, care should be taken to assure the driver and transmission system (couplings, gear, etc.) have sufficient power capability for the maximum speed operating point based on the process system head required curve. This is because a 5% increase in speed and compressor flow will require a 16% increase in driver horsepower.

In Figures 12.5 and 12.6, change of shaft speed provided a means to vary the turbo-compressor characteristic curve and thus meet the system resistance demands exactly without excess energy loss. Essentially, two means of achieving this objective are available to us: variable speed control as previously discussed and inlet guide vane control. Speed control varies the amount of turbo-compressor input energy by changing the impeller tip speed and therefore the flow and head. Adjustable speed prime movers available are:

- Steam turbines
- Gas turbines
- D.C. and variable frequency motors
- Variable speed drive devices

It should be mentioned that variable frequency motors are becoming more popular and will gain in popularity in the near future. However, where the alternative of variable speed is not available, adjustable guide vanes can be used. Adjustable guide vanes vary the turbo-compressor characteristic curve by changing the velocity of the gas through the impeller. This is achieved by changing the inlet angle of the gas and therefore altering the velocity relative to the blades. The higher value of the velocity relative to the blades, the less head (energy) will be produced. This results from reduction in exit tangential velocity due to the higher velocity relative to the blades.

Adjustable guide vanes are limited in their application and are effective only for one stage of compression since they vary the inlet gas flow to the impeller. Axial compressors can accommodate multiple stage inlet guide vane configurations and are frequently used. For multi-stage centrifugal compressors however, adjustable guide vanes are not usually employed since their effectiveness is limited only to the first stage. One exception is in multi-stage integral gear centrifugal compressors. In this design, accessibility to the inlet of each stage for guide vanes is available and they can be employed. Recently, compressor vendors have also added the option of adjustable discharge diffuser vanes to this design. The combined effect of adjustable inlet and adjustable diffuser vanes produces a compressor curve with maximum areas of optimum efficiency.

Before leaving this discussion, it must be mentioned that adjustable guide vanes, due to mechanical complexity, must be maintained in order to minimize guide vane friction and wear which can result in lack of guide vane movement. It is advisable also to consider purges to the guide vane actuator and control arm assemblies to minimize the possibility of excessive assembly friction. Remember, the guide vane assembly is a system and must be treated as such.

In the application of constant speed multi-stage centrifugal compressors, adjustable guide vanes cannot efficiently be utilized. Therefore, for these applications suction throttling is usually used to vary the operating point of the compressor. It must be understood that suction throttling is a system resistance change while adjustable guide vanes do not introduce and appreciable pressure drop into the suction of the compressor. As previously explained they only alter the angle of the inlet gas to the impeller and thus vary the velocity of the gas relative to the blades.

Figure 12.7 presents details concerning the function and operation of inlet guide vanes.

Turbo-compressor protection

The function of any protection system is to prevent damage to the turbo-compressor unit by taking immediate action to remove the cause

Adjustable Inlet Guide Vanes

Inlet guide vanes change the gas velocity into the impeller thus varying the velocity relative to the blades and the head (energy) produced.

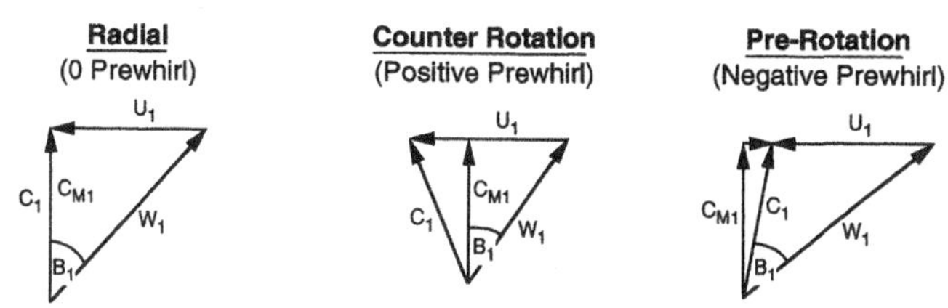

Constant B_1 and U_1 Where:

B_1 = Inlet Blade Angle
W = Inlet Relative Velocity
C_{M1} = Axial Velocity
C_1 = Velocity Off Guide Vane
U_1 = Inlet Blade Tip Speed

Figure 12.7 Adjustable inlet guide vanes

of potential damage. Generally, protection systems fall into the categories shown in Figure 12.8.

Turbo-compressor protection systems

- Overspeed
- Condition (vibration, temperature, pressure, differential pressure)
- Surge

Figure 12.8 Turbo-compressor protection systems

All variable speed driven turbo-compressor trains are protected against excessive overspeed. The industry accepted practice is to trip (shutdown) the unit if the operating speed exceeds 110% of the train maximum continuous speed.

All turbo-compressor trains are protected against operating conditions that can cause damage to any components in the train (rotors, seals, bearings, couplings, etc.). Upon reaching the condition value determined by the equipment vendor, the turbo-compressor train will be immediately shutdown (tripped). The usual condition protection systems are:

- High shaft vibration
- Excessive axial displacement
- Low lube oil pressure
- Low control oil pressure
- Low seal oil differential pressure
- High bearing pad temperature
- High process gas temperature

In present day systems most condition protection systems require a 2 out of 3 vote to permit shutdown of the equipment train. This is to prevent spurious trips of the train resulting from instrument failure. In these systems, all condition protection devices must be supplied in triplicate (3 individual devices required). The one exception is usually shaft vibration since the level of shaft vibration varies around the orbit of the shaft in the bearing. If this approach were used to 'vote' on an excessive orbit, six (6) individual vibration probes would be required.

If any possibility exists that will cause the head required by the process system to exceed the head produced by the turbo-compressor, surge protection must be employed. A surge protection system prevents surge by immediately reducing the head required by the process system. As previously discussed, every turbo-compressor curve has the characteristic of increasing flow throughput for decreasing head required by the process system. The cause of surge is low flow throughput. Therefore, rapidly increasing the turbo-compressors flow throughput will prevent surge.

The fastest way to reduce the head required by any process system is to reduce the discharge pressure (P_2) since,

$$hd_{required} = \frac{1545}{MW} = \times T_1 \times \frac{n}{n-1} \times Z \times \left[\left(\frac{p_2}{p_1}\right)^{\frac{(n-1)}{n}} - 1\right]$$

This is accomplished by rapidly opening (in 1–2 seconds) the bypass or surge control valve. Figure 12.9 shows a typical schematic for a surge protection system.

The surge controller 'C' shown functions as follows:

- The input is the operating point (usually defined by flow and P_2–P_1)
- The set point is the surge control line

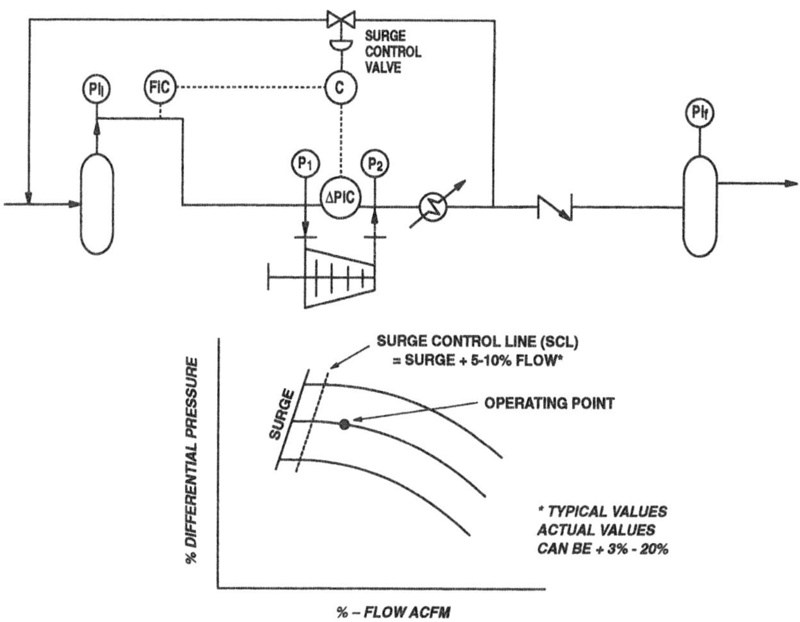

Figure 12.9 Protecting the system

- The output to immediately open the surge control valve is activated if: Operating point ≤ SCL

The surge control system will be examined in detail in the next chapter. At this point it is important to remember 'SYSTEM'. No matter how accurate or fast the controller, the system reliability depends on the reliability of **each** component in the system. A flow meter out of calibration, a plugged pressure tap, a slow surge control valve, etc., will affect the surge control system's responsiveness.

13 Surge control systems

- Introduction
- System objectives
- Available options
- System design considerations
- A history of system types

Introduction

Having examined the cause of surge in dynamic compressors, we will now devote this chapter to the system necessary to protect dynamic compressors against surge and corresponding damage. In this section we will define the objective of the surge control system, the available options to protect compressors, surge system design considerations and a history of surge system types.

System objectives

The surge system objectives are to assure that the gas velocity in each impeller stage in any dynamic compressor exceeds the critical velocity that will cause stall. Refer to Figure 13.1.

The surge system objective

- Assure the relative gas velocity in each impeller stage exceeds the critical velocity that will cause stall

Figure 13.1 The surge system objective

Available options

In order to assure that impeller velocity always exceeds the stated critical velocity, a number of options are available. These options mainly fall into three (3) categories. Refer to Figure 13.2.

Options to achieve the objective

- Shut down unit before critical flow is reached
- Increase the relative gas velocity
  - By impeller speed increase
  - By adjusting inlet gas angle
- Decrease system resistance
  - Process system decrease
  - Direct gas flow through blow off or recycle line

Figure 13.2 Options to achieve the objective

1. **Shut down the unit before the critical flow is reached**
   Naturally, when one considers that most dynamic compressors are in critical service, that is unspared, this option is not possible.
2. **Increase the relative gas velocity**
   This objective can be achieved by either increasing the speed of the impellers or by adjusting the inlet angle of the gas relative to the impeller vanes. Both alternatives result in the increase of turbo-compressor energy and should provide the higher velocity through the impeller. Note however that this alternative may not be effective under certain system resistance characteristics.
3. **Decrease the system resistance**
   This can be achieved by either directly reducing process system

resistance, fully opening a throttle valve or by directing gas flow through a blow off or a recycle line. Of the two alternatives available, the latter is more efficient since it does not affect the energy in the process system.

Given the applications in which turbo-compressors are employed, it can be readily observed that alternative option three (3) is the most cost effective way to protect turbo-compressors against surge and corresponding mechanical damage.

System design considerations

Accepting the fact that reduced system resistance is the most cost effective way to eliminate surge damage, a number of surge system design considerations should be mentioned. These considerations fall into two categories, compressor considerations and process system considerations. The considerations are presented in Figure 13.3.

System design considerations

- Compressor considerations
  - Fast system response
  - Control line automatic adjustment
  - Ability to adjust control line to gas density and operating point changes
  - Recycle stream cooling
  - Check valve location
- Process system considerations
  - Gradual process flow changes
  - Flow reversals
  - Start-up procedure
  - Shut down procedure

Figure 13.3 System design considerations

A history of system types

Figure 13.4 presents a general chronological listing of surge system history.

Surge system history

- 'Surgeless' applications
- Surge alarm systems
- Single parameter systems – flow
- Biased systems – flow and differential pressure
- Modern systems
  - Compensation
  - System back-ups
  - Controller output options

Figure 13.4 Surge system history

In the writer's experience many so called 'surgeless' applications have been observed. These are applications where the process designer assesses the system and concludes that there is no possibility to surge the machine. Many times this assessment is based on system resistance alone. The conclusion being that since there are not any variable resistance points in the process system (coolers, reactors that can foul or control valves) the turbo-compressor cannot surge. If the molecular weight and gas inlet temperature are constant and the turbo-compressor will not be subjected to mechanical damage or fouling, this assessment is true. If any of the above mentioned facts can occur however, the turbo-compressor in question can be forced into surge as a result of the inability of the turbo-compressor to produce the energy required for throughput flow. As previously discussed, the operating point of any compressor is the equilibrium between required process system energy and the compressor produced energy. Both can change, and do.

The second system encountered has been surge alarm type systems. Again, the process designer assesses a particular system and determines that the possibility of surge is minimal. In addition, the application, hopefully, is such that a shut down can be achieved without significant revenue loss. Therefore, the surge system is simply an alarm or shut down system that will stop compressor operation in the event of surge, thus protecting the machine. It should be mentioned, however, that a full assessment should be made to determine the cost impact of such a shut down and the potential problems incurred on the start-up. A properly installed surge system can eliminate these unnecessary shut downs.

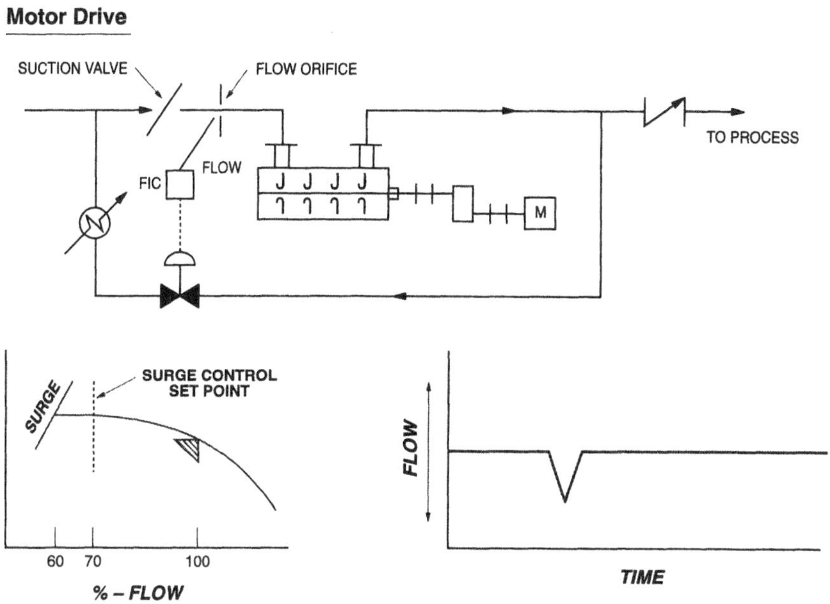

Figure 13.5 A single parameter surge protection system

Before leaving this subject, the reliability of the surge detection system should be mentioned. Frequently, these detection systems are simply differential pressure switches. Depending on the application, surge can be very violent or very light. However, in the case of low pressure applications or low molecular weight applications the effect of surge (flow and pressure pulsations) will be minimal and the sensitivity of the surge sensor device must be investigated to assure that such an event is detected.

Figure 13.5 shows a single parameter surge protection system.

Single parameter surge systems have been employed extensively and very simply consist of a flow measuring device and an anti-surge valve. These systems can work if properly sized and installed. However, it must be understood that a one parameter system will not be economical in that a rather large area of operation as in the cases of suction throttle constant speed devices and variable speed drives will not be allowed. The result will be unnecessary re-circulation and a loss of product revenue. The next system shown in Figure 13.6 is the standard two parameter or biased system that has been employed almost exclusively in the industry over the years.

Turbine Drive

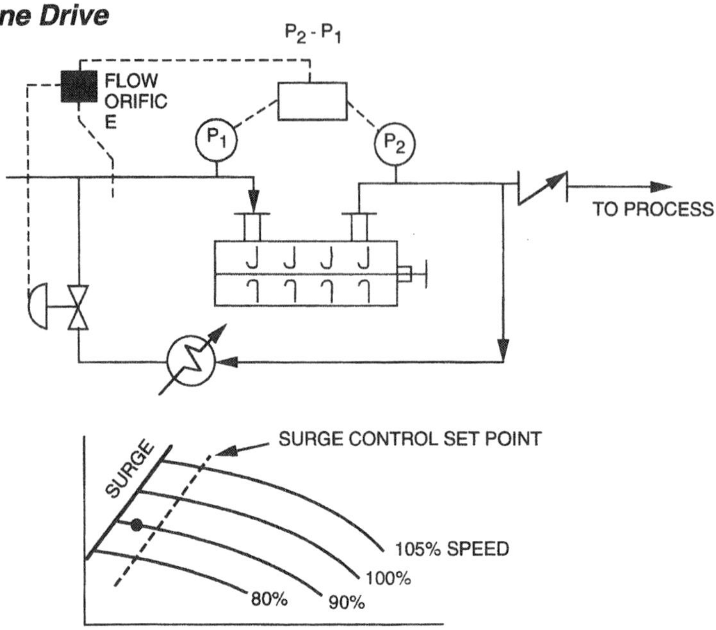

Figure 13.6 The standard two parameter or biased system

This system consists of flow measurement and differential pressure measurement across the compressor and results in a surge control line that roughly parallels the actual surge line of the compressor. It allows operation across a wider range than the single parameter system and if properly sized, can adequately protect the compressor.

The final system in the evolution of surge controls is the modern microprocessor system which affords many options to fully protect the turbo-compressor against surge occurrences. These systems usually employ compressor inlet and discharge temperature inputs in addition to flow and ΔP for each compressor section. Examine the relationship for polytropic head and you will see that this approach yields a more accurate approximation of the operating point. This is because input of T_1 and T_2 allow calculation of:

$$\frac{N-1}{N}, T_1, Z_1 \quad \text{and M.W.}$$

The other available options include compensation for gas composition changes, system backups that will protect the compressor in the event of a surge and controller output options to enable the surge system to be soft in relation to the process system. That is, operation of the surge control system will not adversely affect the process system.

14 Turbo-compressor mechanical design overview

- Introduction
- The casing
- The inlet guide vanes
- The rotor
- Diaphragms
- Inter-stage seals
- Shaft end seals
- Journal bearings
- Thrust bearings

Introduction

In this chapter we will define the functions of each major component of a turbo-compressor. That is, what the purpose of each component is or "What it Does". By understanding what each component is supposed to do, you will be in a better position to know if it is performing its duty correctly. We will present each major component starting with the casing, state its function, operating limits and discuss the problems that can occur.

This chapter presents an overview of the mechanical components in a turbo-compressor. Figure 14.1 defines the major components of a typical turbo-compressor.

Function of major components of a turbo-compressor

Gas path components

- Casing
- Guidevanes (stators)
- Rotor (impeller (blades) and shaft)
- Diaphragms (diffuser and return channel)
- Stage seals (impeller and shaft)

Mechanical components

- Shaft end seals
- Journal bearings
- Thrust bearing
- Balance drum

Figure 14.1 Function of major components of a turbo-compressor

Note that these components can be grouped into either gas path or mechanical. That is, their primary function is either to guide the gas or change its velocity or to seal the gas or support the shaft. As we shall see, most components have more than one function.

Figure 14.2A and B shows all of these components in the two major types of Dynamic Compressors – Axial and Centrifugal.

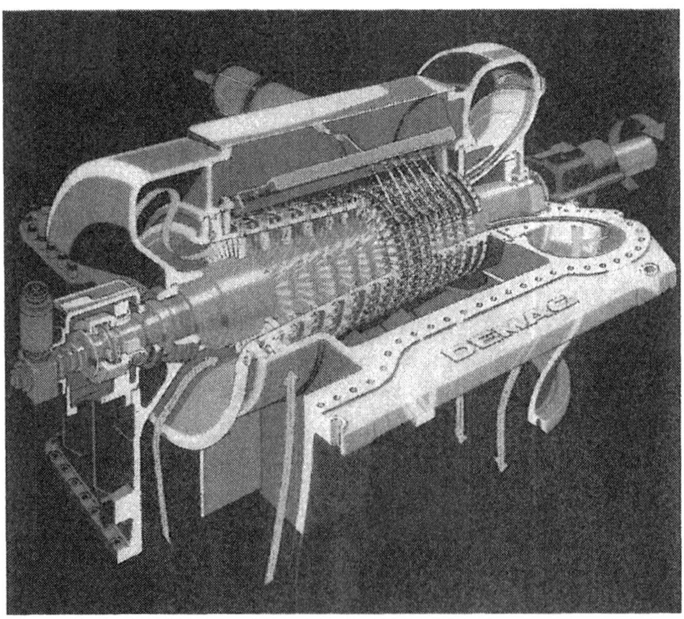

Figure 14.2A Dynamic compressors basic types – Axial (Courtesy of Mannesmann Demag)

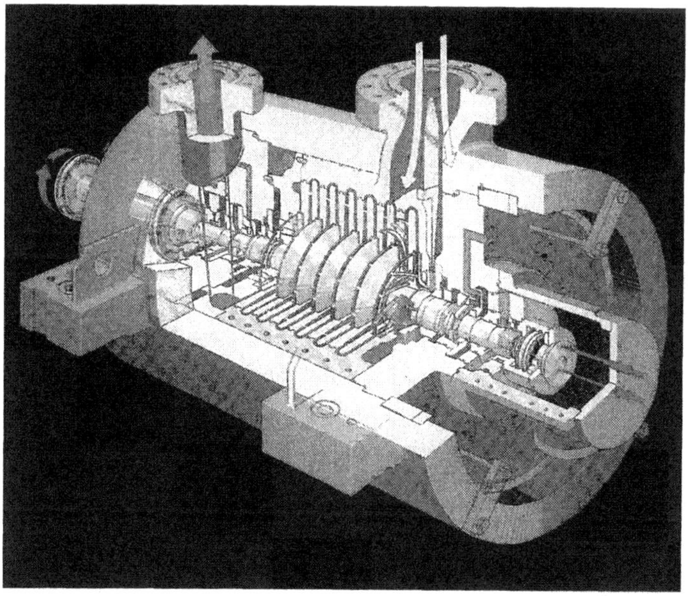

Figure 14.2B Dynamic compressors basic types – Centrifugal (Courtesy of Mannesmann Demag)

The casing

The function of the turbo-compressor casing is defined in Figure 14.3.

Casings can be either horizontally or vertically (barrel type) split. Generally, barrel type casings are used above 600 Psi (40 Bar) pressure or for mixed gases with a high hydrogen content. From a maintenance standpoint horizontally split casings with bottom connections are preferred.

The major problem connected with casing design is distortion of the casing caused by excessive piping or foundation forces. During construction or whenever piping is removed from the casing, dial indicators mounted external to the casing should be positioned on the shaft in the horizontal and vertical planes. Movement of more than 0.002 inch (0.05mm) when piping is bolted up to the casing should require correction to the piping system (pipe flange alignment or spring support modification). The only way for the turbo-compressor shaft to move is if the piping or foundation cause excessive casing strain. A check for excessive casing strain caused by the foundation is commonly known as "soft foot". To check for soft foot, a dial indicator is mounted external to each casing support. If the support moves more than 0.002 inch (0.05mm) when the support foot is unbolted, soft foot exists. Soft foot should be checked with process piping disconnected. The cause of soft foot is that the plane of the support is not parallel to the plane of

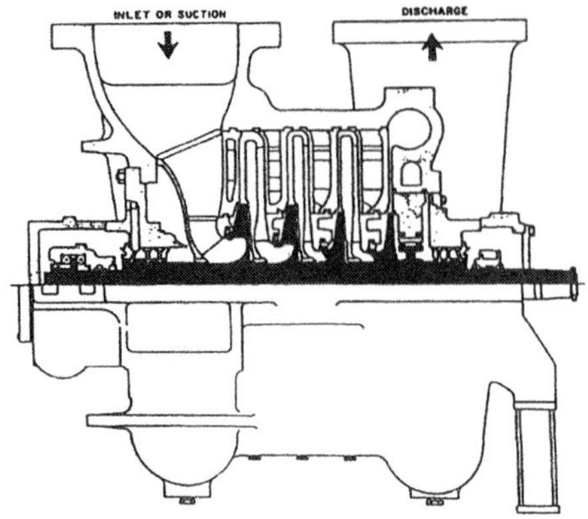

FUNCTION: THE CASING CONTAINS THE PROCESS GAS AND RIGIDLY SUPPORTS ALL STATIONARY INTERNAL PARTS AND THE BEARINGS.

Figure 14.3 The compressor casing

the foundation and/or a gap exists between the foundation and support foot.

The inlet guide vanes

The function of compressor inlet guide vanes is defined in Figure 14.4.

Centrifugal compressor guide vanes are usually fixed. Plant and instrument air compressors can be supplied with adjustable guide vanes. Axial compressors are frequently supplied with adjustable guides vanes.

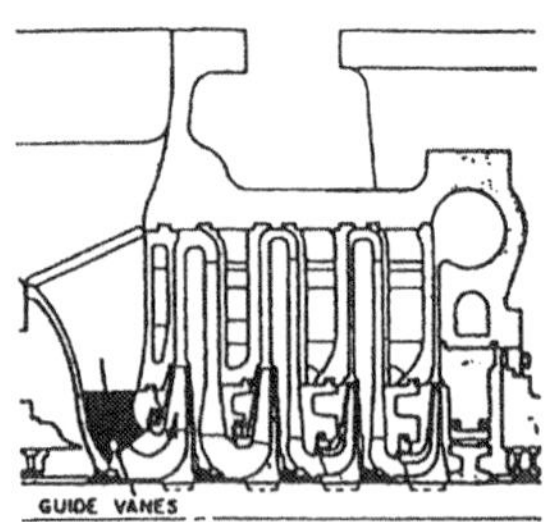

FUNCTION: DIRECTS AND DISTRIBUTES THE GAS TO EACH IMPELLER. CENTRIFUGAL COMPRESSORS GUIDE VANES ARE FIXED. AXIAL COMPRESSOR GUIDE VANES CAN BE ADJUSTABLE WHILE OPERATING.

Figure 14.4 The inlet guide vanes

Adjustable guide vanes alter the dynamic compressor performance curve by changing the velocity and direction of the gas relative to the compressor impeller blades. Impact of the gas velocity vector with the blades (counter rotation) will increase the gas tangential velocity and produce greater energy (head). Leading the gas velocity vector (pre-rotation) will decrease the gas tangential velocity and produce less energy (head).

Guide vanes, stationary or adjustable, positioned at an incorrect angle can alter compressor stage head and efficiency. This problem will be discovered with stationary guide vanes during an initial shop performance test (if required by the End User) and corrected. However, adjustable guide vanes can be improperly set (often known as "scheduled") if the individual guide vanes, bushings or linkages are replaced. Be sure to consult the instruction book and consult the equipment vendor if necessary when replacing any components in an adjustable guide vane assembly.

The rotor

The components and function of a dynamic compressor rotor are presented in Figure 14.5.

Dynamic compressor rotors can contain axial and/or centrifugal impellers. Centrifugal impellers can handle flows from 100 to over 100,000 CFM. Axial impellers (blading rows) are usually used above 100,000 CFM. The head produced by a centrifugal stage can vary from approximately 5,000 to 25,000 FT/LB_F/LB_M. An axial stage usually produces approximately 3,000 FT/LB_F/LB_M of head. The most

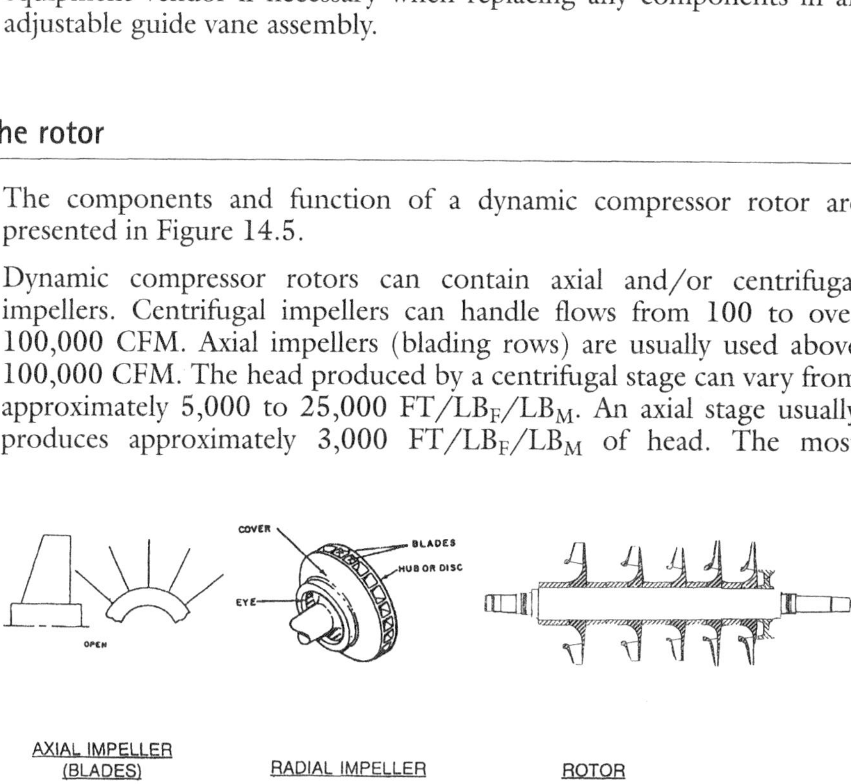

FUNCTION: THE ROTOR WHICH CONSISTS OF THE SHAFT, IMPELLERS, THRUST DISC AND BALANCE DRUM ROTATES THUS INCREASING THE ENERGY OF THE (GAS)

Figure 14.5 The rotor (impeller and shaft)

common problems associated with rotors are; component unbalance, rotor unbalance caused by improper impeller assembly, excessive blade or impeller stress and rotor system natural frequencies. These topics will be discussed in detail later in the book.

Diaphragms

The definition of a diaphragm and its components are presented in Figure 14.6.

Diaphragms can be manufactured of either cast iron or cast steel. Both materials are acceptable. However, in the event of severe axial rubs caused by excessive continuous surge, cast iron diaphragms can crack and will require replacement. Since centrifugal diaphragms are rarely spared, failure will cause long periods of downtime and loss of product revenue.

Diffusers are usually vaneless (parallel walled) but can be vaned or scrolled to increase head produced per stage. Older designs with vaned or scrolled diffusers often encountered problems with impeller side wall breakage caused by fatigue of the impeller side plates resulting form alternating pressures at the impeller discharge. This phenomena was caused by the pressure build up and decay resulting from blade passing interference between the impeller and diffuser vanes. The classic fix is to "scroll" the impeller wheel between each vane. A modified impeller is called a "cabbage cutter wheel" due to its appearance. The idea behind this modification is to remove the effect (side plate cracks) by removing the affected material. This modification has proven successful and does not significantly effect impeller efficiency. Modern "vaned diffuser"

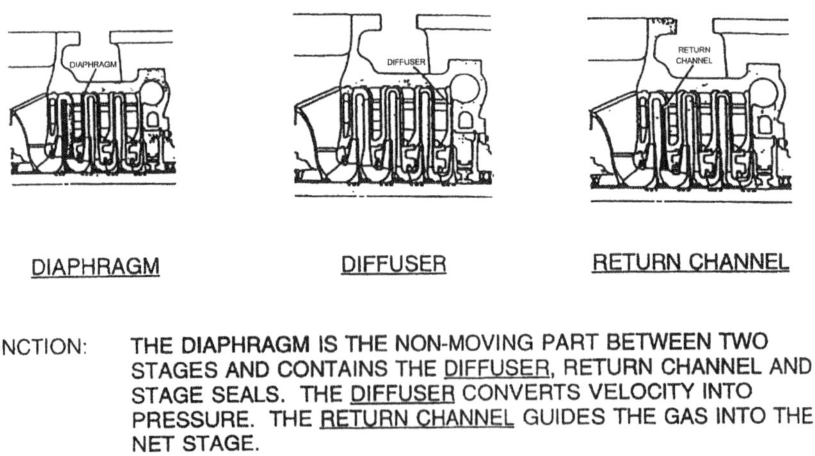

Figure 14.6 Diaphragms (diffuser and return channel)

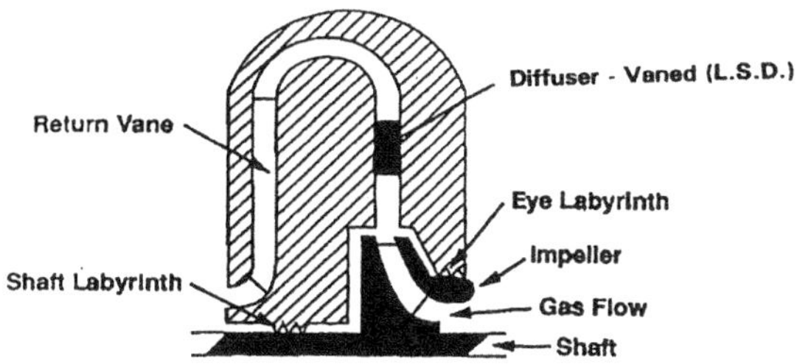

FUNCTION: BOTH IMPELLER AND SHAFT STAGE SEALS PREVENT LEAKAGE FROM RECYCLING FROM THE DISCHARGE OF THE IMPELLER. INCREASED "CLEARANCE" RESULTS IN DECREASED THROUGHPUT.

Figure 14.7 The stage seals (impeller and shaft)

designs like the low solidity diffuser (L.S.D.) move the vane up the diffuser out of the immediate effect of the impeller vanes thus eliminating significant pressure fluctuation. An example of an L.S.D. diffuser is shown in Figure 14.7.

Inter-stage seals

The function of inter-stage seals are defined in Figure 14.7.

Each impeller diaphragm contains one eye seal and one shaft seal. These seals are designed to minimize leakage from the diffuser to the inlet of the impeller (eye seal) and to the inlet of the next stage (shaft seal). Interstage seals are usually manufactured of aluminum to assure that contact with the rotor does not result in an external excitation to cause vibration. However, certain gases and saturated gas mixtures require that alternative materials be used.

Since the early 1970's various abradable material interstage seals have been used. Abradable seals allow for contact with the rotor without causing excitation to the rotor system. This type of interstage sealing system had it's origin in the aircraft engine industry. The advantage of an abradable seal is increased efficiency. Increases in stage efficiency as high as 3% have been obtained. Frequently used abradable seal materials are fluorisint (mica impregnated teflon) and honeycomb arrangements of hastelloy. The temperature limits of these materials are approximately 350°F and 1100°F respectively. Usually, the labyrinth (knife edges) are positioned on the rotors whenever abradable seals are used. It has proven cost effective not to attempt to machine knife edges in the abradable material. One note of caution regarding fluorisint labyrinths

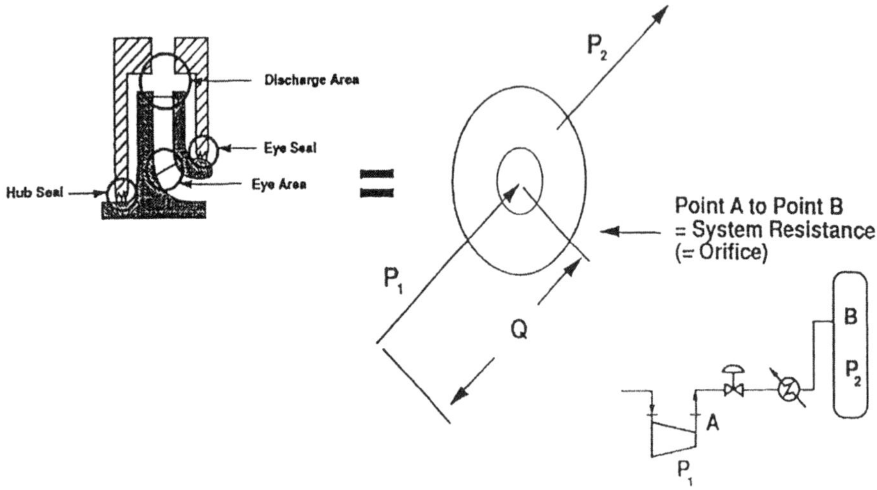

Figure 14.8 Reduce it to an equivalent orifice

... they should be pressure balanced. Pressure balancing of each interstage labyrinth assures that the high pressure side of the seal will not deflect or force the material into the rotor surface. The lack of pressure balance has resulted in premature wear of labyrinth seals. Pressure balance is easily achieved by assuring the pressure behind the seals is less than or equal to the pressure between the labyrinth surface and rotor surface.

Deterioration of labyrinth seals in a centrifugal compressor is one of the major causes of reduced performance. The eye seal is effected to a much greater extent than the shaft seal since the ΔP across this seal is greater. The causes of labyrinth seal wear are: rotor vibration, excessive moisture and fouling. If centrifugal compressor head and efficiency fall off greater that 10%, the compressor should be inspected at the next opportunity if the compressor is known not to be fouled.

Refer to Figure 14.8 and observe how this conclusion can be made by thinking of the interstage labyrinth as equivalent orifices.

Shaft end seals

Compressor shaft end seals perform the function of containing the process fluid by minimizing external leakage and directing the leakage to a safe location. The types of shaft end seals used in compressor applications are shown opposite.

Seal type	Typical applications	Limitations Speed	Limitations Pressure	Seal system type
Labyrinth	■ Inert Gases ■ Hydrocarbon	None	None	■ Eductor
Mechanical Contact (oil)	■ Hydrocarbon	Appx.* 12,000 RPM	None	■ ΔP (Oil to Gas) Control
Liquid film (Bushing)	■ Hydrocarbon	None	None	■ Level (Oil to Gas) Control
Dry Gas	■ All Applications (since 1990)	Appx.* 15,000 RPM	Appx.* 1000/ seal	■ ΔP(Buffer Gas)

*Consult vendor's experience list

With the exception of labyrinth seals used for inert gas service, all compressor seals are configured as double seals. That is, the sealing fluid is supplied at a pressure greater than the process fluid and therefore requires a seal between the process and the sealing fluid and between atmosphere and the sealing fluid. Figure 14.9 shows a liquid film bushing seal used in hydrocarbon service.

Follow the path of the sealant and observe in Figure 14.9 that this seal is in a double seal configuration. The inner bushing seal will leak a small amount (typically less than 5 gallons/day) into the seal oil drainers. The outer seal, frequently called the atmospheric bushing directs oil flow back to seal oil reservoir. Typical values vary from 1–10 gallons/ minute. The seal shown in Figure 14.9 also has a "flow-thru" feature. This feature provides additional cooling to the seal and is an advantage in applications that start up with low process gas pressure but normally operate at high process gas pressures.

Figure 14.10 presents the seal oil system for the seal shown in Figure 14.9.

Trace the seal oil supply from the reservoir to the seal. Note the seal oil, which is maintained at approximately 3 psi above reference gas, is distributed into three different paths: the gas side bushing, the atmospheric bushing and the flow through oil. Compressor seals and sealing systems will be covered in separate modules later in the course. Before leaving this subject, all seal systems are designed to automatically maintain a constant seal fluid to process fluid differential across the inner seal. **However**, the differential pressure across the outer seal varies directly with the process gas pressure. This fact presents a

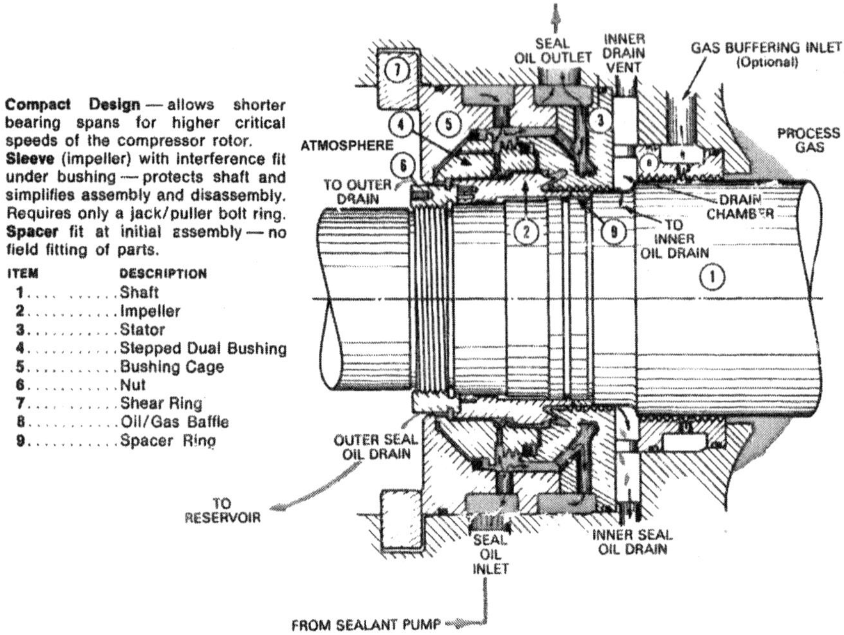

Figure 14.9 Shaft end oil seals (Courtesy of AC Compressor Company)

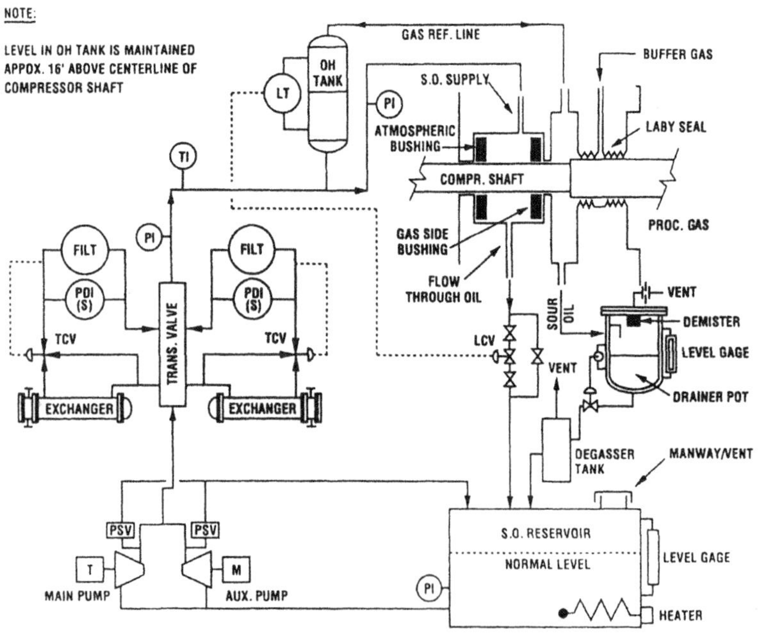

FUNCTION: THE SEAL OIL SYSTEM CONTINUOUSLY SUPPLIES CLEAN, COOL, OIL TO THE SEALS AT THE PROPER DIFFERENTIAL (OIL TO GAS) PRESSURE AND FLOW RATE. IT REMOVES THE HEAT OF FRICTION FROM THE SEALS.

Figure 14.10 The seal oil system (Courtesy of M. E. Crane, Consultant)

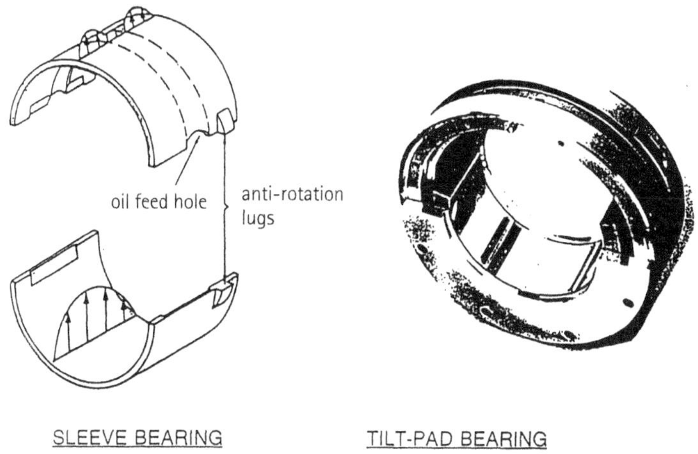

FUNCTION: JOURNAL BEARINGS REGARDLESS OF TYPE CONTINUOUSLY SUPPORT THE ROTOR WITH AN OIL FILM OF LESS THAN ONE THOUSANDTH OF AN INCH.

Figure 14.11 Journal bearings

problem during start-up with low process gas pressure since the seal fluid supply pressure is low and the temperature in the outer seal can become excessive causing seal damage.

Journal bearings

The function of a journal bearing is defined in Figure 14.11.

Most compressors use hydrodynamic bearings as shown. A hydrodynamic bearing supports the rotor with a thin film oil wedge. Small compressors, fans or blowers sometimes use anti-friction (ball or roller) bearings. Regardless of the type of bearing, the basic principle in bearing design is as follows:

$$P = \frac{F}{A}$$

Where P is the pressure in psi on the bearing oil film or component (ball or roller) that results from all the rotor forces (static and dynamic) acting on the bearing area. If the forces are too large or the bearing area is too small, the pressure load will become excessive and the bearing will fail.

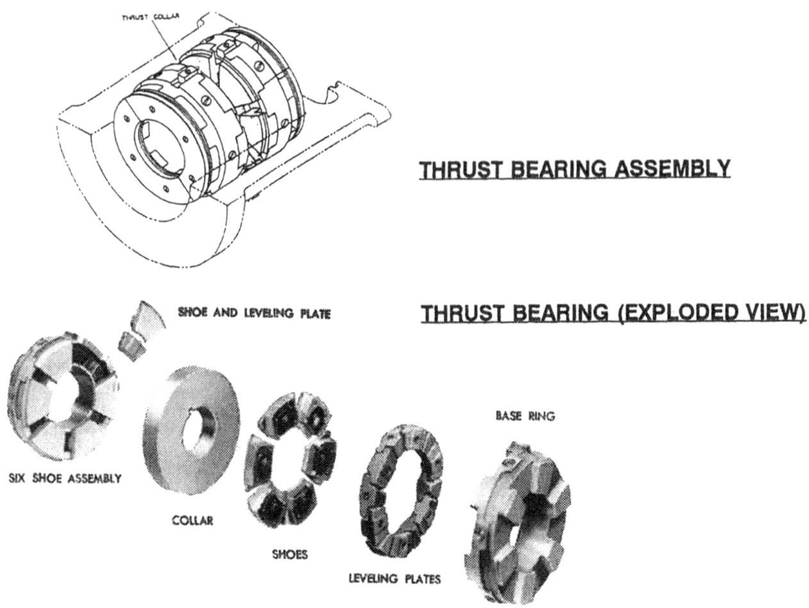

FUNCTION: THE THRUST BEARING POSITIONS THE ROTOR AXIALLY IN THE CASING AND TAKES THE LOAD GENERATED BY THE BLADES AND BALANCE DRUM*

*NOTE: ONLY INSTALLED ON REACTION TURBINES

Figure 14.12 Thrust bearing (Courtesy of Kingsbury, Inc)

Thrust bearings

Figure 14.12 shows a typical hydrodynamic double acting, self equalizing thrust bearing.

The same principle applies in the design of thrust bearing. That is,

$$P = \frac{F}{A}$$

The thrust bearing supports all compressor axial loads. For any hydrodynamic bearing, journal or thrust, the approximate pressure that would cause a bearing failure is 500 psi. Most industry specifications limit the load pressure to 50% of this value or approximately 250 psi.

It should be noted that there are exceptions to this rule. Gearboxes and compressors with integral gearboxes usually have design loads in the 300–400 psi range. This is because major source of bearing load in a gear box is torque. The designer must consider the entire torque range. That is, start up conditions (low torque) and maximum torque conditions. As a result, during start-up the load pressure P will be relatively small on the bearings. This is the reason that many gears are unstable (vibrate) at low loads. The only way to correct this is to

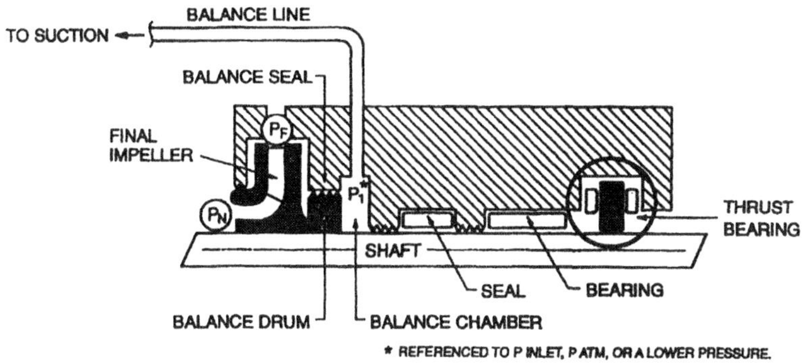

Total Impeller Thrust (LB) = Σ Individual Impeller Thrust
Balance Drum Thrust (LB) = (P_F - P_1) x (Balance Drum Area)
Thrust Bearing Load (LB) = Total Impeller Thrust - Balance Drum Thrust

FUNCTION: The Balance Drum and Seal Reduce Thrust (Force) on the Thrust Bearing by Maintaining Design Clearance Between the Drum and Seal

Figure 14.13 Balance drum and seal

increase the bearing load pressure at low loads at the expense of having a larger value of load pressure at maximum loads.

In many thrust bearing applications, it is not possible to provide sufficient bearing area to maintain the load pressure at 250 psi. In these cases, a thrust balancing system is required to reduce the thrust load. Please refer to Figure 14.13.

The balance system reduces thrust by providing a thrust force which counteracts the total impeller thrust force thus reducing the thrust bearing load P (P.S.I.) This is accomplished by designing the face dimensions of the balance drum for sufficient area (A) to provide the desired force (F) for the differential pressure (P_2–P_1) across the balance drum. Therefore the balance force is calculated by:

$$F_{BALANCE} = (\Delta P)(A_{BALANCE\ DRUM})$$

The balance line, which usually runs from the balance chamber to the compressor suction, maintains the balance chamber pressure at P_1 + Balance Line ΔP. An effective method of balance drum condition monitoring is to monitor balance drum ΔP. An increasing ΔP, at the same operating point, will indicate increased balance drum clearance. Often, thrust bearing failures are caused by balance seal wear. Monitoring balance line ΔP will indicate the root cause of a thrust bearing failure in this case. Other important facts concerning thrust balance are:

- Thrust can be balanced to any value (0 thrust, towards suction or towards discharge)

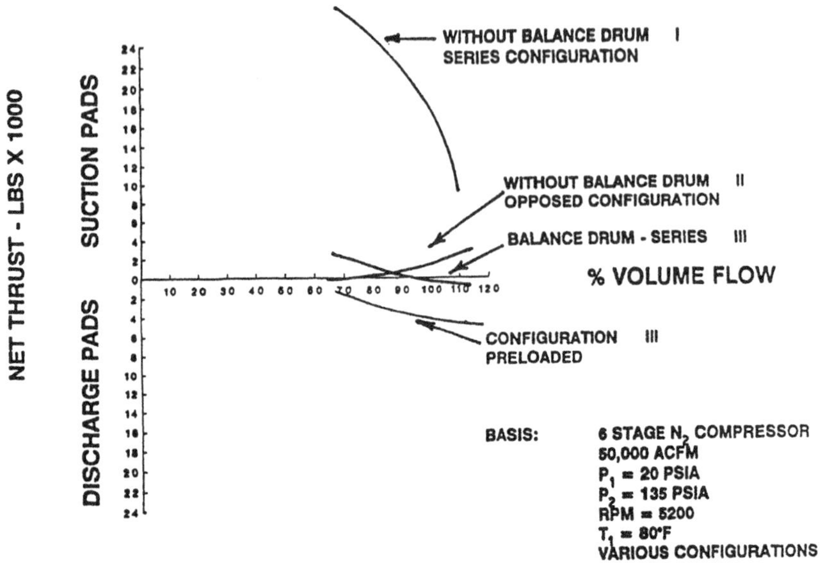

Figure 14.14 Net thrust load vs. % flow

- Active thrust direction is the direction the rotor thrusts at rated conditions
- Thrust position is a function of performance and will change with flow rate, speed and gas density.

Figure 14.14 shows how thrust load can change with balance system design and compressor performance.

Note that a compressor thrust vs flow curve without a balance drum (I) actually follows the shape of a performance curve. A single stage compressor without a balance drum will have its highest operating thrust at surge. An example would be a circulator compressor application. Curves II and III show various other thrust balancing arrangements. Note that the thrust balance can be designed in many different ways! Every custom designed compressor has a unique thrust vs flow curve based on the design of the thrust balancing system.

Figure 14.15 shows a typical lubrication system and its function.

Such a system would continuously provide lubrication to the journal and thrust bearings present in a compressor train.

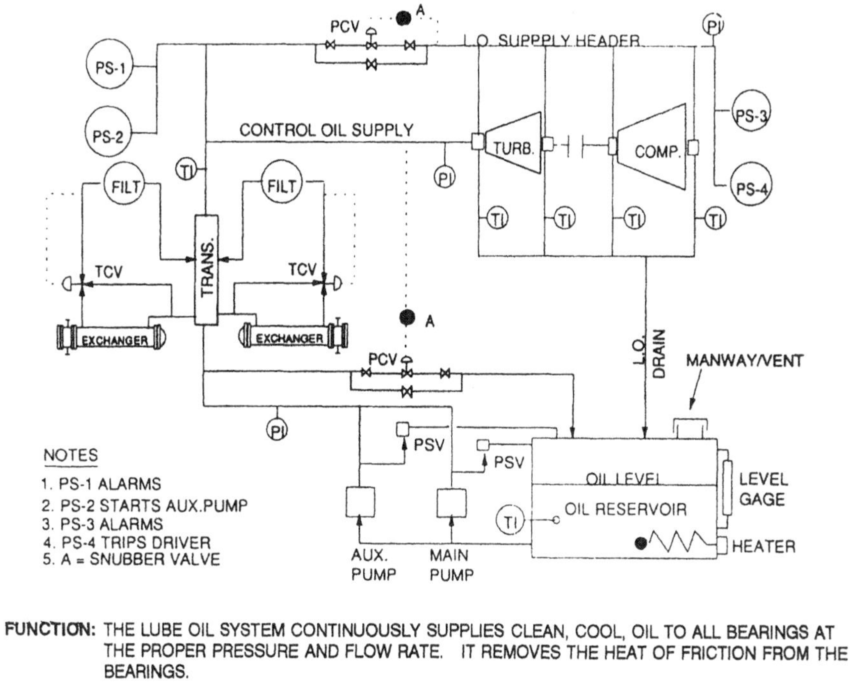

Figure 14.15 The lube oil system (Courtesy of M. E. Crane, Consultant)

15 Compressor casings and stationary internals

- Introduction
- Casing functions
- Casing types
- Casing fabrication options
- Casing stresses and deflections

Introduction

In this chapter we will devote our attention to the compressor case and the stationary internals. We will first define the functions of the casing as follows:

- Flow guidance
- Positive sealing
- Full rotor support

The different types of turbo-compressor cases will be defined:

- Single stage
- Multi-stage
- Horizontal split
- Radial
- Integral gear

Nozzle orientation will also be covered. Emphasis will be on minimal disturbance to the process piping during compressor disassembly. Casing fabrication types will be discussed highlighting cast and fabricated casings. Casing stresses and deflections due to internal forces and external loads will be covered.

Casing functions

The major functions of any compressor case, regardless of type are:

- To provide support of all bearing seals and stationary internals.
- To provide positive containment of the process fluid.
- To provide efficient guidance of the process fluid.

Often, the support function of the casing is not mentioned when stating the major casing functions. Excessive casing forces caused by the process piping or foundation can cause bearing, seal and rotor failures. Positive containment of the process fluid is accomplished by the design and manufacture of leak tight joints. All casings are hydrostatically tested at 1.5 × maximum working pressure prior to the shipment. Gases containing more than 50% H_2 require a gas test at rated pressure in addition to the hydrostatic test.

Casing joints are either horizontal or vertical (radial). Horizontal joints are usually metal to metal as opposed to vertical joints which always incorporate an 'O' ring. Some manufacturers do incorporate an 'O' ring in their horizontal design. In the writer's opinion, metal to metal horizontal joints are preferred to assure that a leak will not develop if the 'O' ring is damaged.

An effective design feature or modification for horizontal joints is to 'relieve' the joint. That is, to minimize the joint contact area and therefore increase the joint contact stress by machining the horizontal faces in certain areas of the casing flange. The machined areas are usually 0.032–0.062 deep. A plan view of a typical relieved casing is shown in Figure 15.1. The relief can be in either the top or bottom half of the casing joint.

Barrel or radially split casings are used for gases with high H_2 content or for pressures above approximately 600 psi. The advantages of the barrel type compressor are:

- Minimum casing joint area
- Removal of internals with removal of process piping

The disadvantages are:

- Special tools are required for disassembly

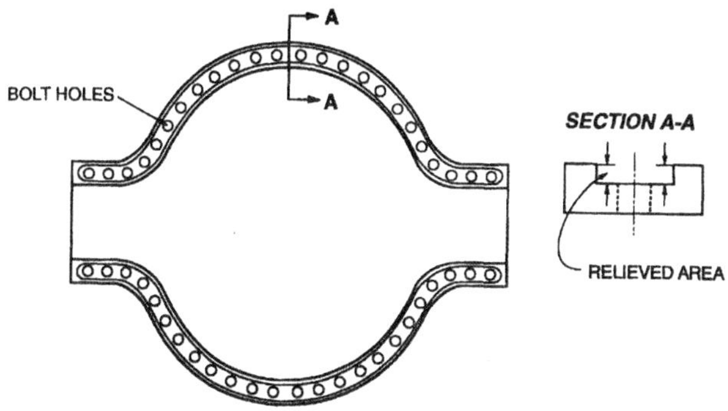

Figure 15.1 Relieved horizontal joint – horizontal split casing – plan view

- Additional maintenance space is required for removal of internals

When a two-case application uses barrel type compressors, it is advisable to design the driver with double shaft extensions for ease of maintenance.

Casing types

The available types of turbo-compressor casings are:

- Single stage radial split
- Single state integral gear
- Multistage radial split
- Multistage horizontal split
- Multistage integral gear

Figure 15.2 is an example of a single stage radially split casing.

This type of casing is typically used for high pressure applications and for gases with high hydrogen content. Its design is similar to that of a single stage overhung pump. The design shown is 'front pullout' that is, the impeller must be removed for seal and/or shaft removal. Other designs allow back pullout. This type offers the advantage of removing the internals without having to remove the process piping.

A single stage casing for an integral gear centrifugal compressor is shown in Figure 15.3 (Sundyne Compressor).

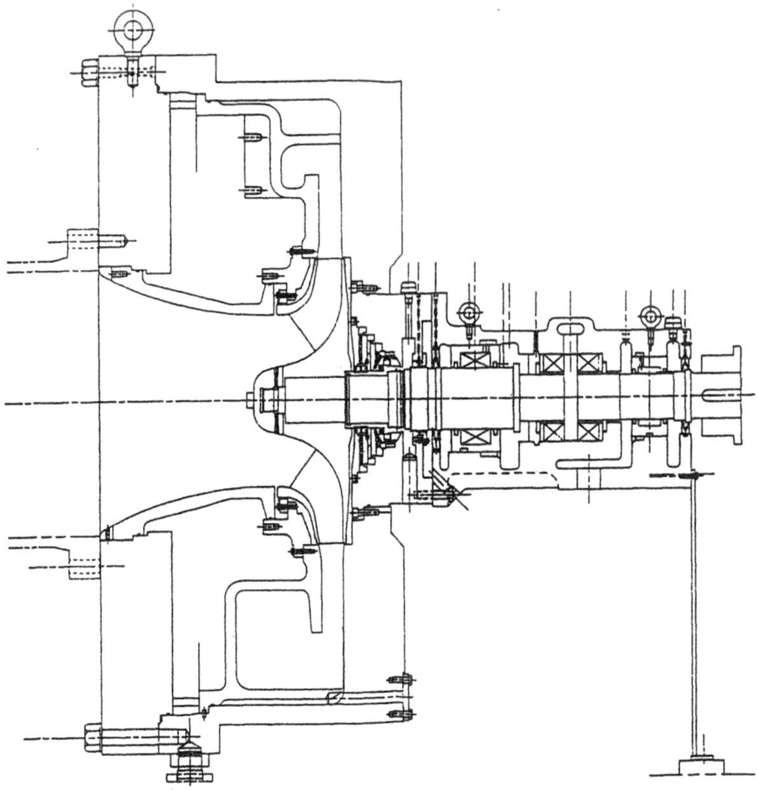

Figure 15.2 Single stage overhung turbo-compressor (Courtesy of A-C Compressor Corp.)

This type of compressor is used in low flow, high head applications. The integral gearbox (not shown) enables this type of compressor to operate at speeds above 30,000 rpm. The casing is cast and is also a 'black pullout' design which allows removal of internals without affecting the process piping.

An example of a multistage, radially split or barrel compressor is shown in Figure 15.4.

Figure 15.5 depicts a multistage, horizontally split centrifugal compressor casing. Most multistage compressors are horizontally split and should be used whenever process pressures and/or gas compositions permit. The advantage of a horizontally split casing should be ease of removal of internal parts. The example shown meets this requirement since all process piping is located on the lower half of the casing. When process piping is connected to the top half of the compressor casing, the maintenance advantage is lost.

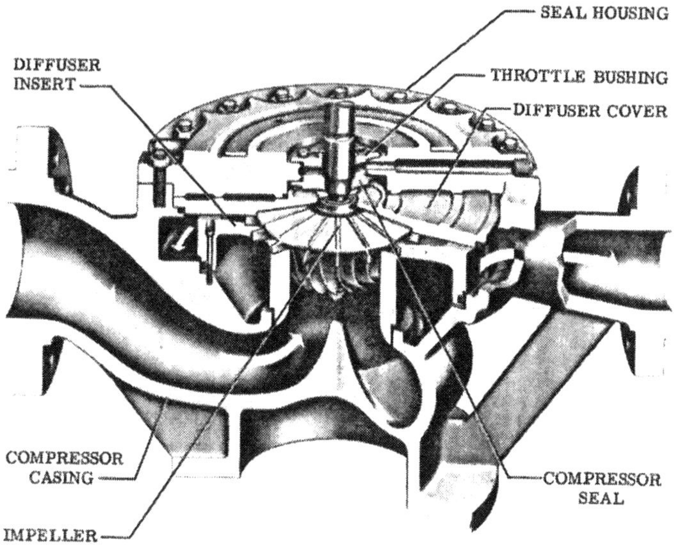

Figure 15.3 Single stage high-speed compressor (gear box and motor removed) (Courtesy of Sundstrand Corp.)

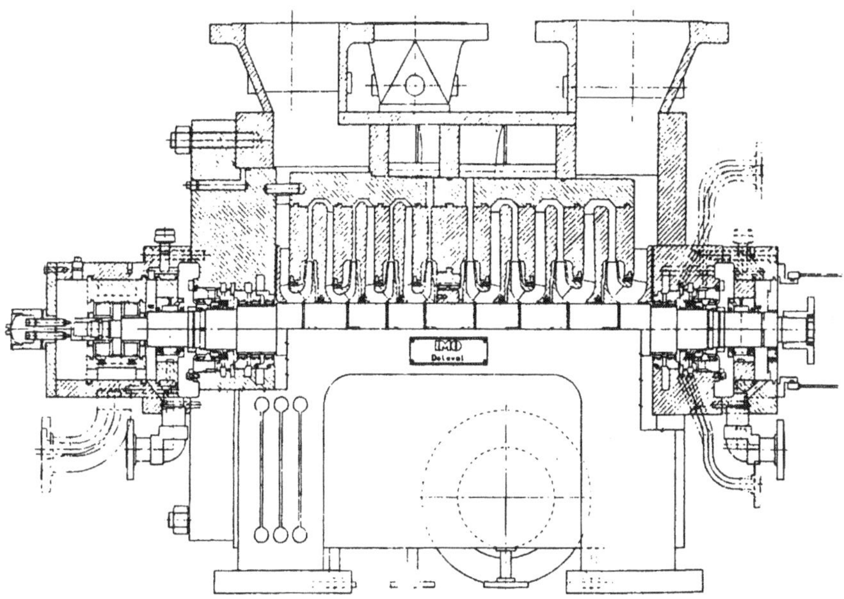

Figure 15.4 Multistage barrel compressor (Courtesy of IMO Industries, Inc.)

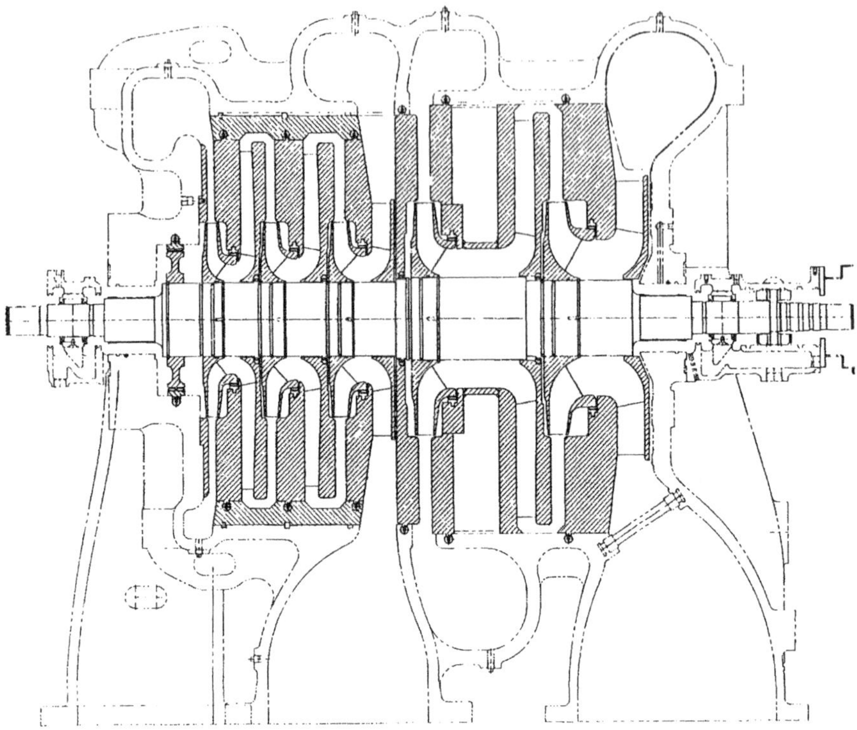

Figure 15.5 Multistage intercooled turbo-compressor (Courtesy of IMO Industries, Inc.)

Readers are advised to become involved in the initial process plant design to assure bottom connected horizontal compressor cases are installed if at all possible. Have a cost analysis performed to determine the additional cost for bottom connections. Note: compressor vendors require five (5) pipe diameters straight runs to process connections. When performing a cost analysis, be sure to include the cost penalty for the additional removal and assembly time for top mounted piping. Also be sure to include the revenue lost per day of additional downtime if the compressor is unspared. Typical additional times for piping removal and assembly are 2–3 days or longer. If the cost is prohibitive, another option would be a barrel compressor with top connections.

A multistage, integral gear compressor casing is shown in Figure 15.6.

Typical applications are plant and instrument air and inert gas compressors. The casing consists of a cast or fabricated gear case and bolted on individual stage cast casings. Some designs allow rotor removal without having to disconnect the process piping. The advantages of an integral gear compressor are reduced number of stages required and higher efficiency. This is because all stages are intercooled

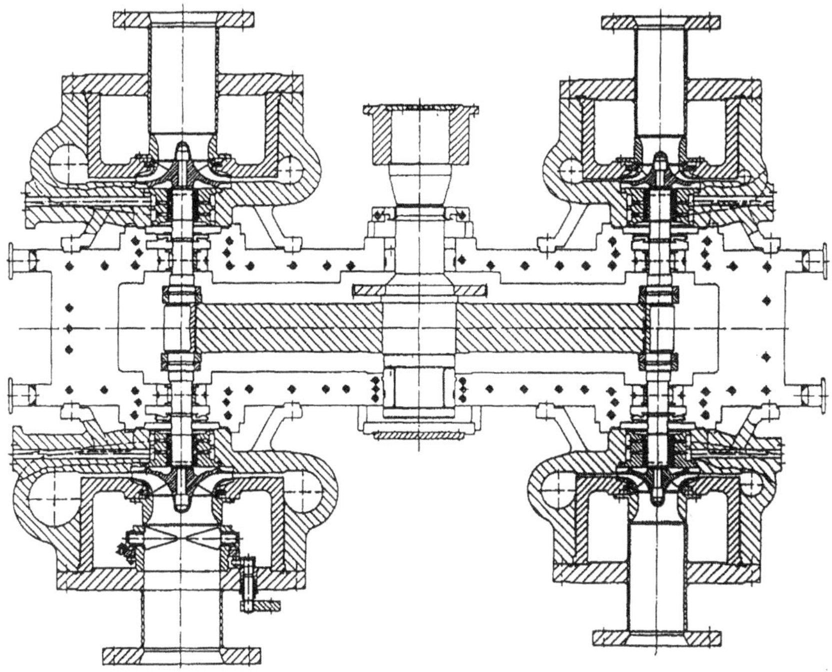

Figure 15.6 4 stage internal gear compressor (Courtesy of Demag Corp.)

and the integral speed increasing gear allows all stages to operate in a higher specific speed.

Multistage, integral gear compressor manufacturers have begun to offer these units for hydrocarbon service. Readers are advised to exercise caution and confirm satisfactory field operating experience on similar gases before purchase. The requirements for gas or oil seals, varying gas composition and flow rates and the possibility of fouling can significantly reduce the reliability of this type of compressor opposed to air or inert gas applications.

Casing fabrication options

Various casing fabrication options exist. Casings can be:

- Cast
- Forged
- Welded

Older designs were generally cast. However, foundry quality has significantly decreased and today, many casings are welded from steel plate. Vendors have existing qualified procedures and welders to fabricate these casings. Figures 15.2, 15.3, 15.5 and 15.6 are examples of cast casings. Figure 15.4 shows a fabricated barrel compressor casing. In high pressure or high H_2 gas applications, forged barrel compressors are required in this design, the entire barrel casing is on continuous forging. However, cast nozzles are welded onto the casing. The compressor shown in Figure 15.4 could be supplied with a forged barrel. The common approach used today is to accept either a fabricated or cast casing as long as the compressor vendor has proven experience with the offered design. Regardless of the type, steel casings are required for all hyrdrocarbon services. Casings are used exclusively for air and some inert gas services.

Casing stresses and deflections

As previously mentioned in this module, casing stresses and deflections caused by excessive piping and/or foundation forces can cause damage to bearings, seals, etc. There have been established standard checks during compressor field construction and disassembly to assure that the values of all external forces and moments on compressor cases are within prescribed vendor limits.

During the compressor design phase the vendor issues an 'allowable piping forces and moments diagram' which states the maximum values of external forces and moments to which the casing can be subjected. The piping systems are then designed to comply with these limits. However, field piping fabrication methods, field foundation building, improper spring support sizing, etc. can result in excessive case stresses even though the systems were properly designed!

Therefore, field checks of proper piping connections and foundation fabrication are required. The basic philosophy of these checks is to assure the casing is not deflected. This is accomplished by checking the movement of the rotor relative to a fixed external point and the movement of the compressor baseplate relative to a fixed external point.

Prior to conducting field checks, it is advisable to make a preliminary inspection of external piping system condition. Figure 15.7 presents guidelines for this inspection.

Preliminary piping connection guidelines

- 'Float' process piping to compressor connections and confirm:
- Bolting is free without use of 'come-alongs'
- Machined faces of flanges are parallel within 0.010–0.020" at four quadrant locations

Figure 15.7 Preliminary piping connection guidelines

If the conditions specified in Figure 15.7 are not met, take corrective action before proceeding with the field checks. Possible corrective action can be:

- Loosening piping flange connections away from the compressor to meet requirements, then re-tighten while performing field check
- Correction of spring support or spring hanger sizing and/or installation
- Heating piping to meet requirements
- Cutting and re-welding piping

The field check to assure that the foundation is not exerting excessive force on the baseplate is known as the 'soft foot check'. The guidelines are shown in Figure 15.8.

Soft foot guidelines

- Perform check with all process piping unbolted, and not exerting force on the casing
- Install dial indicator external to each hold down bolt location on the baseplate
- Loosen bolting, one foot at a time and observe dial indicator does not move more than 0.002"
- If dial indicator moves more than 0.002" at any location, insert stainless steel shims to correct the problem
- If problem persists, foundation must be checked to assure it is continuous

Figure 15.8 Soft foot guidelines

The field check guidelines to assure that the process piping is not exerting excessive force on the compressor are shown in Figure 15.9.

External piping connection guidelines

- Install two (2) dial indicators on exposed compressor shaft (horizontal and vertical)
- Dial indicators must not be mounted on the compressor case
- 'Float piping' to each connection and bolt up following compressor vendors' instructions
- If either dial indicator moves more than 0.002", take corrective action (see 'corrective action' in this chapter)

Figure 15.9 External piping connection guidelines

Readers are advised that excessive casing forces from process piping and foundations are one of the major sources of misalignment. It is a good practice to be aware of any work on process piping systems of foundations and to prepare guidelines to control and monitor this work.

Centrifugal compressor rotor design

- Introduction
- Shaft end design
- Rotor stiffness
- Basic rotor configurations
- Rotor assembly
- Rotor balance

Introduction

In this chapter we will present some of the fundamentals concerning rotor system design. The rotor is comprised of the shaft and the components mounted thereon. For a turbo-compressor rotor the components consist of: impeller, seal collars, thrust collars, balance drum and coupling(s).

We will examine shaft design and particularly shaft end design to transmit the maximum required torque.

The concept of rotor stiffness will be discussed. It will be shown to be proportional to shaft diameter and bearing span. Certain limits for sensitive and insensitive rotors will be presented.

The basic rotor configuration or various types of compressors will be discussed. The reliability considerations for each of these configurations will be reviewed. Points to watch in the assembly, disassembly operation of each configuration will be covered.

Rotor assembly stacking will be presented. Practical rotor runout limits

Figure 16.1 Multistage turbo-compressor rotor (Courtesy of IMO Industries, Inc.)

will be noted. A case history will be presented showing the effects of residual stresses on vibration during operation.

Rotor balance will be briefly covered and the subject of low speed vs high speed balance will be discussed.

Shaft end design

A photograph of a five (5) stage centrifugal compressor rotor is shown in Figure 16.1.

The assembly drawing of the same rotor is shown in Figure 16.2. Figure 16.2 also shows the mode shape of the rotor at the first and second rotor system critical speeds (natural frequencies).

Any compressor rotor, regardless of its type consists of various components mounted on a shaft. For centrifugal compressors, these components are:

Impellers, seal collars or sleeves, thrust collars, shaft sleeves, balance drum and coupling(s). Mounting of these components will be discussed later in this chapter.

The rotor shaft end must be designed to transmit the maximum torque provided by the driver. Torque is equal to horsepower per unit speed:

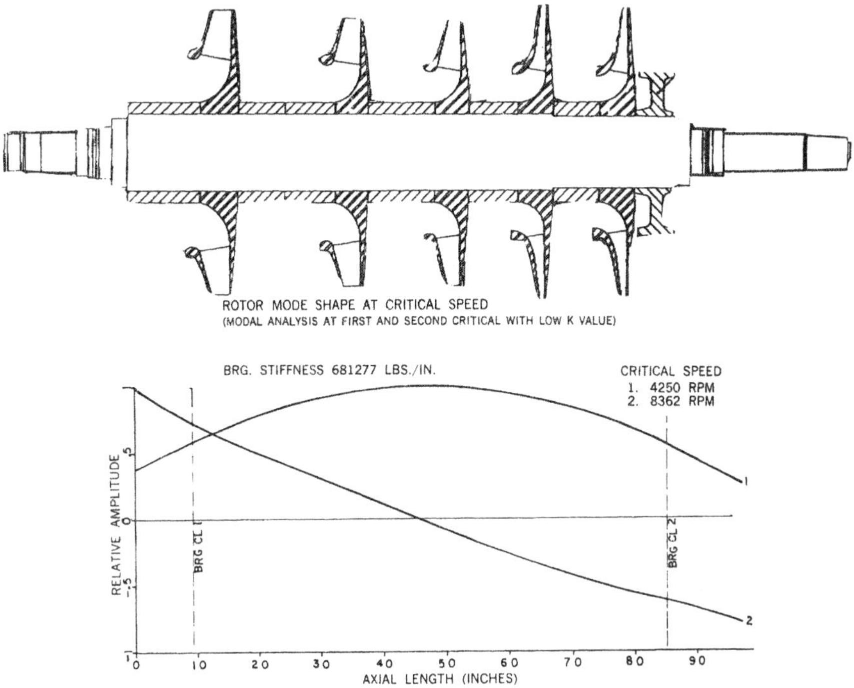

Figure 16.2 Assembly drawing of multistage turbo-compressor rotor (Courtesy of Elliott Company)

$$T = \frac{BHP}{N} \times 63{,}025 \ \frac{\text{in-lbs-rpm}}{\text{H.P.}}$$

Where: T = Torque (in-lbs)
BHP = Brake horsepower
N = Speed RPM

From this relationship, it is obvious that torque will increase with increasing BHP or decreasing N.

All compressor manufacturers have a limit of maximum allowable shear stress for shaft ends. This limit varies between 12,000 psi and 15,000 psi. The most common shaft materials are ANSI 4140 and ANSI 4340 steels which have a shear yield of approximately 36,000 psi. The relationship for shear is:

$$\text{Shaft shear stress} = \frac{16T}{\pi D^3}$$

Where: T = Shaft torque – in-lbs
D = Shaft end diameter

From this relationship, the designer will determine the shaft end diameter. Also note that the shaft diameter at any point along the shaft can also be found from this equation if the torque at that point is known. I have often been asked why the thrust collar is always at the shaft end opposite to the coupling. As we have learned, the objective in bearing design is to provide a maximum bearing surface area 'A'. By positioning the thrust collar at the end of the shaft with lowest torque, the required shaft diameter is the smallest. Therefore, the smaller the shaft bore, the greater the area available for the thrust collar.

Shaft ends can be designed either for hydraulic or keyed coupling fit. Hydraulic fits are usually used for compressors in hydrocarbon service since the coupling must be removed to disassemble the shaft seal and heat required to remove a keyed coupling would present a safety hazard. Coupling fits and mounting procedures will be discussed in the next chapter.

Rotor stiffness

Referring again to Figure 16.2, the rotor natural frequency is directly proportional to shaft stiffness by the relationship:

$$\text{Shaft natural frequency} = \sqrt{\frac{K}{M}}$$

Where: K = Shaft stiffness – in-lbs
M = Rotor mass – lbs

Rotor stiffness is proportional to bearing span (distance between bearing centers) divided by the major shaft diameter (diameter under the impellers).

$$\text{Or } K \propto \frac{\text{Bearing span}}{\text{Diameter under impellers}}$$

Experience has shown that dynamic compressor rotors with stable vibration characteristics usually have a K value ≥ 10. Readers are advised to calculate this value and always ask for references of units operating in the field if K ≥ 10. It is advisable to confirm satisfactory field operation of these units before a purchase is made.

Basic rotor configurations

Figure 16.3 shows schematically the basic rotor configurations.

Usually, single stage overhung and multistage opposed arrangements do not incorporate a balance drum since the rotor thrust values are low enough to be accommodated by a thrust bearing.

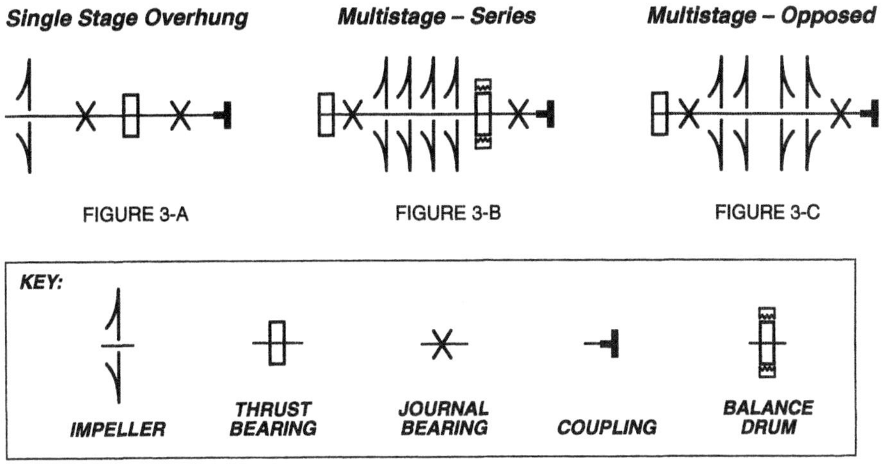

Figure 16.3 Compressor rotor configurations

Multistage opposed designs are usually used in high pressure applications. In these cases, failure of the balance drum seal, if the rotor were a series configuration, (Figure 16.3B) would cause catastrophic damage since the thrust bearing would immediately fail. An opposed rotor design as shown in Figure 16.3C would eliminate the requirement for a balance drum since the impeller thrust forces are opposed. Therefore, the thrust loads will always remain low since the rotor is naturally balanced.

Rotor assembly

Refer to Figure 16.1 of this chapter to observe a typical multistage rotor. All centrifugal compressor rotors are 'built-up' type. That is, the major components are not integral with the shaft and must be shrunk on. Figure 16.4 details the requirements for impeller shrink fits.

Impeller forces and shrink fit

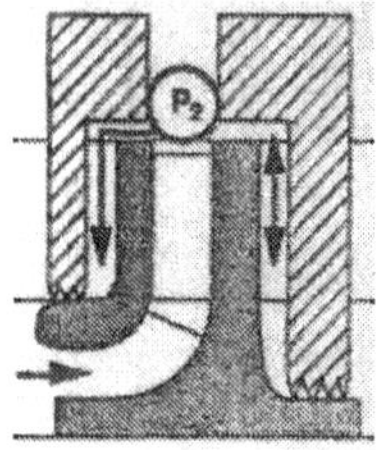

The forces acting on an impeller that create impeller stress and require positioning stress (shrink fit) are:

- Centrifugal force
- Gas load (force from torque)
- Thermal difference between impeller and shaft
- Thrust force

The impeller must be designed such that the shrink fit (shaft diameter – impeller bore diameter) is sufficient to maintain impeller/shaft contact at maximum conditions

Figure 16.4 Impeller forces and shrink fit

It can be seen from Figure 16.4, that the amount of shrink fit required for impellers can become large. Typically, impeller shrink fit values range from 0.002–0.003 inches shrink per inch of shaft diameter. The rotor shown in Figure 16.1 has a shrink fit in excess of 0.040 inch for each impeller. In this case, the impellers would have to be expanded in excess of 0.040 before mounting on the shaft. Typical mounting stresses are 20,000–30,000 psi. Therefore, thermal or hydraulic means are required to expand the components to be shrunk on the shafts. All multistage rotors utilize a thermal means of assembly (furnace or torch). For these rotors, assembly is performed vertically (rotor in a pit) for ease of assembly and to minimize residual shaft stresses.

Single stage rotors usually use hydraulic methods for rotor assembly/ disassembly since most designs require impeller disassembly for seal removal.

Regardless of the method of rotor assembly, residual stresses remaining after assembly can deflect the rotor. Small amounts of rotor deflection (0.001" runout) can cause large amounts of rotor unbalance. If a rotor is low speed balanced in a deflected state, high amounts of vibration due to rotor unbalance will be experienced in the field. This is because the normal forces on the impeller will remove the residual shaft stresses at operating speed resulting in an unbalanced rotor.

In order to prevent this occurrence, all compressor vendors and qualified repair facilities check the rotor deflection (runout) before and after the addition of each impeller to the shaft. If the runout changes by more than 0.001–0.002", the impeller must be removed and reassembled. The usual cause of shaft deflection is the inadequate expansion of the component prior to assembly which results in an uneven mounting on the shaft.

For the reasons stated above, only qualified repair facilities should be used for rotor disassembly and/or assembly. Prior to proceeding, the vendors' procedure should be reviewed and approved by the original equipment manufacturer (O.E.M.). If the O.E.M. is not performing the procedure, consideration should also be given to a required witness of the rotor assembly based on the exposure to high product revenue losses if the rotor must be removed from the compressor in the field and replaced.

Before leaving this subject, it should be mentioned that all other components must also be shrunk on the shaft. However, their shrink fit values per inch of shaft diameter will be considerably lower since they do not carry a torque load. Typical shrink fit values for these components are on the order of 0.0005 inch per inch of shaft diameter. Figure 16.5 defines the relationship used for thermal expansion. This relationship can be used to determine the temperatures components must be heated to for assembly and disassembly.

Rotor/case thermal movement

Since the rotor is in direct contact with the steam and has significantly less mass than the case, a rapid gas temperature increase can yield potentially damaging effects to the rotor and stationary compressor parts.

Both the rotor and case change dimensions by the relationship

ΔL (in) = L (in) x 0.0000065 (in/in-°F) x ΔT°F

But the rotor rate of change will be greater than the casing (stationary parts).

Based on rotor rate of change = 2x case rate of change

ROTOR/CASE CONFIGURATIONS

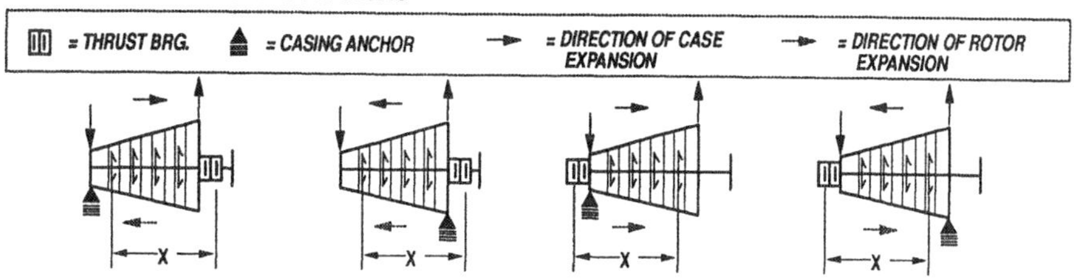

Figure 16.5 Rotor/case thermal movement

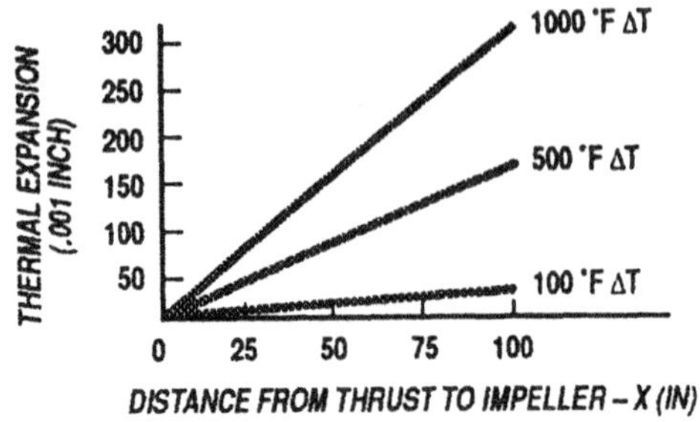

Figure 16.5 Continued

Rotor balance

A rotor is considered to be balanced when the center of rotation (shaft centerline) coincides with the center of rotor mass. All compressor rotors are balanced to a specified limit that will allow the rotors to operate in the field with a minimum of shaft vibration.

The most common criteria for allowable shaft vibration during vendor test is defined by the American Petroleum Institute (A.P.I.) as follows:

$$\text{Allowable shaft vibration} = \sqrt{\frac{12000}{\text{RPM}}}$$

(Mills-Peak to Peak Unfiltered)

Where: RPM = Maximum continuous speed. Examples of allowable shaft vibration are shown below.

RPM	Allowable API value
3,000	2 mils
6,000	1.44 mils
12,000	1.0 mil

In order to obtain these values of allowable vibration on test, vendors use the following A.P.I. guidelines to determine the allowable level of unbalance:

$$\text{Unbalance (oz-in)} = \frac{4W}{N}$$

Where: W = Shaft weight in lbs at the journal bearing location
N = Shaft speed in RPM

Figure 16.6 shows allowable unbalance values as a function of shaft speed for two cases of shaft weight bearing loads.

The standard rotor balance procedure is to progressively balance the rotor to established limits of 4 W/N at low speed (approximately 500

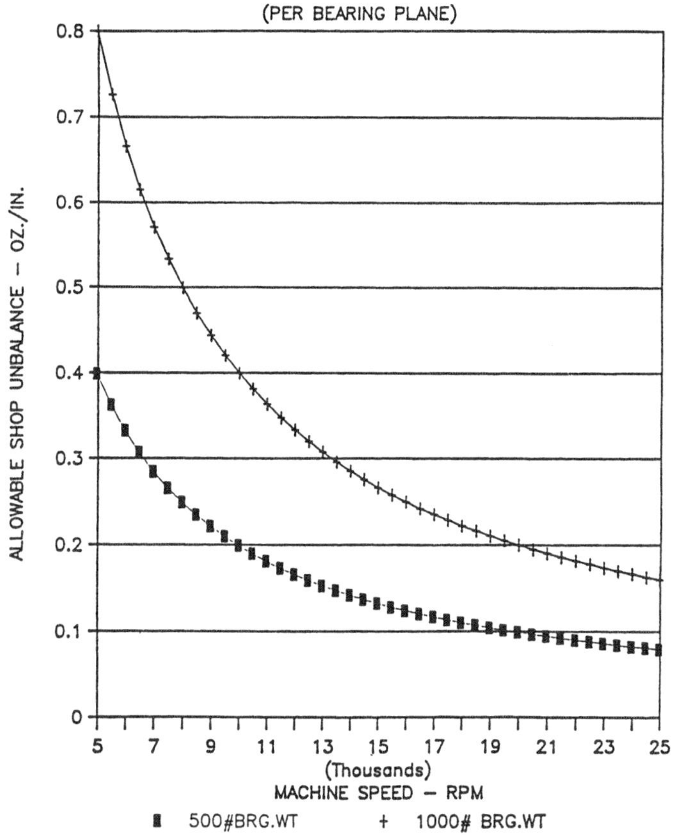

Figure 16.6 Unbalance tolerance (4W/N) (per bearing plane) (Courtesy of M.E. Crane Consultant)

RPM). A progressive balance procedure is based on using previously balanced components and checking the balance against the limits after the assembly of each major component. There are many variations of this procedure. Readers are advised to review all repair shop balance procedures prior to start of work. Many end users require all shop rotor balances to be witnessed. Recently, some vendors have eliminated progressive balance in lieu of high speed balance. This procedure significantly reduces balance time and cost but requires that successive rotor balances must always be performed at high speed since there is no low speed reference balance. More importantly, the contributing source of rotor imbalance is hard to define since a progressive balance was not performed.

I personally always require a low speed progressive rotor balance. In certain cases a high speed 'check balance' should be performed. If corrections are required, it is because the mode shape of the rotor at operating speed is significantly different from the mode shape of the rotor at balance (low) speed.

Over the last 10–15 years, high speed rotor balance has been popular. High speed balance procedures were originally designed for highly sensitive rotors with low values of rotor stiffness (K > 10) that had a history of field vibration problems. High speed balance machines have the capability of balancing the rotor at rated speed. This is achieved by operating in a vacuum and providing sufficient power to operate at this speed.

Applications that can benefit from high speed balance are:

1. Newly designed sensitive rotors (K > 10).
2. Rotors that have been disassembled and will not run in their casings prior to field start-up (new or spare rotors).
3. Use of the high speed balance machine as a test cell for spare rotors if delivery is critical.

As a precautionary measure, it is advisable to confirm the effectiveness of a high speed balance facility by checking past users to confirm that the balance level achieved in the high speed facility was repeated in the field.

Flexible coupling design, installation and operation

- Introduction
- The coupling function
- Types
- The coupling system
- Coupling installation and removal
- Enclosed coupling guards
- Field retrofits from lubricated to dry couplings

Introduction

In this chapter, the subject of couplings or constant speed ratio transmission devices will be discussed. Every equipment train contains one or more couplings. The function of couplings will be defined. The types of couplings most frequently used will be presented and their limitations discussed.

In the writers' experience, the root cause of most coupling failures has not been the coupling itself, but the 'coupling system' (shaft mounting, coupling guard design, etc). The coupling system will be discussed with emphasis on coupling installation, removal and coupling guard design.

The coupling function

The function of a flexible coupling is to transmit torque from the driver to the driven machine while making allowances for minor shaft misalignment and shaft end position changes between the two

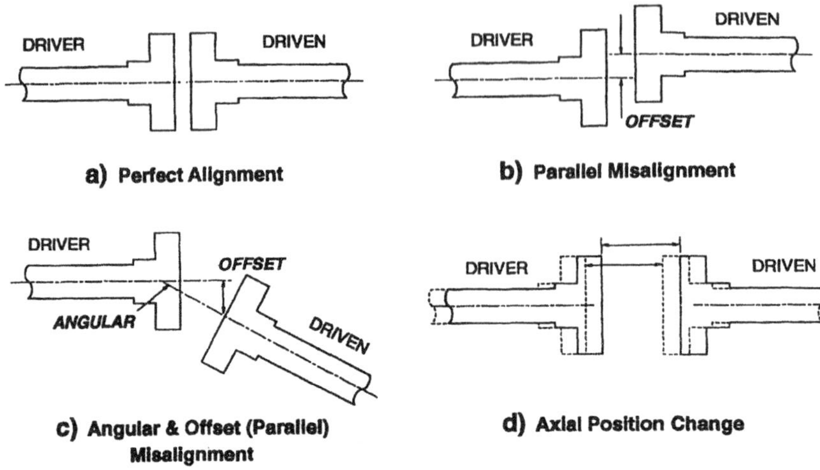

Figure 17.1 Shaft misalignment and axial position

machines. The design of the coupling should provide for transmission of the required torque at the required speed with a minimum of extraneous forces and perturbations exerted on either the driver or driven shaft. Shaft misalignment exists when the centerlines of two shafts joined by a coupling do not coincide. Figure 17.1 shows the various types of misalignment and shaft end position changes that can occur.

Each coupling type has a maximum tolerance of misalignment and axial position change that is noted on the coupling drawing. Regardless of coupling type, misalignment tolerance is stated in degrees and is usually 1/4°. Axial position change tolerance varies with coupling type. Gear type couplings have a large axial position change tolerance compared to flexible element types.

Types

The following is a list of various types of flexible couplings:

I Gear couplings
   A. Continuous lubrication
   B. Grease packed

II Flexible membrane or flexible disc couplings
   A. Single membrane type
   B. Multiple membrane or multiple disc type

III Couplings with elastomer insert flexible drive members.

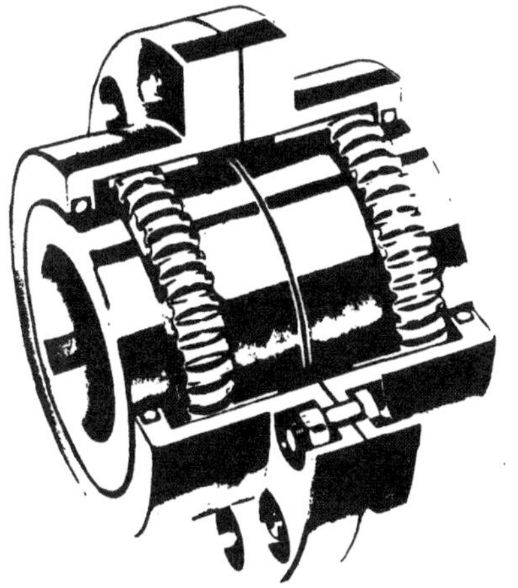

Figure 17.2 Gear tooth coupling (grease packed) (Courtesy of Zurn Industries)

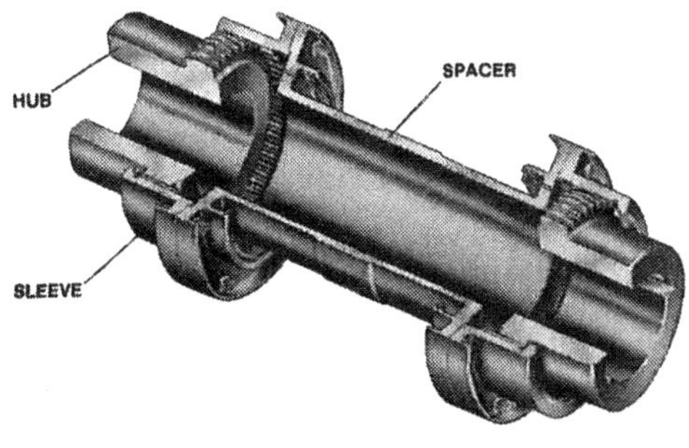

Figure 17.3 Continuously lubricated gear type coupling with spacer (Courtesy of Zurn Industries)

Gear couplings

Gear type couplings are shown in Figures 17.2 and 17.3.

Gear couplings usually include two separate gear mesh units. Each gear mesh unit consists of an external gear which fits closely into an internal gear. The internal gear can either be part of the coupling hub assemblies or mounted on each end of the coupling spacer assembly. If the internal gears are hub mounted, then the external gears are spacer mounted and vice-versa.

Grease pack couplings (see Figure 17.2) are normally designed with hub mounted external gears and the internal gears are part of a sleeve type spacer which serves as a retainer for the grease lubrication. The flange joint of the sleeve is either precision ground to avoid lubrication leaks or has a gasket between the two flange faces. The sleeve ends are fitted with 'O' ring seals to keep dust out and lubrication in.

In recent years, flexible element couplings have been used almost exclusively. However, many gear type couplings are still in use. They are the most compact coupling for a given amount of torque transmission of all the coupling designs. For this reason, they also have the least overhung weight. In addition, the gear coupling can adapt more readily to requirements for axial growth of the driver and driven shafts. Axial position change tolerances are on the order of 1/2" or greater.

There is a common disadvantage in all gear type flexible couplings. Any gear mesh has a break-away friction factor in the axial direction. This is caused by the high contact force between the two sets of gear teeth. The result is that the forces imposed on the driver and driven shafts are not totally predictable and are sometimes higher than desired due to the quality of the tooth machine surfaces and the inevitable build up of sludge or foreign material in the tooth mesh during extended service. These forces are detrimental to the ability of the coupling to make the required corrections for misalignment but, more importantly, can have a disastrous effect on the ability of the coupling to correct for thermal or thrust force changes between the driver and driven machines.

Both coupling manufacturers and user have long been aware of this problem and have used many methods to minimize the effect. Some of these methods are:

- Reduction of the forces between the gear teeth by increasing the pitch diameter of the gear mesh. This is often self defeating in that it results in increased size of the coupling and the coupling weight.
- Reduction of the break-away friction factor by the use of higher quality gear tooth finish and better tooth geometry and fit.
- Reduction of sludge and foreign material build up in the gear mesh by finer filtration of the coupling lubricant.
- Reduction of sludge and foreign material build up in the gear mesh by incorporating self flushing passages and ports in the coupling to allow any contaminants to pass through in the lubricant without being trapped in the gear mesh area.

These steps have been only partially successful and the problem still exists in many applications.

Coupling manufacturers are asked to quote the design break-away friction factor of their coupling as built and shipped from the factory. Machinery train designers then use this figure to calculate the maximum axial force that the coupling would be expected to exert on the connected shafts. From this information, the designers can decide if the thrust bearings adjacent to the coupling are adequate to handle the axial loads within the machine plus the possible load from the coupling resistance to any external forces.

There has been much discussion and some disagreement regarding the friction factor to be used when calculating possible thrust forces which can be transmitted by the coupling. When the coupling is in reasonably good condition, factors from .15 to .30 have been considered reasonable. Since the factor reflects the total force relationship, the coupling design can have a significant effect on the factor used. The factor is a function of the number of teeth in contact and the contact areas of each tooth plus the quality of the tooth contact surface. If we assume that the factor to be used is .30, then the axial force which must be exerted in order to allow the coupling to correct for axial spacing changes can be calculated as:

$$F_a = \frac{0.30 \times T}{Dp/2}$$

Where: F_a = Required axial force in pounds
T = Design torque in in/pounds
Dp = Pitch diameter of gear mesh in inches

We can assume then, that if we use a coupling with a six inch pitch diameter gear mesh transmitting 25,000 in/pounds of torque and a break-away friction factor of .30, the axial force required to move the gear mesh to a new axial position would be 2,500 pounds. Adjacent thrust bearings must be capable of handling this force in addition to the machine's normal calculated thrust forces. Machinery train designers and users must be aware of this and make provisions for it in the built-in safety factors of thrust bearings and machinery mounting design.

The machinery user must know that the same phenomena has an effect on machinery vibration when machinery is operated with excessive misalignment. The gear mesh position must change with each revolution of the shaft to correct for the misalignment. This results in counter axial forces on a cyclic basis since the mesh is moving in opposite directions at each side of the coupling. Vibration detection and monitoring instrumentation will show that the resulting vibration will occur at twice the running frequency of the shafts. Although the primary force generated is axial, the resultant can show up as a radial vibration due to the lever arm forces required on the coupling spacer to make the gear meshes act as ball and socket connections. **Axial or radial vibration in rotating machinery which occurs at twice the frequency of the shaft rotational speed will normally be an indication of misalignment between the two machines.**

Figure 17.3 shows a continuously lubricated, spacer gear type coupling. Spacers are usually required for component removal (seals, etc). They also provide greater tolerance to shaft misalignment. A common spacer size used for unspared (critical) equipment is 18 inches.

Flexible membrane or flexible disc couplings

Couplings in these categories do not have moving parts and derive their flexibility from controlled flexure of specially designed diaphragms or discs. They do not require lubrication and are commonly known as 'dry couplings'. The diaphragms or discs transmit torque from one shaft to the other just as do the gear meshes in a gear coupling.

The following features are common to all flexible disc or flexible membrane type couplings:

1. None require lubrication.

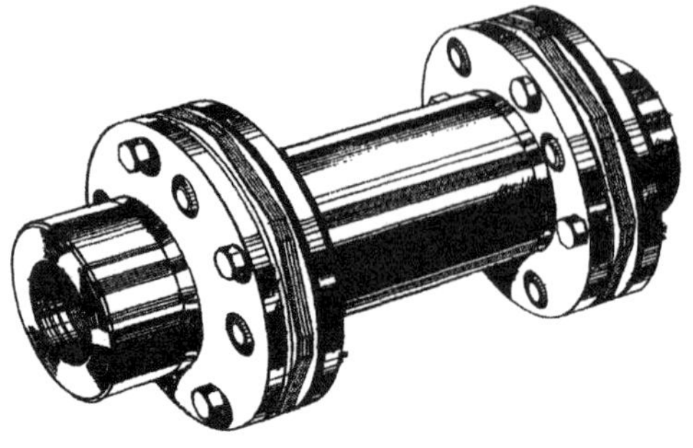

Figure 17.4 Flexible disc spacer coupling (Courtesy of Rexnord)

2. All provide a predictable thrust force curve for a given axial displacement range.
3. Properly applied, operated and maintained, none are subject to wear and have an infinite life span.
4. All provide smooth, predictable response to cyclic correction for minor misalignment.

It should be noted that **none** of the above comments can be applied across the board to gear type flexible couplings. For this reason, more and more special purpose machinery trains are being supplied with flexible metallic element couplings in their design. Many users do not allow the use of gear type coupling for critical (unspared) applications.

The following is a discussion of the various types of 'dry' couplings with comments pertaining to their application ranges and limitations.

Figure 17.4 shows a typical flexible disc coupling.

This is the most common type and is generally used for general purpose applications (pumps, fans, etc). The major consideration with this type of coupling is assuring the shaft end separation (B.S.E.) is within the allowable limits of the couplings. This value is typically only 0.060" for shaft sizes in the 1" – 2" range. At shaft sizes over 4" the maximum end float can be 0.150". Exceeding the allowable end float will significantly increase the axial load on the thrust bearings of the equipment and can fail the coupling discs. A single diaphragm, spacer type coupling is shown in Figures 17.5 and 17.6. Figure 17.5 is a cutaway view and Figure 17.6 presents a two dimensional assembly drawing.

This type of coupling is commonly used for critical (unspared) applications where axial end float values are less than .125". This limit is based on an approximate axial float of ± 0.062". If end float is greater than 0.125", a convoluted (wavy) diaphragm or multiple type diaphragm must be used. During disassembly, care must be taken when

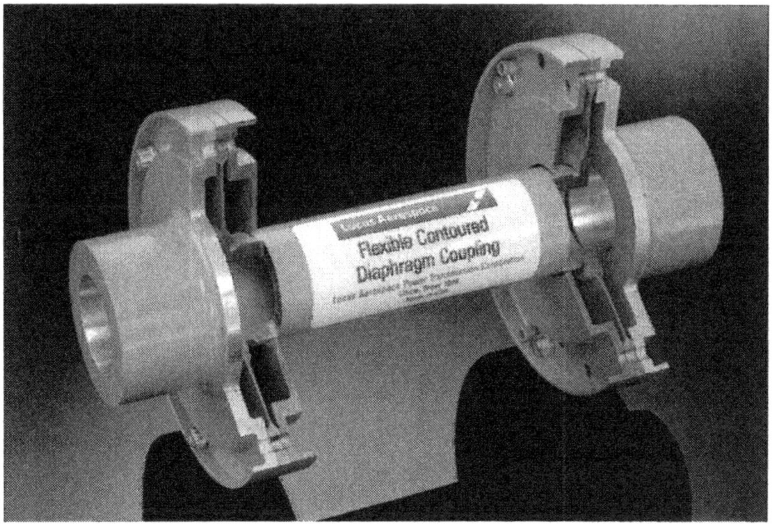

Figure 17.5 Single diaphragm spacer coupling (Courtesy of Lucas Aerospace)

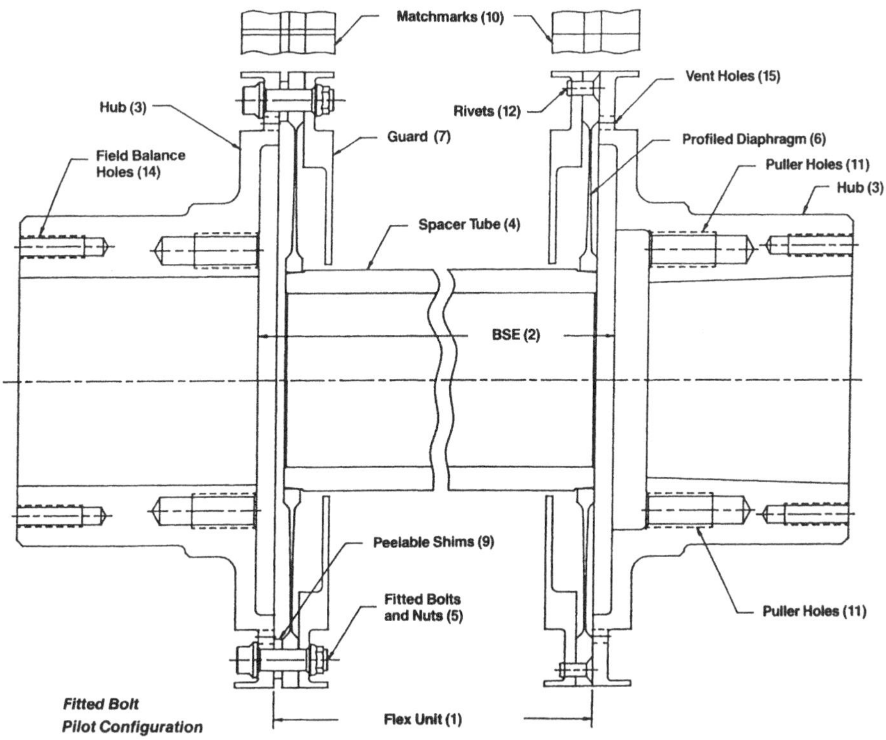

Figure 17.6 Single diaphragm spacer coupling (Courtesy of Lucas Aerospace)

Rotor/Case Thermal Movement – Steam Turbine

Since the Rotor is in direct contact with the Gas and has significantly less mass than the case, a rapid gas temperature increase can yield potentially damaging effects to the rotor and stationary compressor parts.

Both the rotor and case change dimensions by the relationship

$$\Delta L \text{ (IN)} = L \text{ (IN)} \times 0.0000065 \text{ (IN/IN-°F)} \times \Delta T\text{°F}$$

But the rotor rate of change will be greater than the casing (stationary parts).

ROTOR/CASE CONFIGURATIONS

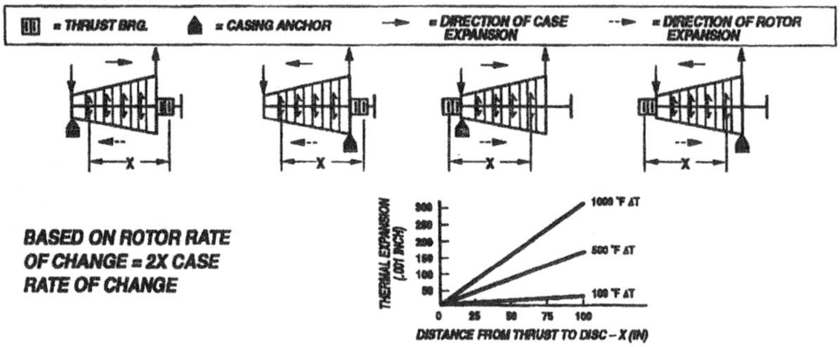

Figure 17.7 Rotor/case thermal movement – steam turbine

removing the spacer to not scratch or dent the diaphragm element. A dent or even a scratch that penetrates the protective coating can cause a diaphragm failure.

Regardless of the type of diaphragm couplings, it is common practice to 'pre-stretch' these couplings to take full advantage of the maximum available end float. Readers are cautioned to always require equipment vendors provide axial shaft movement calculations in order to confirm that the coupling maximum end float is not exceeded. Figure 17.7 graphically displays the various combinations of end shaft movement and the calculation method.

Figure 17.8 is a picture of a multiple, convoluted (wavy) diaphragm spacer coupling.

This type coupling is used whenever large values of axial end float exist. Axial end float values as high as ± 0.875" are attainable with this type of coupling.

As previously mentioned, gear type couplings provide the lowest value of overhung weight (coupling moment) on the bearing. However, a dry type coupling will usually have a higher coupling moment because the flexible assembly is farther from the bearing centerline than the gear teeth in a gear coupling. An excessive coupling moment will reduce the second natural frequency (N_{c2}) of a turbo-compressor and could move it close to or within the operating speed range. A solution in these cases

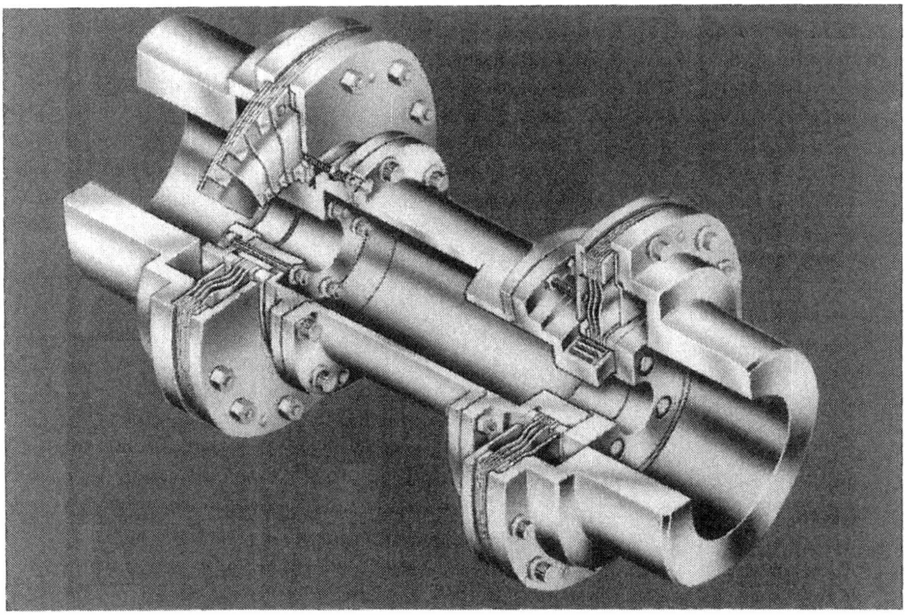

Figure 17.8 Multiple, convoluted diaphragm-spacer coupling (Courtesy of Zurn Industries)

can be to use a reduced moment diaphragm coupling as shown in Figure 17.9.

In this design, the diaphragm is moved to the back of the hub and the flange diameter is reduced thus significantly reducing the coupling moment. The reduced moment coupling approaches the gear coupling in term of coupling moment value.

Couplings with elastomer insert flexible drive members

This type coupling is normally used only for low horsepower, general purpose applications. Their limitations are based primarily on the wear factor and the difficulty in maintaining shape and concentricity of the elastomer insert. These items have a tendency to limit the maximum design speed at which such couplings can be operated. A typical 'Jaw and Spider' type is shown in Figure 17.10.

One exception is a special design used for synchronous motor driven compressor trains. A characteristic of synchronous motors is a variable oscillating torque that decreases linearly in frequency from 2X line frequency (50 HZ or 60 HZ) at 0 RPM to 0 frequency at rated RPM. Figure 17.11 shows a plot of motor RPM vs transient torsional excitation frequency.

The excitation frequency inherent in all synchronous motors will excite all torsional natural frequencies present between 2x line frequency and 0 RPM.

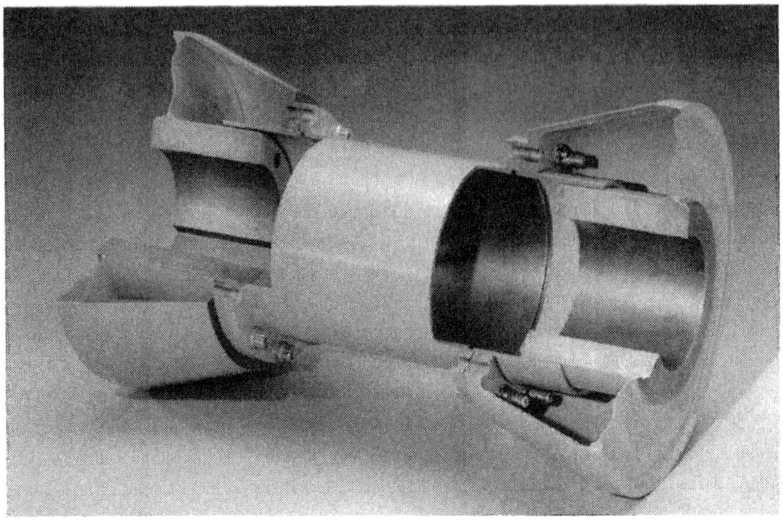

Figure 17.9 Reduced moment convoluted (wavy) diaphragm spacer coupling (Courtesy of Lucas Aerospace)

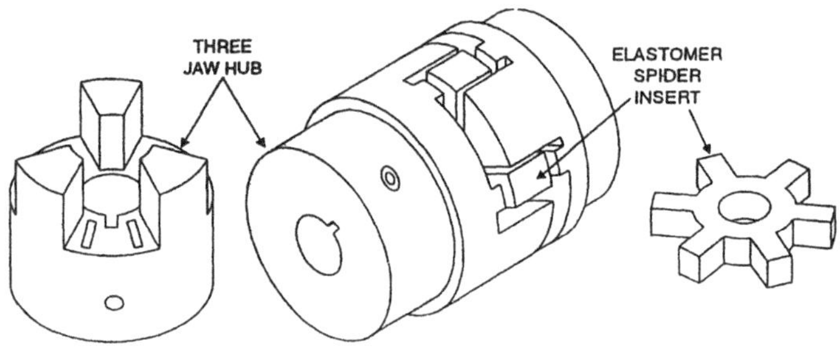

Figure 17.10 Jaw and spider coupling

When the motor torsional excitation frequency briefly coincides with a torsional natural frequency, torque values can amplify to as much as 5 or 6 times full load torque. The 'Holset' or elastomeric coupling shown in Figure 17.12 significantly reduces the torque amplification by dampening out the response in the elastomeric elements. The hardness of these elements is controlled to limit the maximum amplification factor to an acceptable value (usually 2–3 X rated torque).

The coupling system

It has been the writers' experience that if couplings are properly selected, the root cause of failure, if it occurs, is in the coupling system.

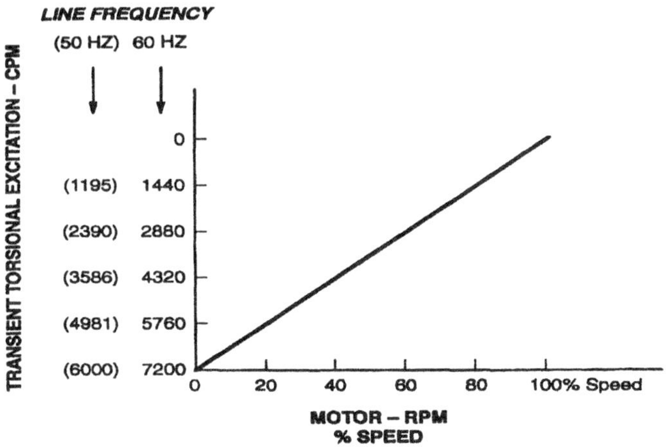

Figure 17.11 Transient torsional excitation - Frequency vs. motor speed

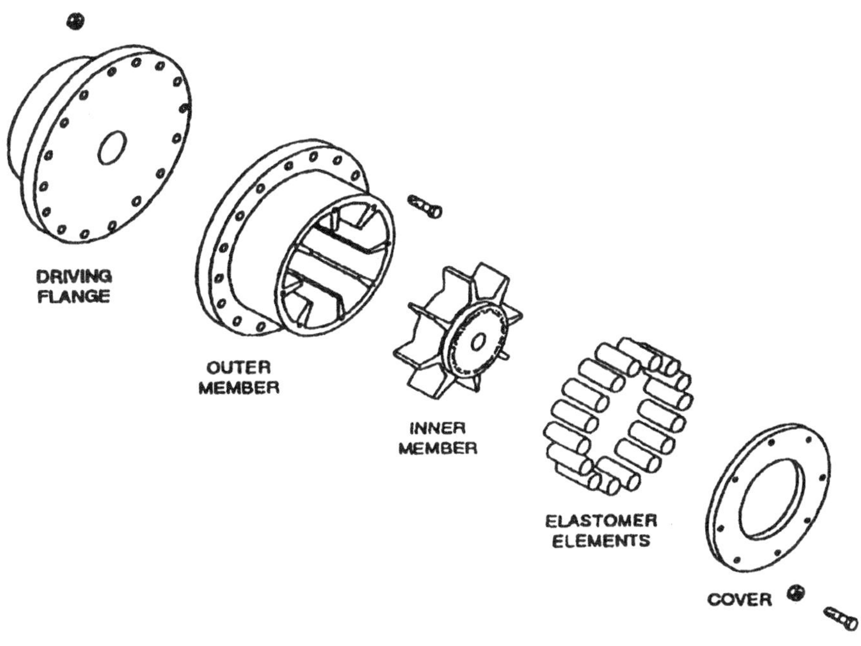

Figure 17.12 Holset coupling (exploded view) non-spacer type

The 'coupling system' must continuously transmit torque safely between the driver and driven equipment and must allow for changes in shaft misalignment and axial movement. The components that make up the shaft system are:

- The driver shaft
- Driver shaft/coupling fit
- Coupling
- Driven shaft/coupling fit
- The driven shaft
- The coupling spacer system
- Lubrication (if required)
- Cooling system (if required)

A schematic of a coupling system is shown in Figure 17.13.

The reliability of the coupling is a function of the coupling system design and assembly. If any of the items noted above are not properly designed or assembled a coupling failure can occur.

Coupling assembly/disassembly errors and enclosed coupling guard design are two important areas that are critical to coupling system reliability.

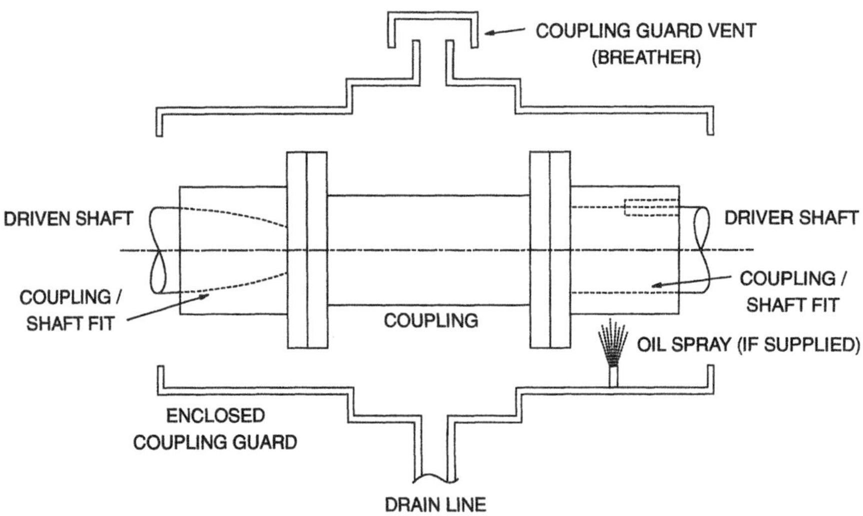

Figure 17.13 The coupling system (Courtesy of M.E. Crane Consultant)

Coupling installation and removal

The most common methods of coupling attachment are:

- Key fit
- Spline fit
- Hydraulic fit

Key fits are used whenever possible. They are the most common method of shaft fit. It is important to assure keys and keyways are properly manufactured to avoid problems with removal or breakage. Key fits will be used on equipment that does not require coupling removal to remove shaft components (seals, bearings, etc). Keyed fits are usually used on motors, gearboxes and most pumps and small steam turbines. Since heat is usually required to remove keyed couplings, they will not be used where removal in the field is necessary.

In these applications either spline or hydraulic fits are used.

Spline fits consist of a male (on the shaft) and female (in coupling hub) finely machined mating gear teeth with line to line fit (no backlash). When assembled on the shaft, the fit is rigid and provides no flexibility. Spline fits are commonly used in the gas turbine industry. They do not usually require heat for removal.

Hydraulic fits are used where heat to remove the coupling hub is either not available or not permitted. Usually, turbo-compressors will utilize hydraulic fits for this reason, since hydrocarbon gas, usually present, requires a flame free environment. Figure 17.14 shows a typical coupling hydraulic shrink fit arrangement. Note that the entire torque load is transmitted by the shrink fit and that no keys are used!

The equipment vendor calculates the required shrink fit based on the shaft and coupling dimensions. Typical values of hydraulic shrink fit are 0.002 inch/inch of shaft diameter.

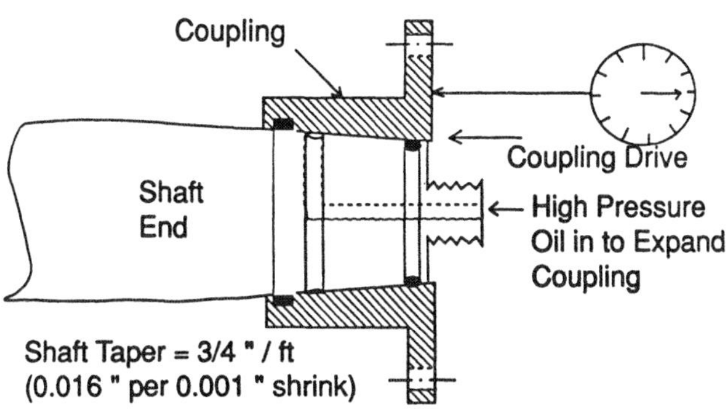

Figure 17.14 Typical coupling hydraulic shrink fit

For ease of hydraulic fit assembly and disassembly, all shafts and coupling hubs are tapered. Different shaft/coupling hub matching tapers are used. The most common are:

- ½° taper
- ½" per foot taper
- ¾" per foot taper (shafts above 4" diameter)

Once the shrink fit is calculated, the value appears on the coupling drawing and is usually expressed as 'drive' or 'push' based on the shaft taper. This is the axial distance the coupling must be moved up the shaft. The coupling drive per 0.001" of shrink fit for the most common shaft tapers is noted below:

Shaft taper	Drive per 0.001 shrink
½°	0.057"
½" per foot	0.024"
¾" per foot	0.016"

As an example, a hydraulic fit coupling with a 4" bore requires a 0.008" shrink fit (i.e. the bore diameter is 0.008 less than the shaft). To expand the coupling bore 0.008", what is the drive if the shaft taper is:

Taper	Drive
½°	0.456"
½" per foot	0.192"
¾" per foot	0.128"

Since the load torque is completely transmitted by the shrink fit, one can see the importance of assuring that the correct shrink fit (or drive) is obtained. The shrink fit amount is directly proportional to torque load capability. If the shrink fit is 50% of the specified value, so is the torque capability! However, industry specifications require that the shrink fit at minimum tolerances be a minimum of 125% greater than the driver maximum torque. Observing the calculated drives in the example above it can be seen that the smaller the shaft, the more critical the correct drive becomes for a given shaft taper. The coupling drive is measured by positioning a dial indicator on the coupling hub and measuring the axial distance traveled during coupling assembly.

Figure 17.15 shows a typical hydraulic coupling mounting arrangement used by Dresser-Rand.

All turbo-compressor manufacturers use similar arrangements. There are some slight differences which are:

- STANDARD ON ALL DRESSER-RAND COMPRESSORS WITH CYLINDRICAL SHAFTS
- SIMPLE, RELIABLE SHRINK FIT
- TORQUE TRANSMITTED WITHOUT KEYS

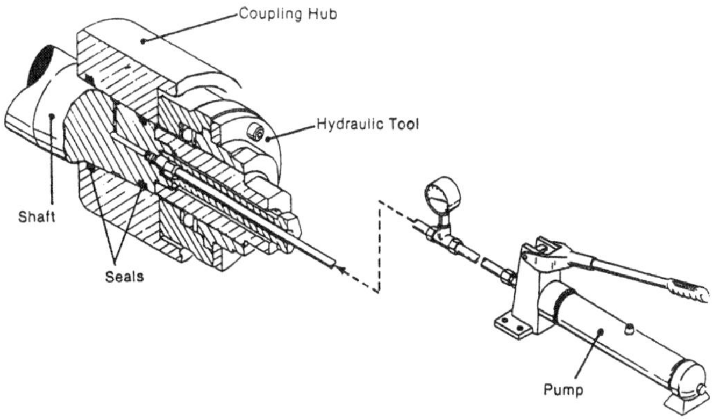

Figure 17.15 Hydraulic fit coupling (Courtesy of Dresser-Rand Corp)

- Hydraulic oil enters the coupling hub and not the shaft.
- An additional pump is used to move the hydraulic tool axially (Figure 17.15 shows a nut which is manually turned to push the coupling axially).

The basic coupling mounting procedure is as follows: (readers must refer to the specific vendors' instruction book for the exact procedure)

1. Clean shaft end and coupling bore with light oil.
2. Remove all 'O' rings from shaft end and coupling.
3. Lightly blue the coupling hub.
4. Push coupling on shaft **without 'O' rings** and tap hub with wood to assure tight fit.
5. While hub is on shaft, index coupling hub axial position relative to a machined surface on shaft (usually shaft end).
6. Remove hub and confirm contact area of blue is a minimum of 85%. If not, correct as required.
7. When coupling contact of 85% is confirmed, clean shaft and coupling hub and install shaft and coupling 'O' rings.
8. Hand push coupling on shaft to **indexed position in step 5. NOTE: It may be necessary to use pump since 'O' rings can provide significant resistance to movement.**
9. **With hub at indexed (zero drive) position**, use hand pump to push coupling axially to value noted on coupling drawing.

Coupling drive must be within tolerances noted. **NOTE: Pump pressures will be high. Be extremely careful when connecting pump and tubing. Be sure to secure pump so that hand jacking cannot break tubing.** Pressures typically required range from 15,000–30,000 PSI depending on shaft dimensions, coupling dimensions and shrink fit.

10. When coupling is on shaft correct amount, do not remove dial indicator but reduce pump pressure to zero and **back off hydraulic tool slightly. Observe that dial indicator does not move before removing tool.**
11. Promptly assemble shaft end coupling nut.
12. Measure between shaft end dimension (B.S.E.) to assure it is as stated on coupling drawing before assembling coupling spacer. If this dimension is not correct, consult instruction book and O.E.M. if necessary before taking corrective action. **Under no circumstances should coupling spacers be added unless allowed by the coupling manufacturer or should equipment axial shaft position be changed without O.E.M. consent.**
13. When coupling is properly assembled check alignment using 'reverse dial indicator procedure'. **NOTE: For coupling removal, consult vendor's instruction book. Under no circumstances should coupling be pulled or heated. Usually, hydraulic pressure required for removal will be higher (5–10%) than that required for assembly. If the value required exceeds 35,000 PSI, do not proceed until consulting O.E.M. for additional options concerning removal.**

Incorrectly mounting a hydraulic coupling can cause catastrophic coupling and/or shaft end failure.

Enclosed coupling guards

Most turbo-compressor couplings are completely enclosed by a spark proof (usually aluminum) coupling guard. This is because the couplings are continuously lubricated gear type or to prevent oil siphoned from the bearing brackets by the windage action of the dry couplings. In either case, proper design of the coupling guard is essential to maintaining coupling reliability. Many coupling failures have resulted from high coupling enclosure temperatures, enclosures full with oil and debris educted into the coupling guard from the atmosphere.

As a minimum the following must be checked by the O.E.M. and coupling vendor during equipment design or field coupling retrofit from gear to dry type:

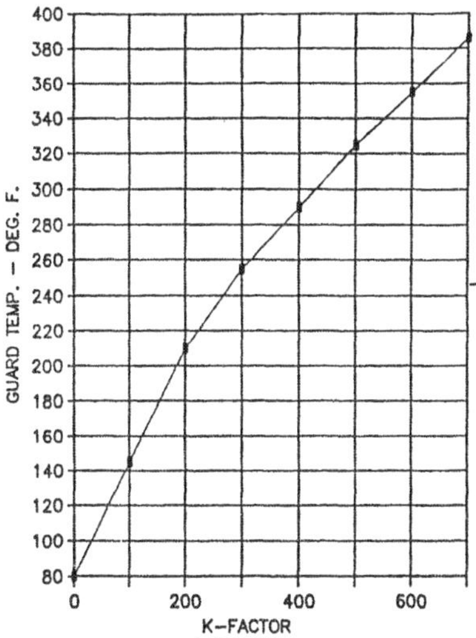

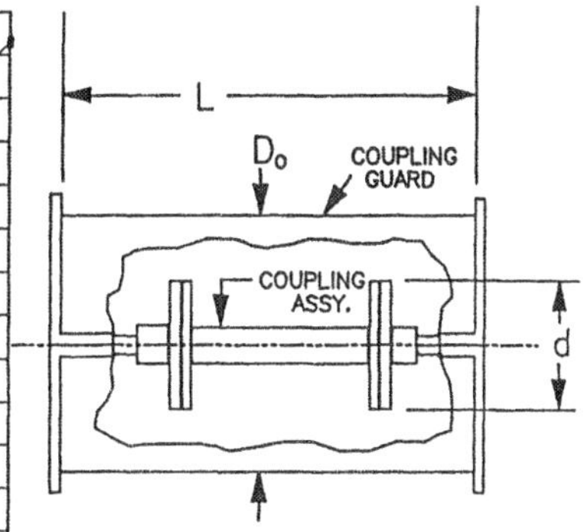

$$K = \frac{1.7 \times 10^{-13} \times d^5 \times N^3}{D_o \times L}$$

d = diaphragm O.D. – inches
N = speed – RPM
D_o = guard O.D. – inches
L = guard length – inches

COUPLING GUARD
TEMPERATURE CALCULATIONS

Figure 17.16 (Courtesy of M.E. Crane Consultant)

- Proper coupling O.D. to guard and/or bearing bracket I.D. clearance.
- Proper coupling guard baffle design to allow proper drainage. **NOTE: All enclosed coupling guards must be supplied with vent and drain.**
- Proper vent breather sizing and design.

Figure 17.16 presents coupling guard dimensional design criteria for dry type couplings operating in enclosed coupling guards. Note that in some designs, D_O may be the I.D. of the bearing bracket.

Recommend coupling guard skin operating temperatures should be below 200°F to avoid coupling and coupling guard leakage problems. Under no circumstances should coupling guard skin temperatures approach the flash point of lubricating oil (400°F for new mineral oil).

Field retrofits from lubricated to dry couplings

Considering the advantages of dry couplings, many users are retrofitting their older style lubricated couplings to dry couplings. Whenever considering a retrofit, the following action should be taken to maintain or increase coupling system reliability.

1. Consult equipment or coupling vendor for proper selection of new coupling.

2. Consult equipment O.E.M.(s) (each affected vendor) to confirm:
   - Critical speeds will not be affected
   - Coupling guard design is acceptable
3. Advise coupling vendor or any environmental considerations that may affect dry flexible element life (environmental gases, temperatures, excessive dust, etc).

Radial bearing design

- Introduction
- Anti-friction bearings
- Hydrodynamic bearings
- Hydrodynamic bearing types
- Condition monitoring
- Vibration instabilities

Introduction

Radial bearings fall into two major categories:

- Anti-friction
- Hydrodynamic

Anti-friction bearings rely on rolling elements to carry the load of the equipment and reduce the power losses or friction.

Hydrodynamic bearings rely on a liquid film, usually lubricating oil, to carry the load of the equipment and minimize friction. In general, anti-friction bearings are used for equipment of low horsepower. Hydrodynamic bearings are generally used for all rotating equipment above approximately 500 horsepower. During this section we will concentrate on the subject of hydrodynamic bearings since they are the principle bearing type used in turbo-compressor operation.

A bearing in the radial position is responsible for mainly supporting the weight of the shaft and any dynamic loads that are present in the system. It can be generally stated that the dynamic loads are in the order of 20% of the static loads on the bearing journal. In the case of gears however, the radial load component is made up principally of the meshing force of the gear teeth and the load will vary from zero to maximum torque. One must be careful in this design to assure that the various load angles occur in zones of the bearing that can support these loads.

We will examine loads on hydrodynamic bearings and present an example of the bearing sizing based on static and dynamic loads of a rotor system. We will see that there are specific oil film pressure limits which dictate the bearing dimensions (length and diameter).

The types of hydrodynamic bearings will be reviewed; plain, stationary anti-whirl types and tilt pads.

We will conclude this section by discussing the condition monitoring requirements for radial bearings and briefly discuss shaft vibration and vibration troubleshooting (diagnostics).

Anti-friction bearings

Anti-friction radial bearings support the rotor using rolling elements to reduce friction losses. They are used in low horsepower applications (below 500 H.P.) and in aero-derivative gas turbines. Examples of roller and ball type anti-friction bearings are shown in Figure 18.1.

As previously mentioned, all bearings are designed on the basis of sufficient bearing area to support all the forces acting on the bearing.

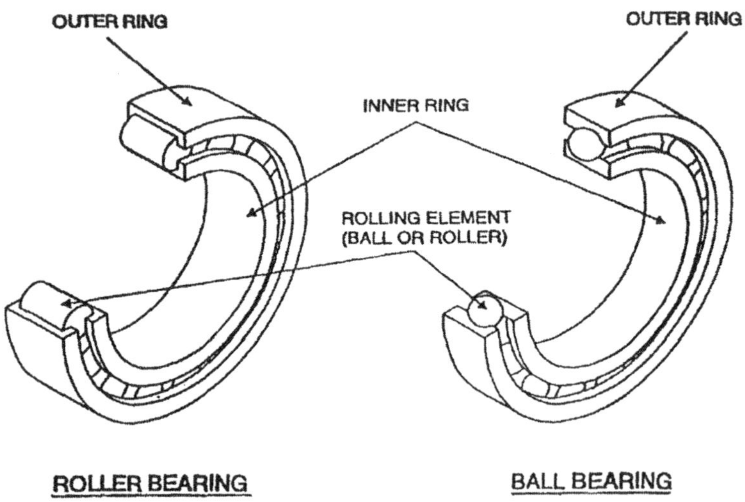

Figure 18.1 Anti-friction bearings

That is:

$$P = \frac{F}{A}$$

Where: P = Pressure on the bearing elements – P.S.I.
F = The total of all static and dynamic forces acting on the bearing
A = Contact area

For anti-friction bearing applications, the pressure, P is the point contact or 'Hertzian' stress on the bearing elements and rings or 'races'. For an anti-friction bearing to be properly designed, its D-N number and bearing life must be determined. Figure 18.2 presents the definition of D-N number and its uses.

D-N Number

- Is a measure of the rotational speed of the anti-friction bearing elements
- D-N number = bearing bore (millimeters) x speed (RPM)
- Approximate lubrication ranges

Lubrication type	D-N Range
Sealed	below 100,000
Regreaseable	100,000–300,000
Oil lube (unpressurized)	300,000–500,000
Oil lube (pressurized)	above 500,000

Figure 18.2 D-N number

Each type of anti-friction bearing has a maximum operating D-N number. If this value is exceeded, rapid bearing failure can occur. In addition, D-N numbers are typically used to determine the type of lubrication required for bearings. A common practice in the turbomachinery industry has been to use hydrodynamic bearings when the D-N number exceeds approximately 500,000.

The exception to this rule is aircraft gas turbines which can have D-N numbers in excess of 2,000,000. In these applications, hydrodynamic bearings are not used since the size and weight of the required lubrication system would be prohibitive.

Anti-friction bearings possess a finite life which is usually specified as 'B-10' or 'L-10' life as defined in Figure 18.3.

'B' OR 'L' – 10 Life

'B' or 'L' – 10 Life is defined as the life in hours that 9 out of 10 randomly selected bearings would exceed in a specific application.

$$\text{'B' or 'L' – 10 Life} = \frac{16700}{N}\left[\frac{C}{F}\right]^3$$

Where: N = RPM
C = Load in LBS that will result in a bearing element life of 1,000,000 revolutions.
F= Actual load in LBS

Figure 18.3 'B' or 'L' – 10 life

An important fact to note is that the life of any anti-friction bearing is inversely proportional to cube of the bearing loads. As a result, a small change in the journal bearing loads can significantly reduce the bearing life. When anti-friction bearings suddenly start failing where they did not in the past, investigate all possible sources of increased bearing loads (piping forces, foundation forces, misalignment, unbalance, etc.). Anti-friction bearings are usually designed for a minimum life of 25,000 hours continuous operation.

Hydrodynamic bearings

Hydrodynamic bearings support the rotor using a liquid wedge formed by the motion of the shaft (see Figure 18.4).

Oil enters the bearing at supply pressure values of typically 15–20 psig. The shaft acts like a pump which increases the support pressure to form a wedge. The pressure of the support liquid (usually mineral oil) is determined by the area of the bearing by the relationship:

$$P = \frac{F}{A}$$

Where: P = Wedge support pressure (P.S.I.)
F = Total bearing loads (static and dynamic)
A = Projected bearing area ($A_{PROJECTED}$)

$$A_{PROJECTED} = L \times d$$

Where: L = Bearing axial length
d = Bearing diameter

As an example, a 4" diameter bearing with an axial length of 2" (L/d = .5) would have $A_{PROJECTED}$ = 8 in^2.

If the total of static and dynamic forces acting on the bearing are 1600 lbs. force, the pressure of the support wedge is:

$$P = \frac{1600\ Lb_{FORCE}}{8\ in^2} = 200\ P.S.I.$$

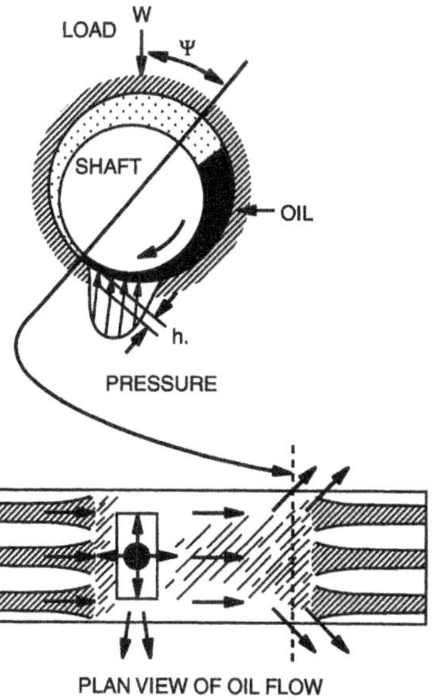

Figure 18.4 Hydrodynamic Lubrication (Courtesy of Bently Nevada Corp.)

The maximum desired design wedge pressure for oil is approximately 500 psi. However, it has been common practice to limit hydrodynamic bearing loads to approximately 250 psi in compressor applications. Figure 18.5 is a side view of a simple hydrodynamic bearing showing the dynamic load forces.

The primary force is the load which acts in the vertical direction for horizontal bearings. However, the fluid tangential force can become

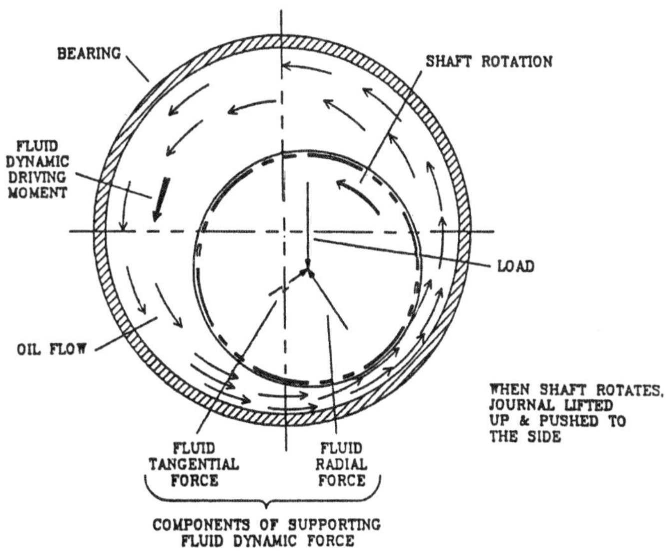

Figure 18.5 Shaft/bearing dynamics (Courtesy of Bently Nevada Corp.)

large at high shaft speeds. The bearing load vector then is the resultant of the load force and fluid tangential force. The fluid radial force opposes the load vector and thus supports the shaft. It has been demonstrated that the average velocity of the oil flow is approximately 47–52% of the shaft velocity. The fluid tangential force is proportional to the journal oil flow velocity. If the fluid tangential force exceeds the load force, the shaft will become unstable and will be moved around the bearing shell. This phenomena is known as oil whirl.

Hydrodynamic bearing types

Regardless of the type of hydrodynamic bearing, all bearing surfaces are lined with a soft, surface material made of a composition of tin and lead. This material is known as Babbitt. Its melting temperature is above 400°F, but under load will begin to deform at approximately 320°F. Typical thickness of Babbitt over steel is 0.060 (1.5mm). Bearing embedded temperature probes are a most effective means of measuring bearing load point temperature and are inserted just below the Babbitt surface. RTD's or thermocouples can be used. There are many modifications available to increase the load effectiveness of hydrodynamic bearings. Among the methods available are:

- Copper backed Babbitt or 'Trimetal' – to aid in heat removal
- Back pad cooling – used on tilt pad bearings to remove heat
- Direct cooling – directing cool oil to maximum load points

A typical straight sleeve hydrodynamic journal bearing is shown in Figure 18.6.

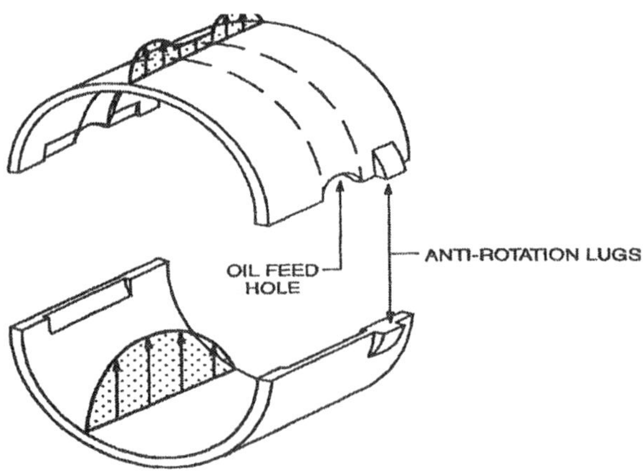

Figure 18.6 Straight sleeve bearing liner (Courtesy of Elliott Co.)

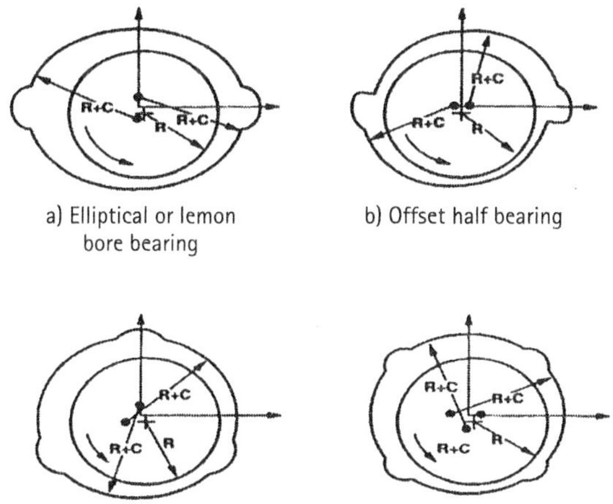

a) Elliptical or lemon bore bearing

b) Offset half bearing

c) Three lobe bearing

d) Four lobe bearing

Figure 18.7 Prevention of rotor instabilities

Straight sleeve bearings are used for low shaft speeds (less than 5,000 RPM) or for older turbo-compressor designs. Frequently, they are modified to incorporate a pressure dam, in the direction of rotation. The pressure dam must be positioned in the top half of the bearing to increase the load vector (see Figure 18.5). This action assures that the tangential force vector will be small relative to the load vector thus preventing shaft instability. It should be noted that incorrectly assembling the pressure dam in the lower half of the bearing would render this type of bearing unstable. When shaft speed is high, other alternatives to prevent rotor instabilities are noted in Figure 18.7.

Shown are examples of anti-whirl bearings. The most common types of these bearings are the 3 and 4 lobe design. Elliptical and offset bearing designs do prevent instabilities but tend to increase shaft vibration if the load vector passes through the major axis of the bearing. These types of bearings may have to be rotated in the bearing brackets to prevent this occupance.

The most common hydrodynamic bearing for higher speed applications is the tilt pad journal bearing shown in Figure 18.8.

A tilting pad bearing offers the advantage of increased contact area since the individual pads conform to the shaft orbit. In addition, this type is also a highly effective anti-whirl bearing since the spaces between the pads prevent oil whirl.

Most end users specify tilt pad radial and thrust bearings for turbo-compressor applications.

Figure 18.9 shows the mechanical frictional losses and oil flow requirements for a tilt pad journal bearing as a function of shaft speed.

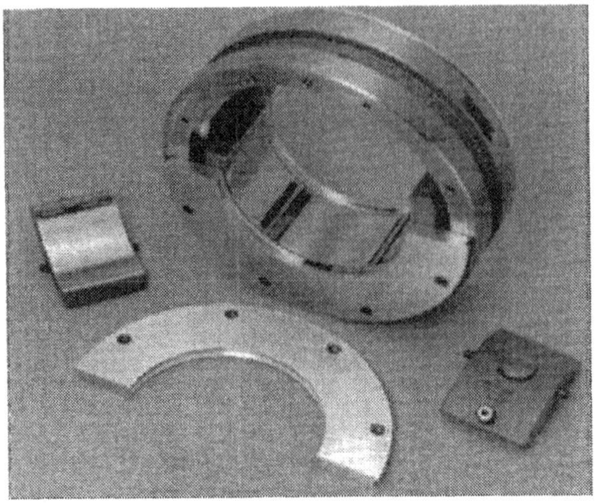

Figure 18.8 Tilting pad journal bearing assembly (Courtesy of Kingsbury, Inc.)

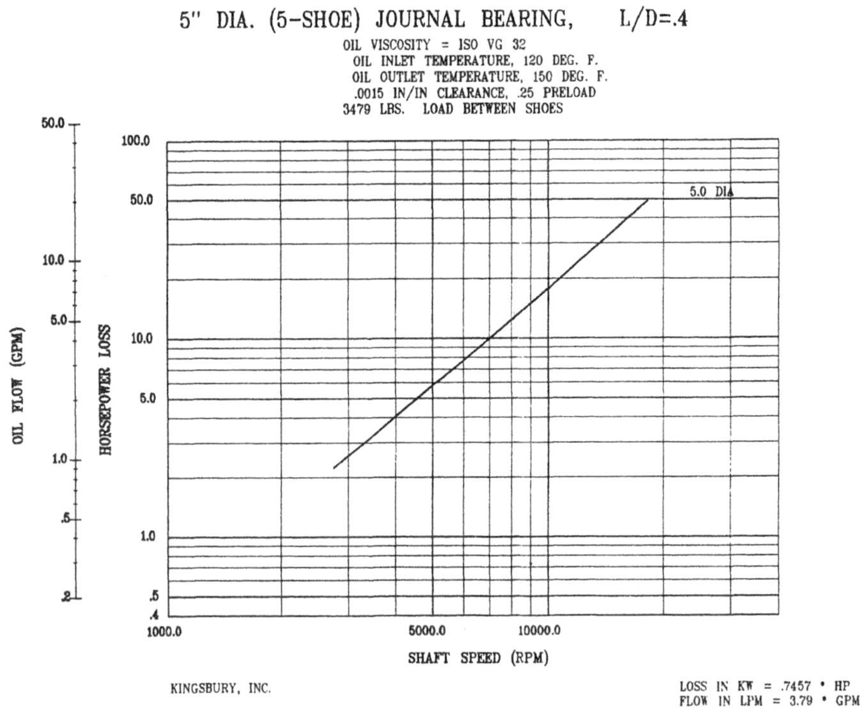

Figure 18.9 Typical journal bearing selection curve (Courtesy of Kingsbury, Inc.)

Note that the basis for horsepower loss and oil flow is an oil temperature rise of 30°F. This is the normal design ΔT for all hydrodynamic bearings. Also given in this figure is the data necessary to calculate bearing pressure at the load point.

As an exercise calculate the following for this bearing:

- Projected Area

$A_{PROJECTED}$ = 5" x 2"

= 10 square inches

- Pressure

= 3479 Lb force ÷ 10 square inches

= 347.9 psi on the oil film at load point

Condition monitoring

In order to determine the condition of any journal bearing, all the parameters that determine its condition must be monitored.

Figure 18.10 presents the eight parameters that determine the condition of a hydrodynamic journal bearing along with typical limits. Attendees are advised to consult the manufacturers instruction book for vendor recommended limits.

Parameter		Limits
1	Radial vibration (peak to peak)	2.5 mils (60 microns)
2	Bearing pad temperature	220°F (108°C)
3	Radial shaft position (except for gearboxes where greater values are normal from unloaded to loaded operation)	>30° change and/or 30% position change
4	Lube oil supply temperature	140°F (60°C)
5	Lube oil drain temperature	190°F (90°C)
6	Lube oil viscosity	Off spec 50%
7	Lube oil flash point	Below 200°F (100°C)
8	Lube oil particle size	Greater than 25 microns

Condition monitoring parameters and their alarm limits according to component:

1. Journal bearing (hydrodynamic)

Figure 18.10 The eight parameters that determine the condition of a hydrodynamic journal bearing along with typical limits

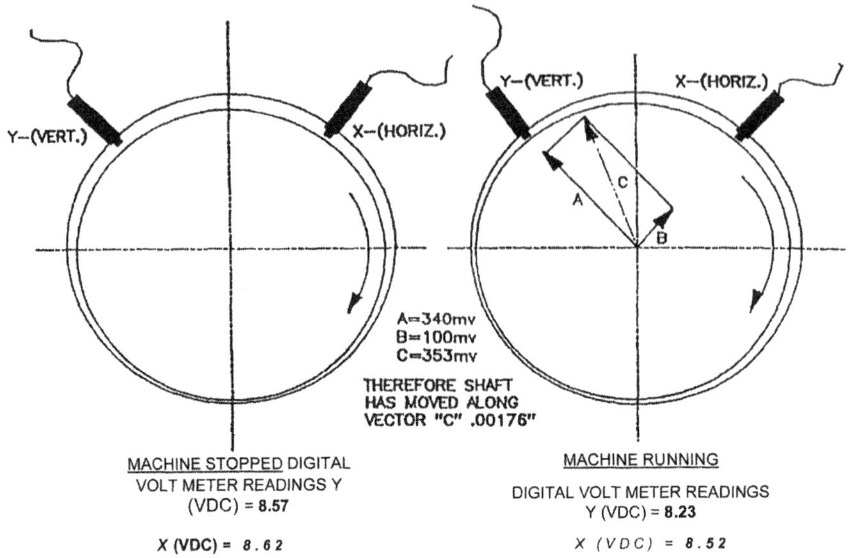

Figure 18.11 Shaft movement analysis (relative to bearing bore) (Courtesy of M.E. Crane Consultant)

One important parameter noted in Figure 18.10 that is frequently overlooked is shaft position. Change of shaft position can only occur if the forces acting on a bearing change or if the bearing surface wears. Figure 18.11 shows how shaft position is determined using standard shaft proximity probes.

Regardless of the parameters that are condition monitored, relative change of condition determines if and when action is required. Therefore, effective condition monitoring requires the following action for each monitored condition.

- Establish baseline condition
- Record condition trend
- Establish condition limit

Figure 18.12 presents these facts for a typical hydrodynamic journal bearing.

Based on the information shown in this trend, the bearing should be inspected at the next scheduled shutdown. A change in parameters during month 6 has resulted in increased shaft position, vibration and bearing pad temperature.

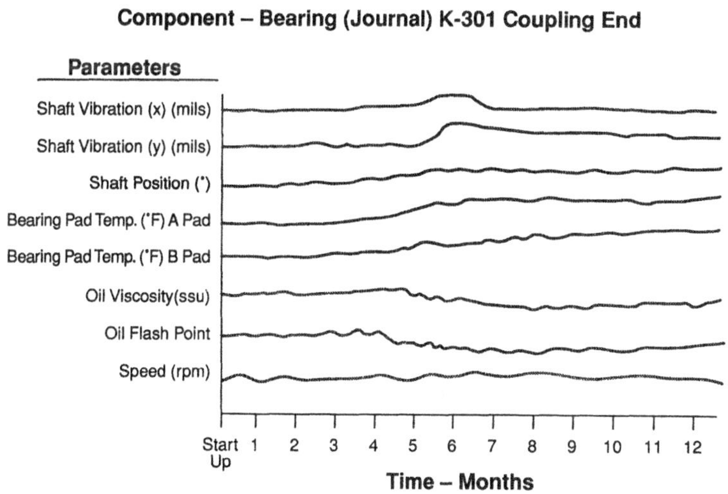

Figure 18.12 Trending data for a typical hydrodynamic journal bearing

Vibration instabilities

Vibration is an important condition associated with journal bearings because it can provide a wealth of diagnostic information valuable in determining the root cause of a problem.

Figure 18.13 presents important information concerning vibration.

Vibration

- Vibration is the result of a system being acted on by an excitation.
- This excitation produces a dynamic force by the relationship:

$$F_{DYNAMIC} = Ma$$

Where: M = Mass (Weight/g)
g = Acceleration due to gravity (386 IN/SEC^2)
a = Acceleration of mass M (IN/SEC^2)

- Vibration can be:
  - Lateral ↕
  - Axial → ←
  - Torsional

Figure 18.13 Vibration

CATEGORY	EXAMPLES	EXCITATION TYPE	VIBRATION TYPE
FORCED VIBRATIONS 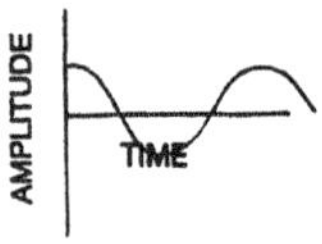	UNBALANCE MISALIGNMENT PULSATION	CONSTANT CONSTANT PERIODIC	LATERAL LATERAL AND AXIAL TORSIONAL
TRANSIENT VIBRATIONS 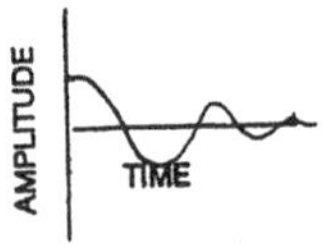	SYNCHRONOUS MOTOR START-UP IMPULSE (SHOCK FORCE)	RANDOM RANDOM	TORSIONAL TORSIONAL, AXIAL, RADIAL
SELF EXCITED	INTERNAL RUB OIL WHIRL GAS WHIRL	RANDOM CONSTANT CONSTANT	LATERAL LATERAL LATERAL

Figure 18.14 Excitation forces with examples

Figure 18.14 defines excitation forces with examples that can cause rotor (shaft) vibration.

Turbo-compressors generally monitor shaft vibration relative to the bearing bracket using a non-contact or 'proximity probe' system as shown in Figure 18.15. The probe generates a D.C. eddy current which continuously measures the change in gap between the probe tip and the shaft. The result is that the peak to peak unfiltered (overall) shaft vibration is read in mils or thousandth of an inch. The D.C. signal is normally calibrated for 200 milli volts per mil. Probe gaps (distance between probe and shaft) are typically 0.040 mils or 8 volts D.C. to assure the calibration curve is in the linear range. It is important to remember that this system measures shaft vibration relative to the bearing bracket and assumes the bearing bracket is fixed. Some systems incorporate an additional bearing bracket vibration monitor and thus record vibration relative to the earth or 'seismic vibration'.

As previously discussed, vibration limits are usually defined by:

$$\text{Vibration(mils p-p)} = \sqrt{\frac{12000}{RPM}}$$

This value represents the allowable shop acceptance level. A.P.I. recommends alarm and trip shaft vibration levels be set as follows:

$$V_{ALARM} = \sqrt{\frac{24000}{RPM}}$$

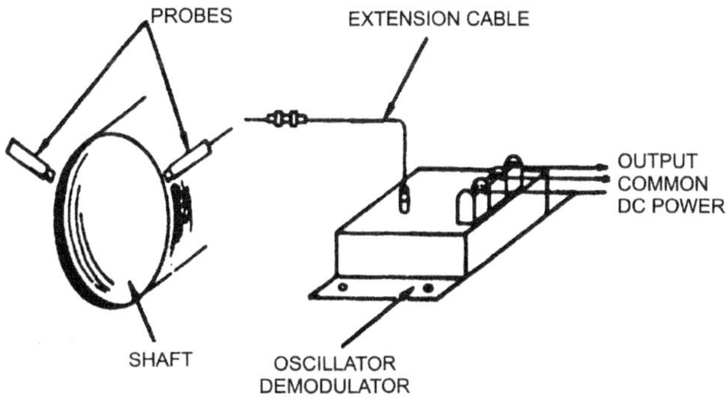

Figure 18.15 Non contact displacement measuring system

$$V_{TRIP} = \sqrt{\frac{36000}{RPM}}$$

In the writers' opinion, shaft vibration alarm and trip levels should be based on the following parameters as a minimum and should be discussed with the machinery vendor prior to establishing levels:

- Application (critical or general purpose)
- Potential loss of revenue
- Application characteristics (prone to fouling, liquid, unbalance, etc.)
- Bearing clearance
- Speed
- Rotor actual response (Bode Plot)
- Rotor mode shapes (at critical and operating speeds)

Figure 18.16 presents a vibration severity chart with recommended action.

A schematic of a shaft vibration and shaft displacement monitor are shown in Figure 18.17.

As mentioned above, vibration is measured unfiltered or presents 'overall vibration'. Figure 18.8 shows a vibration signal in the unfiltered and filtered conditions. All vibration diagnostic work (troubleshooting) relies heavily on filtered vibration to supply valuable information to determine the root cause of the vibration.

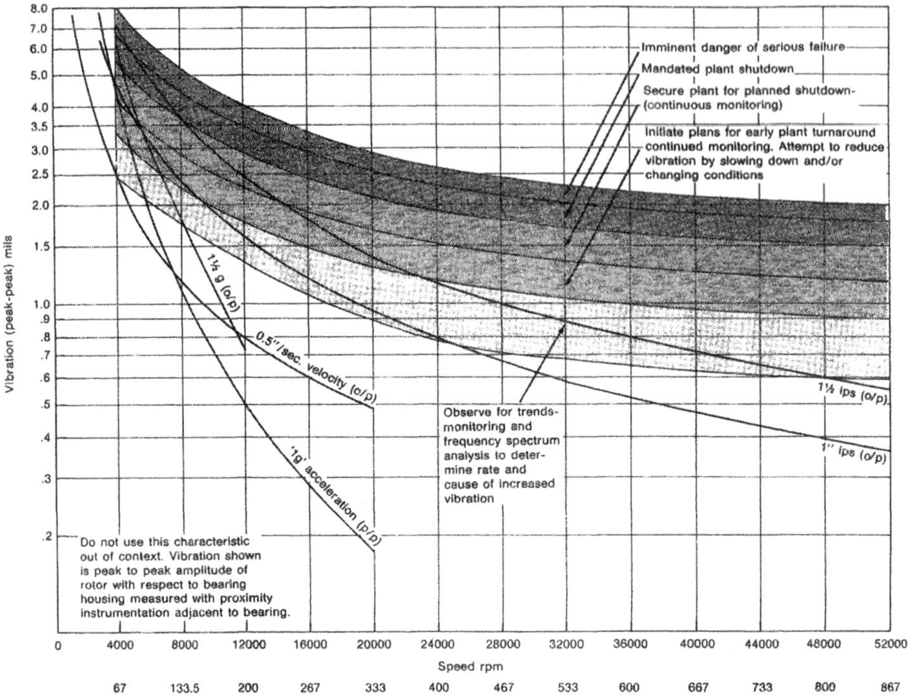

Figure 18.16 Vibration severity chart (Courtesy of Dresser-Rand and C. J. Jackson P.E.)

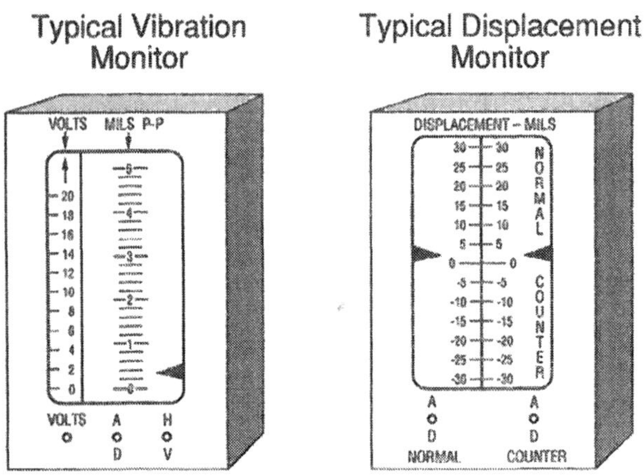

Figure 18.17 Shaft vibration and displacement

ANY VIBRATION SIGNAL IS MADE UP OF ONE OR MORE FREQUENCIES. A TYPICAL UNFILTERED (OVERALL) AND FILTERED SIGNAL ARE SHOWN BELOW:

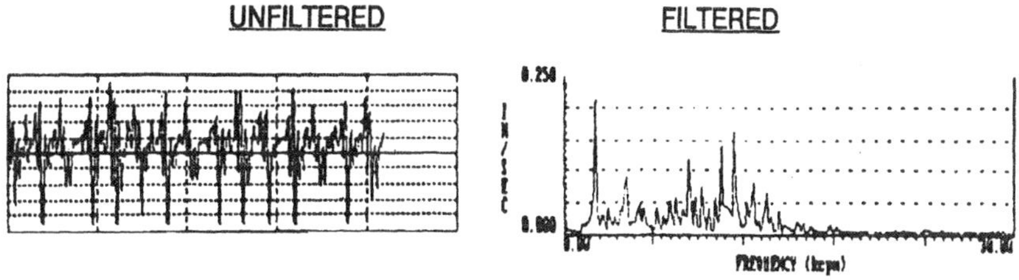

AN ANALOGY TO A FILTERED SIGNAL IS A RADIO. IN A GIVEN LOCALITY, MANY STATIONS ARE TRANSMITTING SIMULTANEOUSLY. ANY GIVEN STATION IS OBTAINED BY ADJUSTING THE TUNER TO THE CORRECT TRANSMISSION FREQUENCY.

Figure 18.18 Vibration frequency

Figure 18.19 presents an example of a radio tuner as an analogy to a filtered vibration signal.

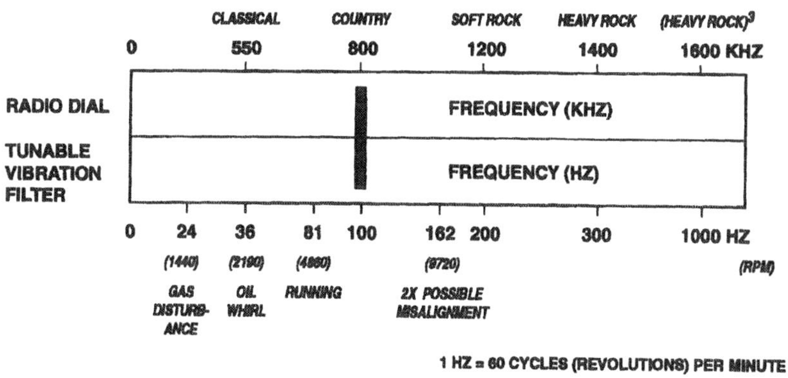

By Adjusting the Tuner (Filter) to a Selected Station (Frequency) the Desired Program (Vibration Frequency Signal) can be Obtained if it Is "On the Air" (Present)

Figure 18.19 Radio tuner/vibration filter analogy

By observing the predominant filtered frequencies in any overall (unfiltered) vibration signal, valuable information can be gained to add in the troubleshooting procedure and thus define the root cause of the problem.

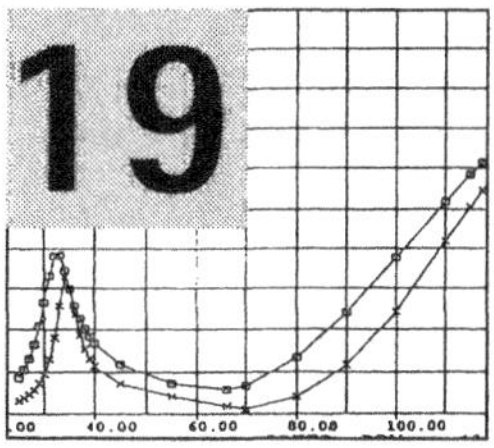

19 Critical speeds and rotor response

- Introduction
- Critical speeds
- The rotor system (input)
- Rotor response (output)
- Measured rotor response

Introduction

The term 'critical speed' is often misunderstood. In nature, all things exhibit a natural frequency. A natural frequency is defined as that frequency at which a body will vibrate if excited by an external force. The natural frequency of any body is a function of the stiffness and the mass of that body. As mentioned, for a body to vibrate, it must be excited. A classical example of natural frequency excitation is the famous bridge 'Galloping Gerty' in the state of Washington. That bridge vibrated to destruction when its natural frequency was excited by prevailing winds.

In the case of turbo-compressor rotors, their natural frequency must be excited by some external force to produce a response that will result in increased amplitude of vibration. One excitation force that could produce this result is the speed of the rotor itself. Thus the term 'critical speeds'. The term 'critical speed' defines the operating speed at which a natural frequency of a rotor system will be excited. All rotor systems have both lateral (horizontal and vertical) and torsional (twist about the central shaft axis) natural frequencies. Only lateral critical speeds will be discussed in this section.

In the early days of rotor design, it was thought that the rotor system consisted primarily of the rotor supported by the bearings. This led to the assumption that only the stiffness of the rotor supported by rigid bearings needed to be considered in the analysis of the natural frequency. Countless machinery problems have proven this assumption to be false over the years. The concept of the 'rotor system' must be thoroughly understood. The rotor system consists of the rotor itself, the characteristics of the oil film that support the rotor, the bearing, the bearing housing, the compressor case that supports the bearing, compressor support (base plate), and the foundation. The stiffness and damping characteristics of all of these components together result in the total rotor system that produce the rotor response to excitation forces.

We will examine a typical rotor response case in this section and note the various assumptions, the procedure modeling, the placement of unbalance, the response calculation output and discuss the correlation of these calculations to actual test results.

Critical speeds

The natural frequency of any object is defined by the relationship:

$$F_{NATURAL} = \sqrt{\frac{K}{M}}$$

Where: K = Stiffness
M = Mass

When excited by an external force, any object will vibrate at its natural frequency. If the frequency of the exciting force is equal to the natural frequency of the object, and no damping is present, the object can vibrate to destruction. Therefore, if the frequency of an exciting force equals the natural frequency of an object, the exciting force is operating at the 'critical frequency'.

Rotor speed is one of the most common external forces in turbo-machinery. When the rotor operates at any rotor system natural frequency, it is said that the rotor is operating at its crucial speed. The critical speed of a rotor is commonly designated as N_C and the corresponding natural frequencies or critical speeds are: N_{C1}, N_{C2}, N_{C3}, etc.

Every turbo-compressor that is designed must have the rotor system critical speeds determined prior to manufacture. In this section, we will follow the procedure for the determination of the necessary parameters to define a rotor systems critical speed. The procedure is commonly known as determination of rotor response. Figure 19.1 is a representation of a critical speed map for a rotor system.

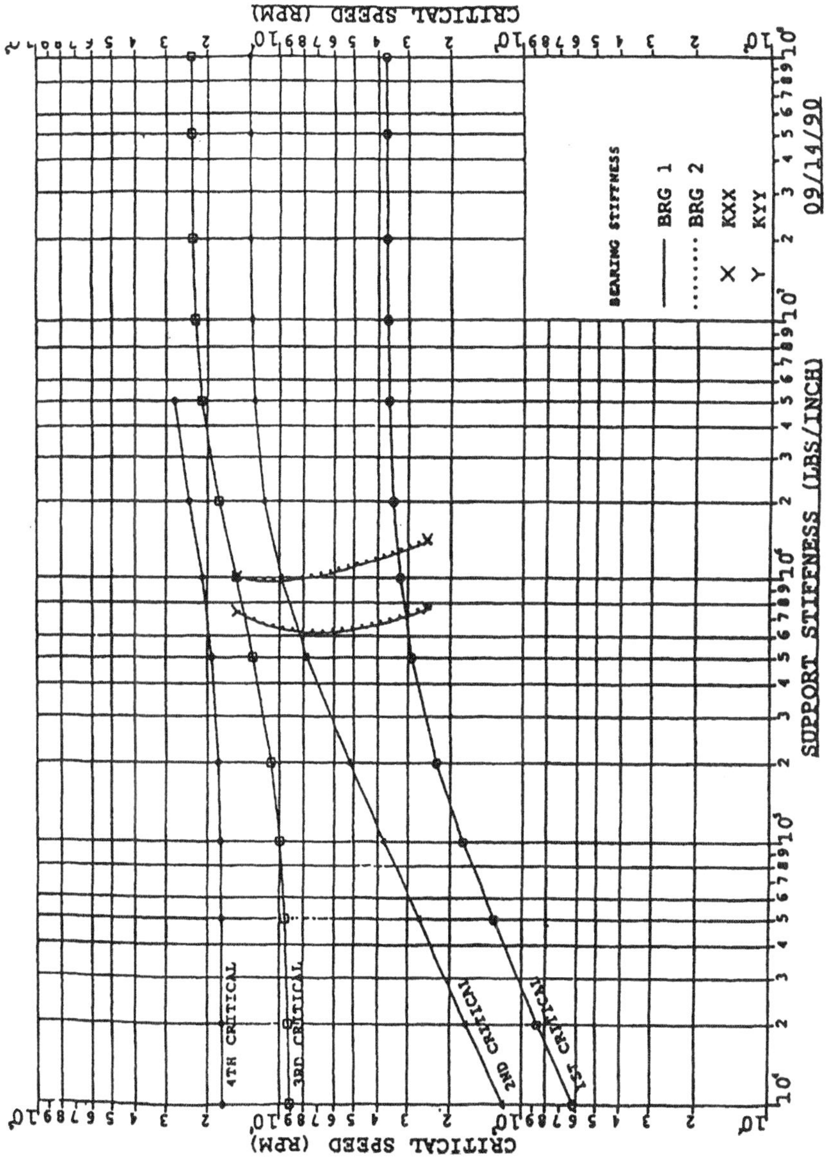

Figure 19.1 Compressor rotor critical speed map – no damping (Courtesy of Elliott Company)

It should be understood that all stiffness values are 'calculated' and will vary under actual conditions. As an exercise, determine N_{C1}, N_{C2} and N_{C3} for the horizontal and vertical directions for each bearing in Figure 19.1 (assume bearing 1 and 2 stiffness are the same)

Critical speed	Horizontal (X)	Vertical (Y)
N_{C1}	3,300 rpm	3,000 rpm
N_{C2}	9,700 rpm	8,000 rpm
N_{C3}	16,000 rpm	15,000 rpm

Based on a separation margin of ± 20% from a critical speed, what would be the maximum allowable speed range between N_{C1} and N_{C2} in Figure 19.1?

- Maximum speed 6,600 rpm
- Minimum speed 4,000 rpm

Remember, changing of any value of support stiffness will change the critical speed. Plotted on the X axis is support stiffness in lbs/inch. The primary components of support stiffness in order of decreasing increasing influence are:

- Oil support stiffness
- Bearing pad or shell
- Bearing housing
- Bearing bracket
- Casing support foot
- Baseplate
- Foundation

Note that this analysis of the critical speed does not include oil film damping. It is common practice to first determine the 'undamped critical speeds' to allow for necessary modifications to the rotor or support system. This is because the effects of stiffness on the location of critical speed are significantly greater than damping. Figure 19.1 shows four (4) distinct critical speeds. Operation within ± 20 of actual critical speeds is to be avoided. Also plotted are the horizontal (x) and vertical (y) bearing stiffness for each bearing. Note that these values vary with speed and are the result of changes in the oil stiffness. Therefore, a change in any of the support stiffness components noted above can change the rotor critical speed. Experience has shown that critical speed values seldom change from ± 5% of their original installed values.

If a turbo-compressor with oil seals experiences a significant change in critical speeds, it is usually an indication of seal lock-up. That is, the seal

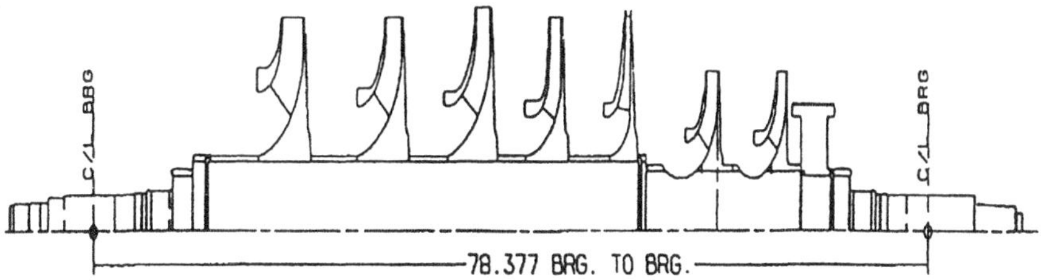

Figure 19.2 Rotor response modeling – rotor (Courtesy of Elliott Co.)

does not have the required degrees of freedom and supports the shaft acting like a bearing. Since the seal span is less than the bearing span, the rotor stiffness 'K' increases and the critical speeds will increase in this case.

The rotor system (input)

Figure 19.2 shows a typical turbo-compressor rotor before modeling for critical speed or rotor response analysis.

Since the natural frequency or critical speed is a function of shaft stiffness and mass, Figure 19.4 presents the rotor in Figure 19.2 modeled for input to the computer rotor response program.

Figure 19.4 is an example of a modeled rotor and only includes the rotor stiffness (K) and mass (M).

In order to accurately calculate the rotor critical speeds, the entire rotor system stiffness, masses and damping must be considered. Figure 19.3 models the oil film stiffness and damping of the journal bearings at different shaft speeds.

4X1.6" tilt 20.5" TB 3.0"		shaftend 7.5–6.5" shaft Bendix coupling	
Static bearing load (lbs)	897	diameter (inches)	4.00
Bearing station	12	length (inches)	1.60
Bearing location	thrust	diam assembly clearance (inches)	5.7487E–03
Bearing type	tilt pad	diam machined clearance (inches)	8.7500E–03
Location of load	between pads	inlet oil temperature (deg F)	120.0
Preload	0.343	type of oil	DTE–light (150SSU @100°F)

Figure 19.3 Typical compressor oil film bearing parameters (Courtesy of Elliott Co.)

Speed (rpm)	50mm No	Fluid film stiffness KXX (lb/in)	KYY (lb/in)	Damping WCXX (lb/in)	WCYY (lb/in)
2500	0.114	1.3871E 06	7.5446E 05	7.7995E 05	4.6249E 05
3000	0.137	1.2984E 06	7.1330E 05	7.8487E 05	4.7587E 05
4000	0.183	1.1769E 06	6.6147E 05	8.0311E 05	5.0825E 05
4500	0.206	1.1341E 06	6.4543E 05	8.1400E 05	5.2564E 05
5500	0.252	1.0703E 06	6.2556E 05	8.3686E 05	5.6116E 05
6613	0.303	1.0230E 06	6.1679E 05	8.6656E 05	6.0354E 05
7000	0.321	1.0109E 06	6.1616E 05	6.7775E 05	6.1885E 05
8000	0.366	9.8751E 05	6.1898E 05	9.0798E 05	6.5935E 05
9000	0.412	9.7305E 05	6.2684E 05	9.4015E 05	7.0111E 05
10000	0.458	9.6556E 05	6.3864E 05	9.7461E 05	7.4430E 05
11000	0.504	9.6360E 05	6.5354E 05	1.0110E 06	7.8878E 05
12000	0.549	9.6610E 05	6.7094E 05	1.0490E 06	8.3434E 05
13000	0.595	9.7225E 05	6.9037E 05	1.0881E 06	8.8080E 05
14000	0.641	9.8144E 05	7.1149E 05	1.1283E 06	9.2801E 05
15000	0.687	9.9317E 05	7.3403E 05	1.1696E 06	9.7586E 05

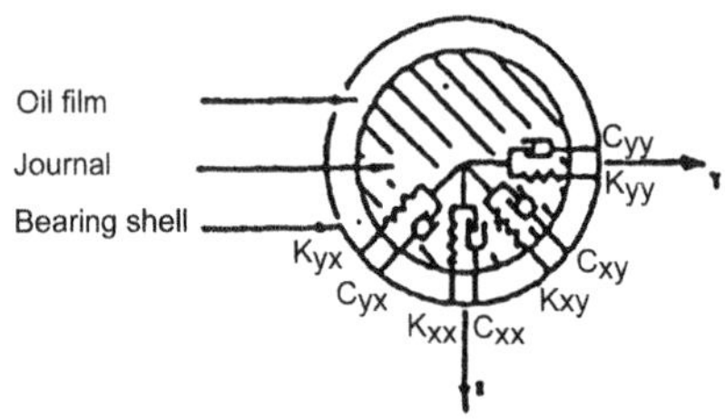

Figure 19.3 Continued – Typical compressor oil film bearing parameters (Courtesy of Elliott Co.)

Note that it is essential that the type of oil to be used in the field (viscosity characteristics) must be known. End users are cautioned to confirm with the OEM before changing oil type as this will affect the rotor response. In addition to modeling of the rotor and bearings, most rotor response calculations also include the following additional inputs:

- Bearing support stiffness
- Oil film seal damping effects

Of all the input parameters, the effects of bearing and seal oil film parameters are the most difficult to calculate and measure. Therefore, a correlation difference will always exist between the predicted and actual values of critical speed. Historically, predicted values of N_{C1} (first critical speed) generally agree within ± 5%. However, wide variations between predicted and actual values above the first critical speed (N_{C1}) exist for N_{C2}, N_{C3}, etc.

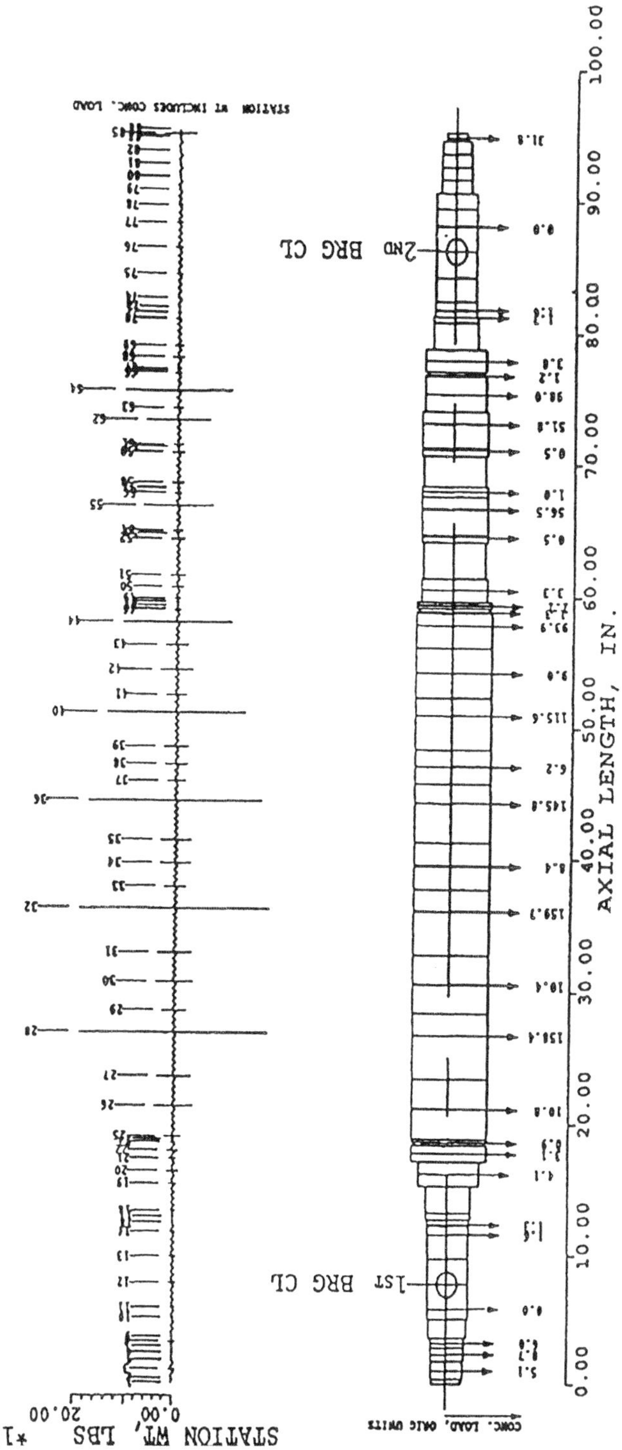

Figure 19.4 Rotor response input data – dimensions, masses and unbalances (Courtesy of Elliott Co.)

When selecting machinery, the best practice is to request specific vendor experience references for installed equipment with similar design parameters as follows:

- Bear span ÷ major shaft diameter
- Speeds
- Bearing design
- Seal design
- Operating conditions (if possible)

Once the rotor system is adequately modeled, the remaining input parameter is the amount and location of unbalance. Since the objective of the rotor response study is to accurately predict the critical speed values and responses, an assumed value and location of unbalances must be defined. Other than bearing and seal parameters, unbalance amount and location is the other parameter with a 'correlation factor'. There is no way to accurately predict the amount and location of residual unbalance on the rotor. Presently, the accepted method is to input a value of 8 X A.P.I. acceptable unbalance limit $\frac{(4W)}{N}$.

This results in a rotor response input unbalance of $\frac{32W}{N}$.

The location of the unbalance is placed to excite the various critical speeds. Typically the unbalances are placed as noted below:

Location	**To excite**
Mid span	N_{C1}
Quarter span (2 identical unbalances)	N_{C2}
At coupling	N_{C2}, N_{C3}

Failure to accurately determine the value and location of residual rotor unbalance is one of the major causes of correlation differences between predicted and actual critical speeds.

Rotor response (output)

The output from the rotor response study yields the following:

- Relative rotor mode shapes
- Rotor response for a given unbalance

Figure 19.5 shows the relative rotor mode shapes for N_{C1}, N_{C2}, N_{C3} and N_{C4}. Usually, the rotor will operate between N_{C1} and N_{C2}.

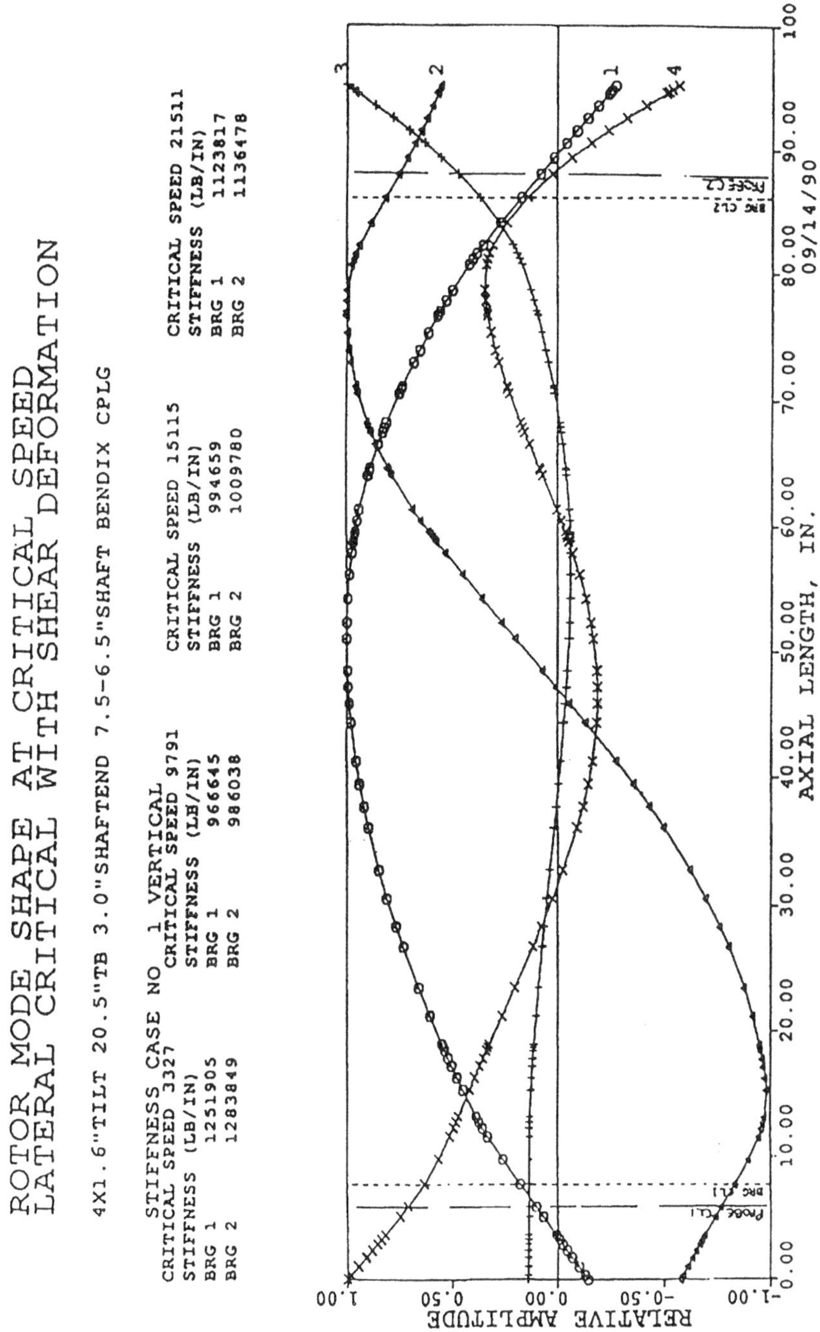

Figure 19.5 Rotor natural frequency mode shapes (Courtesy of Elliott Co.)

Rotor mode shape data is important to the designer because it allows determination of modifications to change critical speed values.

For the end user, this data provides an approximation of the vibration at any point along the shaft as a ratio of the measured vibration data. As an example in Figure 19.5, determine the vibration at the shaft mid span if the vibration measured by the probe C_2 when operating at N_{C1} is 2.00 mills. From Figure 19.5, the vibration at the shaft mid span when operating at the first critical speed of 3327 RPM (50 in location) is:

$\frac{1.00}{.1}$ or 10 x the bearing vibration

Ten (10) times the value at C_2 or 20.0 mils!

Mode shape data should always be referred to when vibration at operating speed starts to increase and your supervisor asks

'When do we have to shut down the unit?'

or

'Can we raise the radial vibration trip setting?'

In this example, the bearing clearance may be 0.006 or 6 mills. And an honest request would be ... 'We'll replace the bearing at the turnaround, please run to 7.0 mils vibration'.

Refer to Figure 19.5 and remember:

- The compressor must go through N_{C1}
- The shaft vibration **increases** at N_{C1} (usually 2X, 3X or more)
- The vibration at center span is approximately **10X** the probe vibration

Therefore:

Vibration at the mid span during the first critical speed will be:

=	(7.0 mils)	x	(2.0)	x	(10)
	Probe value		N_{C1} amplification		Mode shape difference
=	140 Mils!!				

Normal clearance between the rotor and interstage labyrinths is typically 40 mils!! This vibration exposes the diaphragms, which are usually cast iron, to breakage. One final comment ... during shutdown, the rate of rotor speed decrease **CANNOT** be controlled as in the case of start-up. It depends on rotor inertia, load in the compressor, the process system characteristics and the control and protection system. If the vibration at the probe locations is high, the best advice is to stop the compressor fully loaded which will reduce the time in the critical speed range as much as possible. Yes, the compressor will surge, but the short

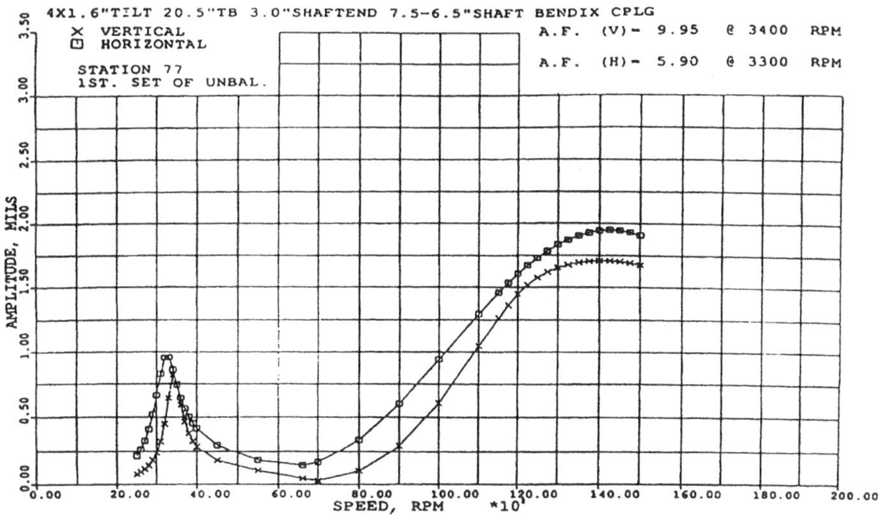

Figure 19.6 Rotor response output at non-drive end bearing (NDE) (Courtesy of Elliott Co.)

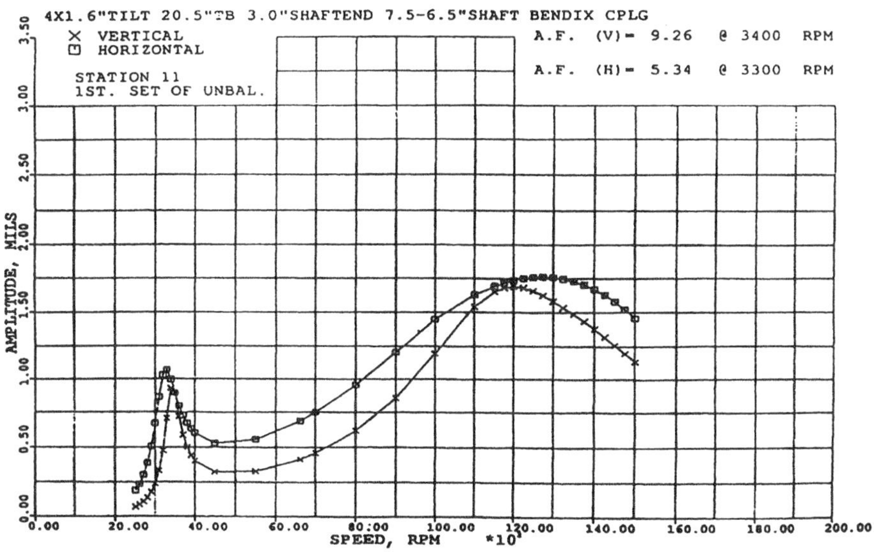

Figure 19.7 Rotor response output drive end bearing (D.E.) (Courtesy of Elliott Co.)

duration will not normally damage the compressor. Figures 19.6 and 19.7 present the primary output of a rotor response study.

Rotor response plots display vibration amplitude, measured at the probes, vs shaft speed for the horizontal and vertical probes. Note that a response curve must be plotted for each set of unbalance locations and unbalance amount.

Figure 19.6 shows the rotor response for the non-drive end (N.D.E.) set of probes with the first set of unbalance. Figure 19.7 shows the rotor response for the drive end set of probes (D.E.). The operating speed range of this example is 6,000–8,000 rpm.

Measured rotor response

During shop test, the rotor response of every turbo-compressor rotor is measured during acceleration to maximum speed and deceleration to minimum speed. Values are plotted on the same coordinates as for the rotor response analysis. The plot of shaft vibration and phase angle of unbalance vs shaft speed is known as a BODE PLOT.

Bode plots represent the actual signature (rotor response) of a rotor for a given condition of unbalance, support stiffness and unbalance. They indicate the location of critical speeds, the change of shaft vibration with speed and the phase angle of unbalance at any speed. A bode plot is a dynamic or transient signature of vibration for a rotor system and is unique to that system for the recorded time frame. Bode plots should be recorded during every planned start-up and shutdown of every turbo-compressor. As discussed in this section, the bode plot will provide valuable information concerning shaft vibration and phase angle at any shaft speed.

Rotor axial (thrust) forces

- Introduction
- The hydrodynamic thrust bearing
- Impeller thrust forces
- Rotor thrust balance
- Thrust condition monitoring

Introduction

In every rotating machine utilizing reaction type blading, a significant thrust is developed across the rotor by the action of the impellers or blades. Also in the case of equipment incorporating higher than atmospheric suction pressure, a thrust force is exerted in the axial direction as a result of the pressure differential between the pressure in the case and atmospheric pressure.

In this section we will cover a specific rotor thrust example and calculate thrust balance for a specific case. We will see the necessity in some applications of employing an axial force balance device known as a balance drum. In many instances, the absence of this device will result in excessive axial (thrust) bearing loadings. For the case of a machine with a balance device, the maintenance of the clearances on this device are of utmost importance. In many older designs the clearances are maintained by a fixed close clearance bushing made out of babbitt which has a melting temperature of approximately 350°F, depending on the pressure differential across the balance drum. If the temperature in this region should exceed this value, the effectiveness of the balance drum would suddenly be lost and catastrophic failures can occur inside the machine. Understanding the function of this device and the potential high axial forces involved in its absence is a very important aspect of condition monitoring of turbo-compressors.

We will also examine various machine configurations including natural balanced (opposed) thrust and see how thrust values change even in the case of a balanced machine as a function of machine flow rate.

Finally, we will examine thrust system condition monitoring and discuss some of the confusion that results with monitoring these machines.

The hydrodynamic thrust bearing

A typical hydrodynamic double acting thrust bearing is pictured in Figure 20.1.

The thrust bearing assembly consists of a thrust collar mounted on the rotor and two sets of thrust pads (usually identical in capacity) supported by a base ring (Michell Type).

The Kingsbury type includes a set of leveling plates between each set of pads and the base ring. This design is shown in Figure 20.2.

Both the Michell and Kingsbury types are used. Figure 20.3 provides a view of the leveling plates providing the self-equalizing feature in the Kingsbury design. The self-equalizing feature allows the thrust pads to lie in a plane parallel to the thrust collar.

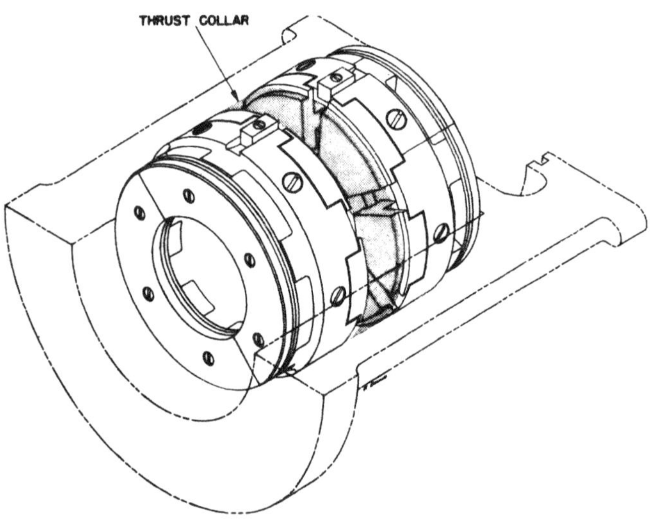

Figure 20.1 Double acting self-equalizing thrust bearing assembly (thrust collar removed) (Courtesy of Elliott Company)

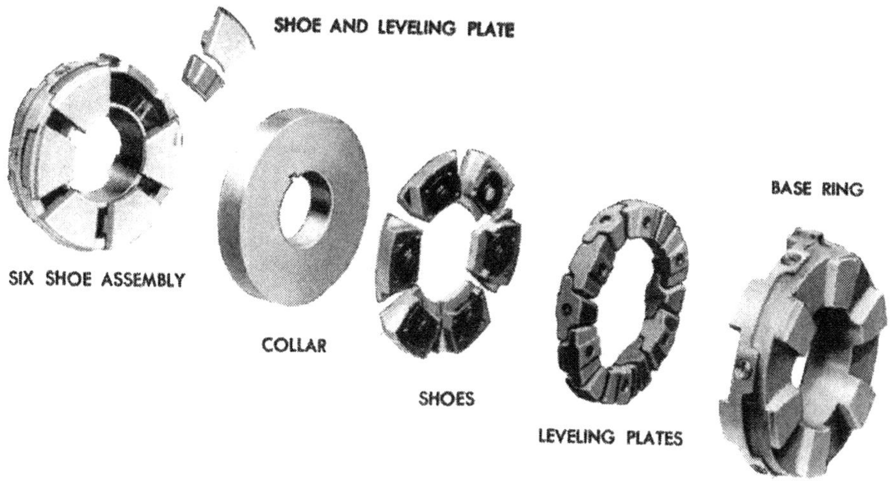

Figure 20.2 Small Kingsbury six-shoe, two direction thrust bearing. Left-hand group assembled, except for one shoe and 'upper' leveling plate. Right-hand group disassembled (Courtesy of Kingsbury, Inc.)

Regardless of the design features, the functions of all thrust bearings are:

- To continuously support all axial loads
- To maintain the axial position of the rotor

The first function is accomplished by designing the thrust bearing to provide sufficient thrust area to absorb all thrust loads without exceeding the support film (oil) pressure limit (approximately 500 psi).

Figure 20.4 shows what occurs when the support film pressure limit is exceeded.

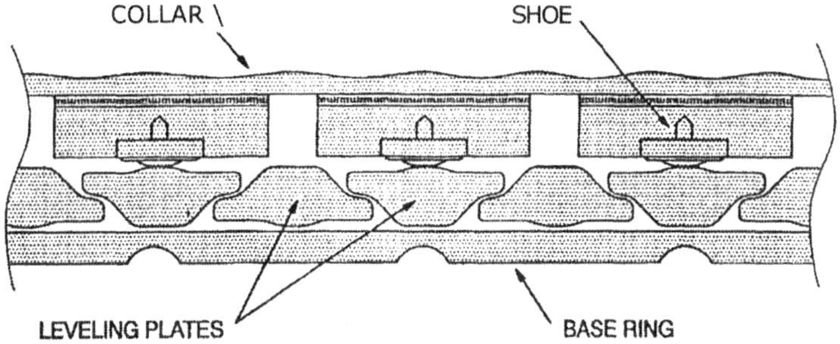

Figure 20.3 Self-equalizing tilt-pad thrust bearing (View – looking down on assembly) (Courtesy of Kingsbury, Inc.)

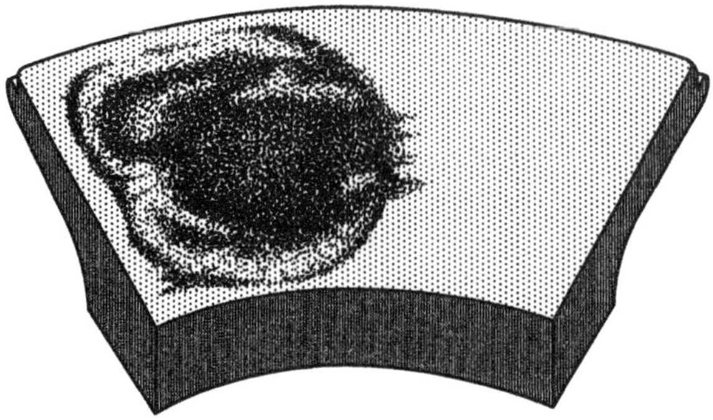

Figure 20.4 Evidence of overload on a tilt-pad self-equalizing thrust bearing pad (Courtesy of Kingsbury Corp.)

The oil film breaks down, thus allowing contact between the steel thrust collar and soft thrust bearing pad overlay (Babbitt). Once this thin layer (1/16") is worn away, steel to steel contact occurs resulting in significant turbo-compressor damage.

Thrust pad temperature sensors, located directly behind the babbitt at the pad maximum load point protect the compressor by tripping the unit before steel to steel contact can occur.

Figure 20.5 presents different Kingsbury bearing size rated capacities as a function of speed.

Figure 20.6 shows how thrust pad temperature and thrust load are related for a given thrust bearing size and shaft speed. Note that the

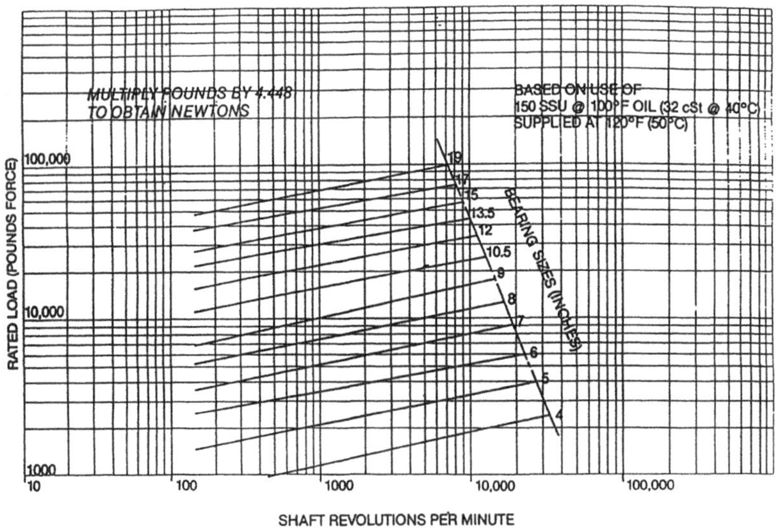

Figure 20.5 Thrust bearing rated load vs. speed (Courtesy of Kingsbury Corp.)

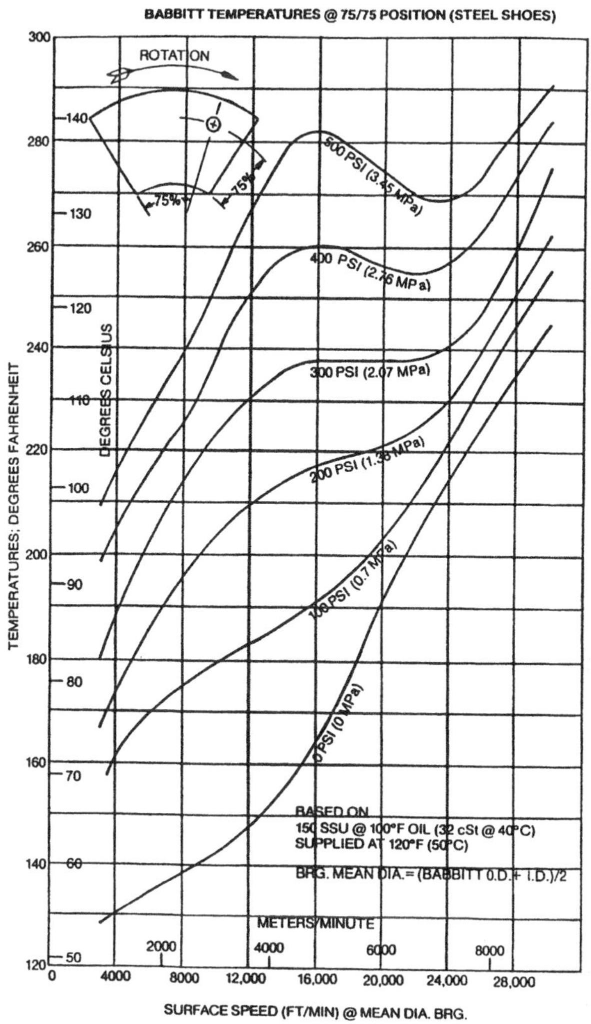

Figure 20.6 The relationship between thrust pad temperature and thrust load (Courtesy of Kingsbury, Inc.)

greater the thrust load (P.S.I.), the smaller the oil film and the greater the effect of oil viscosity on oil flow and heat removal. Based on a maximum load of 500 psi, it can be seen from Figure 20.6 that a turbo-compressor thrust bearing pad temperature trip setting should be between 260° and 270°F.

Other than to support the rotor in an axial direction, the other function of the thrust bearing is to continuously maintain the axial position of the rotor. This is accomplished by locating stainless steel shims between the thrust bearing assembly and compressor axial bearing support plates. The most common thrust assembly clearance with the thrust shims installed is 0.011 – 0.014. These values vary with thrust bearing size. The vendor instruction book must be consulted to determine the proper clearance.

The following procedure is used to assure that the rotor is properly positioned in the axial direction.

1. With thrust shims removed, record total end float by pushing rotor axially in both directions (typically .250"–.500").
2. Position rotor as stated in instruction book.
3. Install minimum number of stainless steel thrust shims to limit end float to specified value.*

*An excessive number of thrust shims act as a spring resulting in a greater than specified axial clearance during full thrust load conditions.

Proper running position of the rotor is critical to obtaining optimum efficiency and preventing axial rubs during transient and upset conditions (start-up, surge, etc.)

Impeller thrust forces

Every reaction type compressor blade set or impeller produces an axial force towards the suction of the blade or impeller. Refer to Figure 20.7.

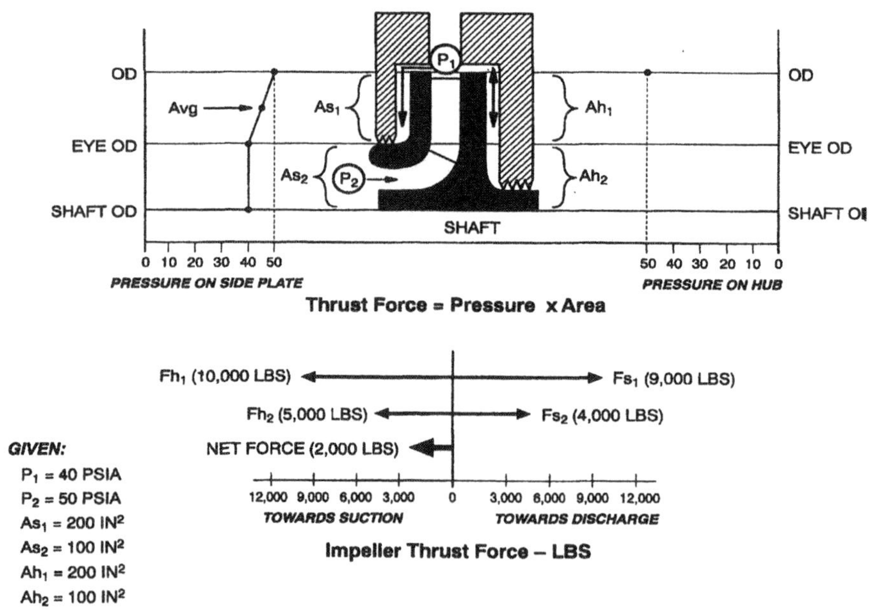

Figure 20.7 Impeller thrust force

In this example, the net force towards the compressor suction is 2,000 psi for the set of conditions noted. Note that the pressure behind the impeller is essentially constant (50 psi), but the pressure on the front side of impeller varies (from 50 psi to 40 psi) because of the pressure drop across the eye labyrinth. Every impeller in a multistage compressor will produce a specific value of axial force towards it's suction at a specific flow rate, speed and gas composition. A change in any or all of these parameters will produce a corresponding change in impeller thrust.

Rotor thrust balance

Figure 20.8 shows how a balance drum or opposed impeller design reduces thrust force. The total impeller force is the sum of the forces from the individual impellers. If the suction side of the impellers is opposed, as noted in Figure 20.8, the thrust force will be significantly reduced and can approach 0. If the suction side of all impellers are the same (in series), the total impeller thrust force can be very high and may exceed the thrust bearing rating. If this is the case, a balance drum must be mounted on the rotor as shown in Figure 20.8. The balance drum face area is varied such that the opposing force generated by the

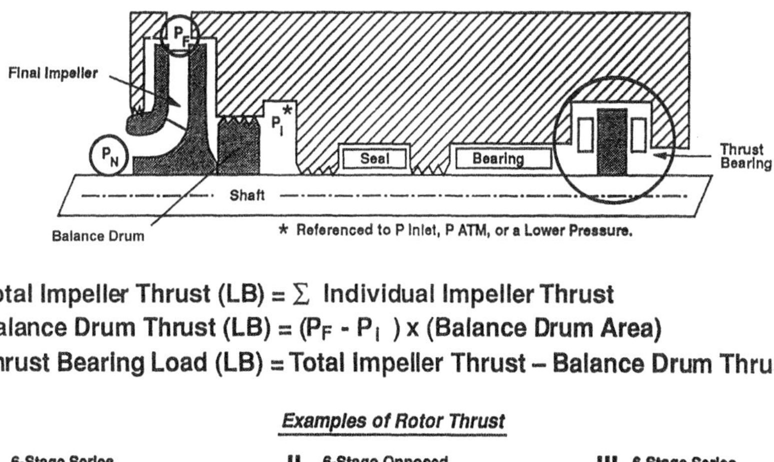

Total Impeller Thrust (LB) = Σ Individual Impeller Thrust
Balance Drum Thrust (LB) = (P_F - P_I) x (Balance Drum Area)
Thrust Bearing Load (LB) = Total Impeller Thrust – Balance Drum Thrust

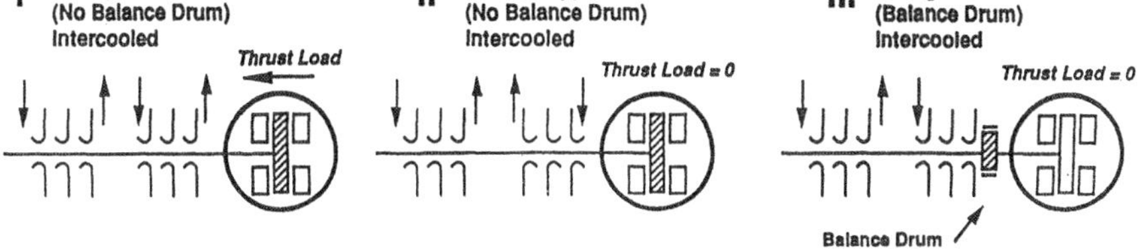

Figure 20.8 Rotor thrust force

balance drum reduces the thrust bearing load to an acceptable value. The opposing thrust force results from the differential between compressor discharge pressure (P_F) and compressor suction pressure (P_1) since the area behind the balance drum is usually referenced to the suction of the compressor. This is accomplished by a pipe that connects this chamber to the compressor suction. This line is typically called the 'balance line'.

It is very important to note that a balance drum is used only where the thrust bearing does not have sufficient capacity to absorb the total compressor axial load. And the effectiveness of the balance drum depends directly on the balance drum seal. Fail the seal, (open clearance significantly) and thrust bearing failure can result.

A common misunderstanding associated with balance drum systems is that a balance drum always reduces the rotor thrust to zero. Refer to Figure 20.9 and observe that this statement may or may not be true depending on the thrust balance system design. And even if it is, the thrust is zero only at one set of operating conditions.

Figure 20.9 shows a rotor system designed four (4) different ways. Note how the thrust **always** changes with the flow rate regardless of the design. Another misconception regarding thrust balance systems is the normal or 'active' direction of thrust. In many cases, the active thrust is assumed to always be towards the suction of the compressor.

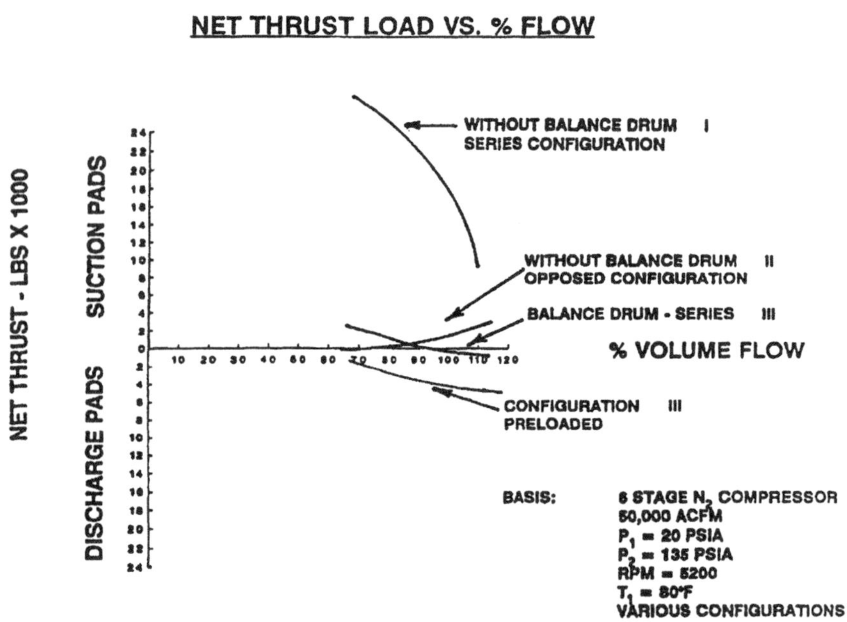

Figure 20.9 Rotor system designed four different ways

Observing Figure 20.9, it is obvious that the 'active' direction can change when the turbo-compressor has a balance drum or is an opposed design. It is recommended that the use of active thrust be avoided where possible and that axial displacement monitors be labeled to allow determination of the thrust direction at all times.

Please refer to Figure 20.10 which shows a typical thrust displacement monitor.

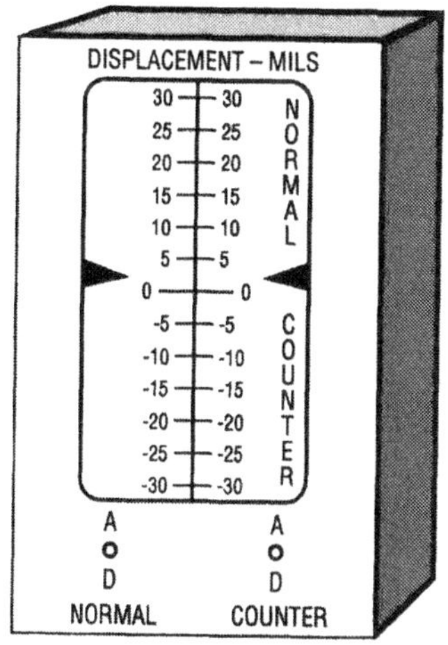

Figure 20.10 Typical axial thrust monitor

These monitors detect thrust position by targeting the shaft end, thrust collar or other collar on the rotor. Usually two or three probes (multiple voting arrangement) are provided to eliminate unnecessary compressor trips. The output of the probes is noted on the monitor as either + (normal) or – (counter). However, this information gives no direct indication of the axial direction of the thrust collar. The following procedure is recommended:

1. With compressor shutdown, push rotor towards the suction and note direction of displacement indicator.
2. Label indicator to show direction towards suction of compressor.

Knowing the actual direction of the thrust can be very useful during troubleshooting exercises in determining the root cause of thrust position changes.

Thrust condition monitoring

Failure of a thrust bearing can cause long term and possibly catastrophic damage to a turbo-compressor. Condition monitoring and trending of critical thrust bearing parameters will optimize turbo-compressor reliability.

The critical thrust bearing condition monitoring parameters are:

- Rotor position
- Thrust pad temperature
- Balance line ΔP

Rotor position is the most common thrust bearing condition parameter and provides useful information regarding the direction of thrust. It also provides an indication of thrust load but does not confirm that thrust load is high. Refer to Figure 20.11.

All axial displacement monitors have pre-set (adjustable) values for alarm and trip in both thrust directions. Typically, the established procedure is to record the thrust clearance (shims installed) during shutdown and set the alarm and trip settings as follows:

$$\text{Alarm} = \frac{\text{Clearance}}{2} + 10 \text{ mils (each direction)}$$

$$\text{Trip} = \text{Alarm Setting} + 5 \text{ mils (each direction)}$$

The above procedure assumes the rotor is in the mid or zero position of the thrust clearance. An alternative method is to hand push the rotor to the assumed active position and add appropriate values for alarm and trip.

The writer personally recommends the first method since an active direction of thrust does not have to be assumed.

As noted, axial displacement monitors only indicate the quantity of thrust load. False indication of alarm or even trip settings can come from:

- Compression of thrust bearing components
- Thermal expansion of probe adaptors or bearing brackets
- Loose probes

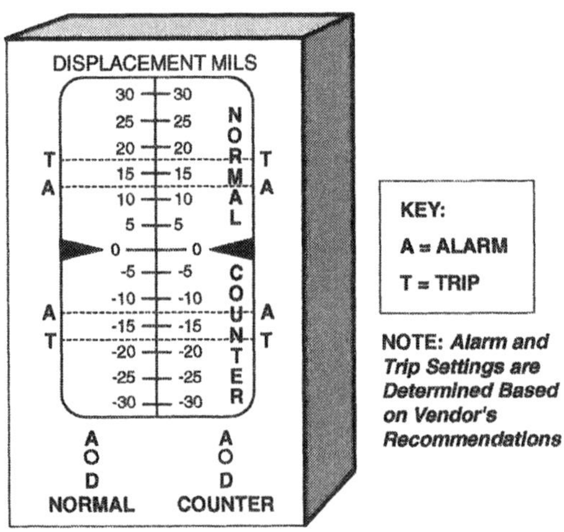

Figure 20.11 Typical axial displacement monitor

It is strongly recommended that any alarm or trip displacement value be confirmed by thrust pad temperature if possible prior to taking action. Please refer back to Figure 20.6 of this chapter and note that the thrust pad temperature in the case of thrust pad overload is approximately 250°F. If an axial displacement alarm or trip signal is activated **observe** the corresponding thrust pad temperature. If it is below 220°F, take the following action:

- Observe thrust pads. If no evidence of high load is observed (pad and back of pad) confirm calibration of thrust monitor and change settings if necessary.

The last condition monitoring parameter for the thrust system is balance line pressure drop. An increase of balance line ΔP will indicate increased balance drum seal leakage and will result in higher thrust bearing load. Noting the baseline ΔP of the balance line and trending this parameter will provide valuable information as to the root cause of a thrust bearing failure. In many field case histories, the end user made many thrust bearing replacements until an excessive balance drum clearance was discovered as the root cause of the thrust bearing failure. It is a good practice to always check the balance line ΔP after reported machine surge. Surging will cause high internal gas temperatures which can damage the balance drum seal.

21 Compressor seal system overview and types

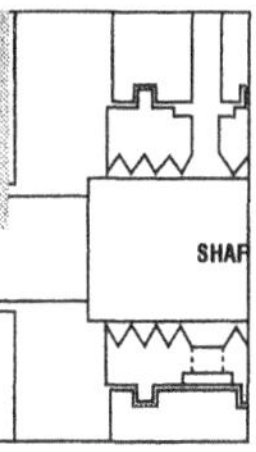

- Introduction
- The supply system
- The seal housing system
- Seal supply systems
- Seal supply system summary
- Seal liquid leakage system

Introduction

There are numerous types of fluid seal systems since the types of seals utilized, sealing fluids and sealing pressures vary widely. Regardless of the type of seal used, the function of a critical equipment seal system is as follows: '*To continuously supply clean fluid to each specified seal interface point at the required differential pressure, temperature and flow rate*'. A typical seal system for a centrifugal compressor is shown in Figure 21.1.

The system shown is for use with clearance bushing seals. Let's examine this figure by proceeding through the system from the seal oil reservoir through the compressor shaft seal and back through the reservoir. As previously discussed, the concept of sub-systems can be useful here. The seal oil system shown can be divided into four major sub-systems:

A The supply system

B The seal housing system

C The atmospheric drain system

D The seal leakage system

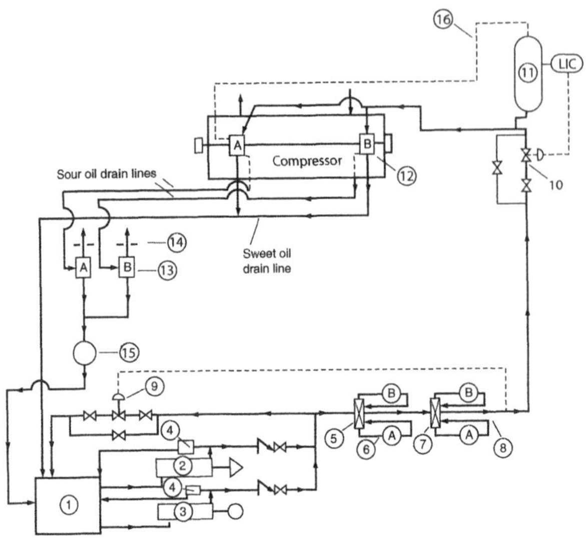

1. Oil reservoir
2. Main screw pump - turbine driven
3. Aux screw pump - motor driven
4. Relief valves
5. Transfer valve
6. Oil coolers (A & B)
7. Transfer valve
8. Oil filters (A & B)
9. Back pressure control valve
10. Level control valve
11. Overhead seal oil tank (typically 15 ft above compressor centerline)
12. ISO sleeve seals (also called liquid bushing seals)
13. Seal oil drainers (A & B)
14. Drainer orifice vents (A & B)
15. Degassing tank
16. Reference gas line to seal oil tank

Note: component condition instrumentation and autostarts not shown

Figure 21.1 API 614 lube/seal oil system for ISO-sleeve seals (Courtesy of Elliott Co.)

A The supply system

This system consists of the reservoir, pumping units, the exchangers, transfer valves, temperature control valves, and filters. The purpose of this sub-system is to continuously supply clean, cool sealing fluid to the seal interfaces at the correct differential pressure.

B The seal housing system

This system is comprised of two different seals. A gas side bushing, and an atmospheric bushing. The purpose of the seal housing system is to positively contain the fluid in the compressor and not allow leakage to the atmosphere. The seal fluid is introduced between both seal interfaces, thus constituting a double seal arrangement. Refer to Figure 21.2 for a closer examination of the seal.

The purpose of the gas side bushing seal is to constantly contain the reference fluid and minimize sour oil leakage. This bushing can be conceived as an equivalent orifice. This concept is similar to bearings previously discussed with the exception that the referenced downstream pressure of the gas side bushing can change. In order to assure a constant flow across this 'orifice,' the differential pressure must be maintained constant. Therefore, every compressor seal system is designed to maintain a constant differential against the gas side seal. The means of obtaining this objective will be discussed as we proceed.

The other seal in the system is the atmospheric bushing whose purpose is to minimize the flow of seal liquid to an amount that will remove

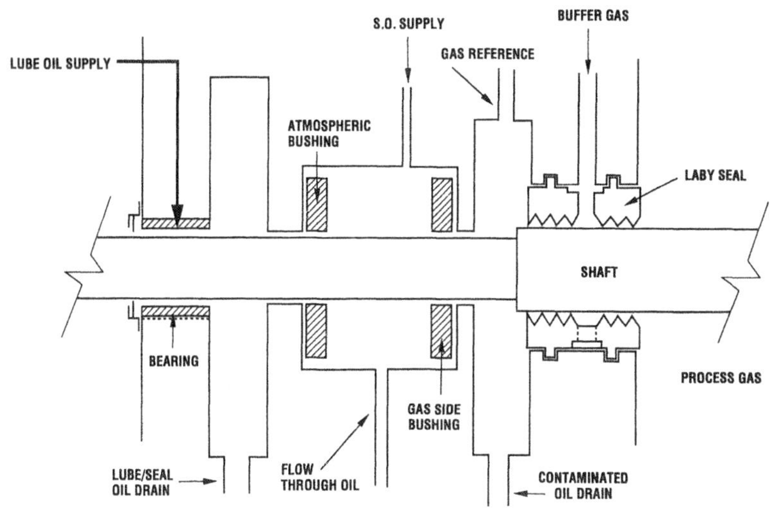

Figure 21.2 Bushing seal schematic (Courtesy of M.E. Crane Consultant)

frictional heat from the seal. This bushing can be conceptualized as a bearing, since the downstream pressure is usually atmospheric pressure. In systems that directly feed into a bearing the atmospheric bushing downstream pressure will be constant (approximately 20 psi). However, the upstream supply pressure will vary with the pressure required by the sealing media in the compressor.

As an example, if a seal system is designed to maintain a constant differential of 5 lbs. per square inch between the compressor process gas and the seal oil supply to the gas side bushing, the supply pressure with 0 PSIG process gas pressure, would be 5 psi to both the gas side bushing and atmospheric bushing. Therefore, gas side bushing and the atmospheric bushing differential would both be equal to 5 psi. If the process gas pressure were increased to 20 psi, the seal oil system would maintain a differential of 5 psi across the gas side seal, and the supply pressure to the gas side bushing and atmospheric bushing would be 25 psi. In this case, the differential across the gas side bushing would remain constant at 5 psi, but the atmospheric bushing differential pressure would increase from 5 to 25 psi. As a result, a primary concern in any seal liquid system is the assurance that the atmospheric bushing receives proper fluid flow under all conditions. After the seal fluid exits the seal chamber, it essentially returns through two additional subsystems.

C The atmospheric draining system

The flow from the atmospheric bushing, if it does not directly enter the bearing system, will return to the seal oil reservoir. In addition, flow from any downstream control valve will also return through the atmospheric drain system to the seal oil reservoir. Both these streams should be gas free since they should not come in contact with the process gas.

D The seal leakage system

The fluid that enters the gas side bushing is controlled to a minimum amount such that it can be either discarded or properly returned to the reservoir after it is degassed. Typically, this amount is limited to less than 20 gallons per day per seal. Since this liquid is in contact with the high speed shaft it is atomized and combines with sealing gas to enter the leakage system. This system consists of:

- An automatic drainer
- A vent system
- Degassing tank (if furnished)

The function of each component is as follows:

The drainer

The drainer contains the oil-gas mixture from the gas side seal. The liquid level under pressure in the drainer, is controlled by an internal float or external level control valve to drain oil back to the reservoir or the de-gassing tank, as required.

The vent system

The function of the seal oil drainer vent system is to assure that all gas side seal oil leakage is directed to the drainer. This is accomplished by referencing the drainer vent to a lower pressure than the pressure present at the gas side seal in the compressor. The drainer vent can be routed back to the compressor suction, suction vessel or a lower pressure source.

The degassing tank

This vessel is usually a heated tank, with ample residence time (72 hours or greater) to sufficiently de-gas all seal oil such that it will be returned to the reservoir and meet the seal oil specification. (Viscosity, flashpoint, dissolved gasses, etc.) These items will be discussed in detail later.

We will now proceed to discuss each of the major sub-systems in detail. Defining the function of each such that the total operation of a seal system can be simplified.

The supply system

Referring to definition of a seal oil system, it can be seen that the function is identical to that of a lube oil system, with one exception. The exception is that the seal fluid must be delivered to the seals at the specified differential pressure. Let's examine this requirement further.

Refer again to Figure 21.2 which shows an equivalent orifice diagram for a typical compressor shaft seal. Notice, that the atmospheric bushing downstream pressure is constant (atmospheric pressure). However, the gas side bushing pressure is referenced to the compressor process pressure. This pressure can and will vary during operation. If it were always constant, the requirement for differential pressure control would not be present in a seal system and would be identical to that of a lubricating system. Another way of visualizing the systems is to understand that the lube system utilizes differential pressure control as well, but the reference pressure (atmospheric pressure) is constant and consequently all control valves need only control lube oil pressure. However seal systems require some means of constant differential pressure control (reference gas pressure to seal oil supply pressure). This objective can be accomplished in many different ways. Referring back to Figure 21.1 it can be seen that the supply system function is identical to that of a lube oil system with the exception that the liquid is referenced to a pressure that can vary and must be controlled to maintain a constant differential between the referenced pressure and the seal system supply pressure. The sizing of the seal oil system components is also identical to that of the lube oil system components. Refer back and observe the heat load and flow required of each seal is determined in a similar way to that of the bearings. Seals are tested at various speeds and a necessary flow is determined to remove the heat of friction under various conditions. The seal oil flow requirements and corresponding heat loads, are then tabulated and pumps exchangers, filters, and control valves are sized accordingly.

The seal oil reservoir is sized exactly the same way as lube oil reservoir in our previous example. The only major difference between the component sizing of a seal and a bearing is that the seal flows across the atmospheric bushing will change with differential pressures. As previously explained, any liquid compressor seal incorporates a double seal arrangement. The gas side seal differential is held constant by system design. The atmospheric side seal differential varies with varying seal reference (process) pressure. Therefore, the total flow to the seals will vary with process pressure and must be specified for maximum and minimum values when sizing seal system components. Remembering the concept of an equivalent orifice, a compressor at atmospheric conditions will require significantly less seal oil flow than it will at high pressure (200 psi) conditions. This is true since the differential across the atmospheric seal and liquid flow will increase from a low value to a significantly higher value, while the gas side bushing differential and liquid flow will remain constant provided seal clearances remain constant.

Many seal system problems have been related to insufficient seal oil flow through the atmospheric bushing at low suction pressure conditions. Close attention to the atmospheric drain cavity temperature is recommended during any off design (low suction pressure condition) operation.

The seal housing system

Regardless of the type, the purpose of any seal is to contain the fluid in the prescribed vessel (pump, compressor, turbine, etc.). Types and designs of seals vary widely. Figure 21.3 shows a typical mechanical seal used for a pump.

Since the contained fluid is a liquid, this seal utilizes that fluid to remove the frictional heat of the seal and vaporize the liquid, thus attaining a perceived perfect seal. A small amount of vaporized liquid constantly exits the pump across the seal face. It is a fact that all seals leak. This is the major reason that many pump applications today are required to utilize seal-less pumps to prevent emission of toxic vapors. The following is a discussion of major types of seal combinations used in centrifugal compressor seal applications.

Gas seals

A typical gas seal is shown in Figure 21.4.

Gas seals have recently drawn attention since their supply systems appear to be much simpler than those of a traditional liquid seal system.

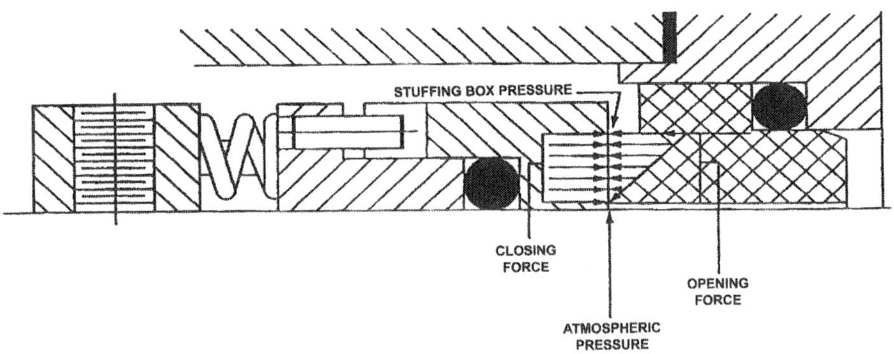

Figure 21.3 Typical pump single mechanical seal

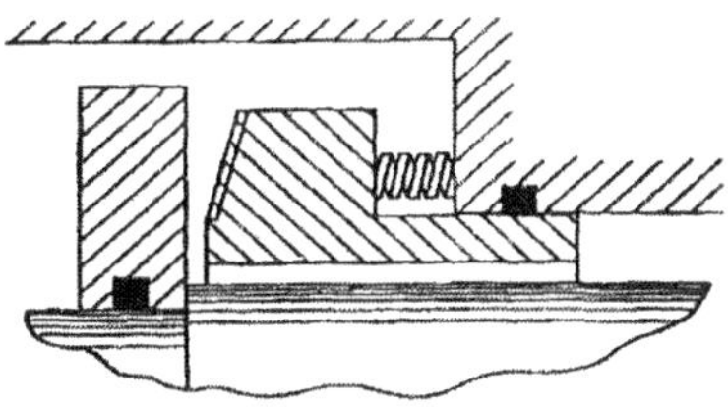

Typical Design For Curved Face — Spiral Groove Non-contact Seal; Curvature May Alternately Be On Rotor

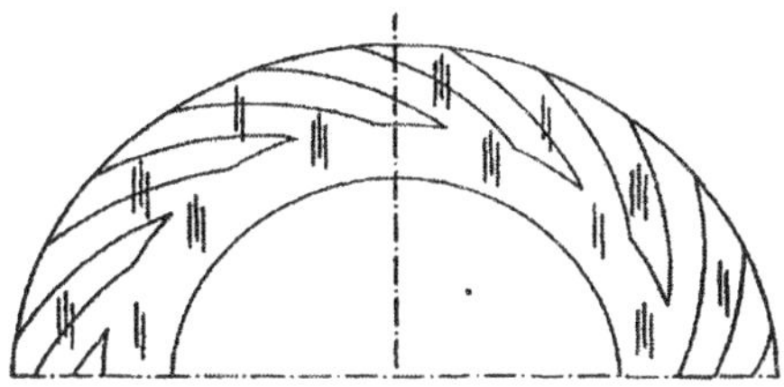

Typical Spiral Groove Pattern On Face Of Seal
Typical Non-contact Gas Seal

Figure 21.4 Typical gas seal (Courtesy of John Crane Co.)

Since gas seals utilize the sealed gas or a clean buffer gas, a liquid seal-system incorporating pumps, a reservoir and other components, is not required. However, one must remember that the sealing fluid still must be supplied at the proper flow rate, temperature and cleanliness. As a result, a highly efficient, reliable source of filtration, cooling, and supply must be furnished. If the system relies upon inert buffer gas for continued operation, the supply source of the buffer gas must be as reliable as the critical equipment itself. Gas seal configurations vary and will be discussed in detail in the next section. They can take the form of single, tandem (series), or multiple seal systems. The principle of operation is to maintain a fixed minimum clearance between the rotating and non-rotating face of the seal. The seal employed is essentially a contact seal with some type of lifting device to maintain a fixed minimum clearance between the rotating faces. It is essential that the gas between these surfaces be clean since any debris will quickly clog areas and reduce the effectiveness of the lifting devices, consequently resulting in rapid damage to the seal faces.

Liquid seals

Traditionally, the type of seal used in compressor service has been a liquid seal. Since the media that we are sealing against is a gas, a liquid must be introduced that will remove the frictional heat of the seal and assure proper sealing. Therefore, all compressor liquid seals take the form of a double seal. That is, they are comprised of two seals with the

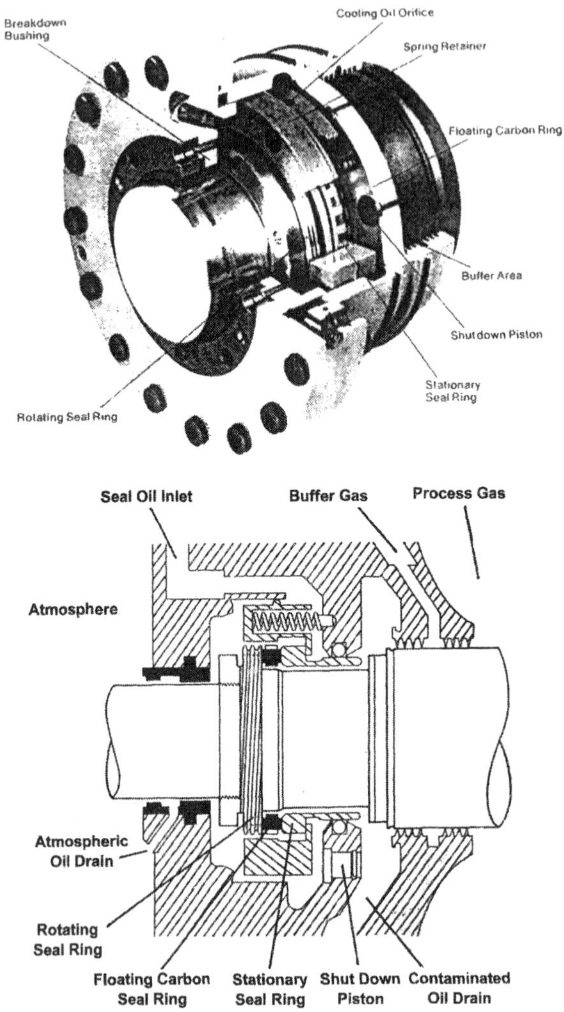

Figure 21.5 Iso carbon seal (Courtesy of Elliott Co.)

sealing liquid introduced between the sealing faces. Refer to Figure 21.5. To assure proper lubrication of both the gas side (inboard) and atmospheric side (outboard) seals, the equivalent 'orifices' of each seal must be properly designed such that the differential pressure present provides sufficient flow through the seal to remove the heat of friction at the maximum operating speed. The type of gas side seal used in Figure 21.5 is a contact seal similar to that used in most pump applications. This seal provides a minimum of leakage (five to ten gallons per day per seal) and provides reliable operation. (Continuous operation for 3+ years.) As will be discussed below, the specific types of seals used in the double seal (liquid) configuration can vary.

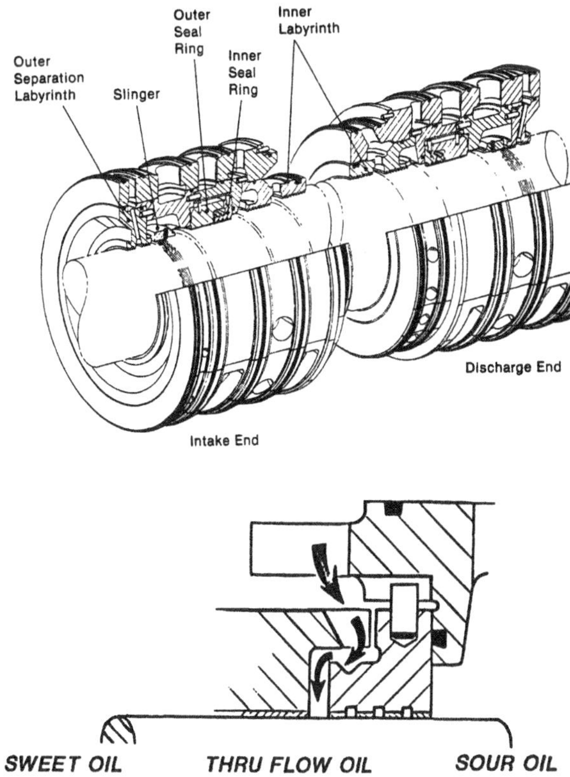

Figure 21.6 Bushing seal – top: oil film seal: bottom: seal oil flow (Courtesy of Dresser-Rand)

Liquid bushing seals

A liquid bushing seal can be used for either a gas side or an atmospheric side seal application. Most seals utilize a liquid bushing seal for an atmospheric bushing application. A typical bushing seal is shown in Figure 21.6.

The principle of a bushing seal is that of an orifice. That is, a minimum clearance between the shaft and the bushing surface to minimize leakage. The bushing seal is designed such that the clearance is sufficient to remove all the frictional heat at the maximum power loss condition of that bushing with the available fluid differential across the bushing. It is important to realize that while acting as a seal, the bushing must not act as a bearing. That is, it must have degrees of freedom (float) to assure that it does not support the load of the rotor. Since its configuration is similar to a bearing, if not allowed freedom of movement, it can act as an equipment bearing and result in a significant change to the dynamic characteristics of equipment with potential to cause damage to the critical equipment. In order to achieve the objectives of a bushing seal, clearances are on the order of 0.0005" diametrical clearance per inch of shaft diameter.

Liquid bushing seals are also used for gas side seals, however, their leakage rate will be significantly larger than that of a contact seal since they are essentially an orifice. When used as a gas side bushing, therefore, the system must be designed to minimize the differential across the bushing. As a result, the differential control system utilized must be accurate enough to maintain the specified oil/gas differential under all operating conditions. The typical design differential across a gas side bushing seal is on the order of five to ten psid. The accurate control of this differential is usually maintained by a level control system.

Referring back to Figure 21.6, one can see that functioning of the bushing seal totally depends on maintaining a liquid interface between the seal and shaft surface. Failure to achieve this results in leakage of gas outward through the seal. It must be fully understood that all bushing seals must continuously maintain this liquid interface to assure proper sealing. All systems incorporating gas side bushing seals must have the seal system in operation whenever pressurized gas is present inside the compressor case. If a liquid interface is not maintained, gas will migrate across the atmospheric bushing seal and proceed through the system returning back to the supply system. There have been cases in such system designs where failures to operate the seal system when the compressor is pressurized have resulted in effectively turning the gas side bushing into a filter for the entire process gas system! This resulted in the supply side of the seal oil system being filled with extensive debris that required lengthy flushing and system cleaning operations prior to putting the unit back into service. Remember, any system incorporating a gas side bushing seal must be designed such that the entrance of process gas into the supply system is prohibited at all time. This can be accomplished by either:

- Continuous buffer gas supply
- A check valve installed as close as possible to seals in the seal oil supply header
- Rapid venting and isolation of the compressor case on seal system failure

In the second and third cases above, supply seal oil piping must be thoroughly checked for debris prior to re-start of the compressor. It is our experience that many bushing seal system problems have resulted from improper attention to the above facts.

Contact seals

Figure 21.7 shows a typical compressor contact seal. As mentioned, these seals are similar in design to pump seals. In order to remove the

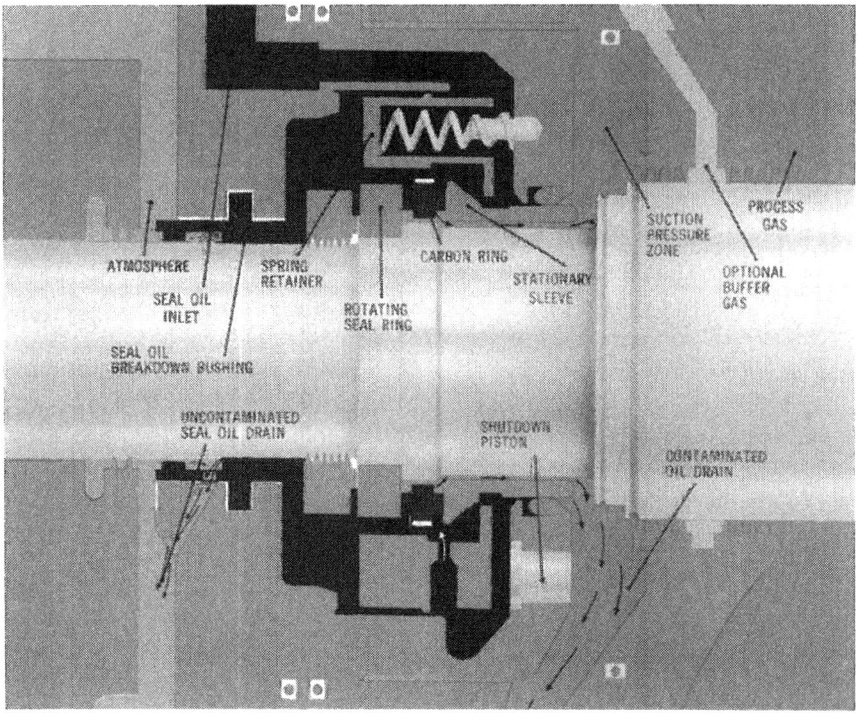

Figure 21.7 Compressor contact seal (Courtesy of Elliott Co.)

heat of friction for this type of seal, a sufficient differential pressure above the reference gas must be maintained. Typical differentials for contact seals vary between 35 and 50 lbs. per square inch differential pressure. Leakage rates with a properly installed seal can be maintained between five to ten gallons per day per seal.

A limitation in the use of contact seals are shaft speed, since the contact seal operates on a surface perpendicular to the axis of rotation, the rubbing speed of the seal surface is critical. As a result, contact seals are speed limited. Typical maximum speeds are approximately 12,000 revolutions per minute. Above those speeds, bushing seals are used, since the sealing surface is maintained at a lower diameter and correspondingly lower rubbing speed. The maximum limit of differential pressure across contact seals is controlled by the materials of construction and is approximately 200 lbs. per square inch differential. As a result, contact seals are usually used for gas side seal applications. They are very seldom utilized for atmospheric seal applications since they are differential pressure limited.

Since the differential pressure required across the seal face is relatively high as compared to a bushing seal, contact seals utilize differential pressure control as opposed to level control for most bushing seals. This fact will be discussed in the next section.

Restricted bushing seals

The last type of seal to be discussed is a restricted bushing type seal. This type of seal is shown in Figure 21.8.

This particular type of restricted bushing seal utilizes a small pumping ring in the opposite direction of bushing liquid flow to compensate for the relatively large leakage experienced with bushing seals by introducing an opposing pumping flow in the opposite direction. Seals of this type can be designed for practically zero flow leakages. However, it must be pointed out that in variable speed applications, the pumping capability of the trapped seal ring must be calculated for both minimum and maximum speeds. Failure to do so can result in the actual pumping of gas from the compressor into the sealing system. It is recommended that such seals be designed to leak a small amount at maximum operating speed. Any retrofits of equipment employing this type of seal should be investigated when higher operating speeds are anticipated. A restricted bushing seal is used exclusively for gas side service.

In summary, the basic types of liquid seals used for compressor applications can be either: open bushing types, contact types, or restricted bushing types. Contact types are used primarily on the gas side. Liquid bushing types are used on either the gas or atmospheric side. Restricted bushing types are used exclusively on the gas side.

Compact Design — allows shorter bearing spans for higher critical speeds of the compressor rotor.
Sleeve (impeller) with interference fit under bushing — protects shaft and simplifies assembly and disassembly. Requires only a jack/puller bolt ring.
Spacer fit at initial assembly — no field fitting of parts.

ITEM	DESCRIPTION
1	Shaft
2	Impeller
3	Stator
4	Stepped Dual Bushing
5	Bushing Cage
6	Nut
7	Shear Ring
8	Oil/Gas Baffle
9	Spacer Ring

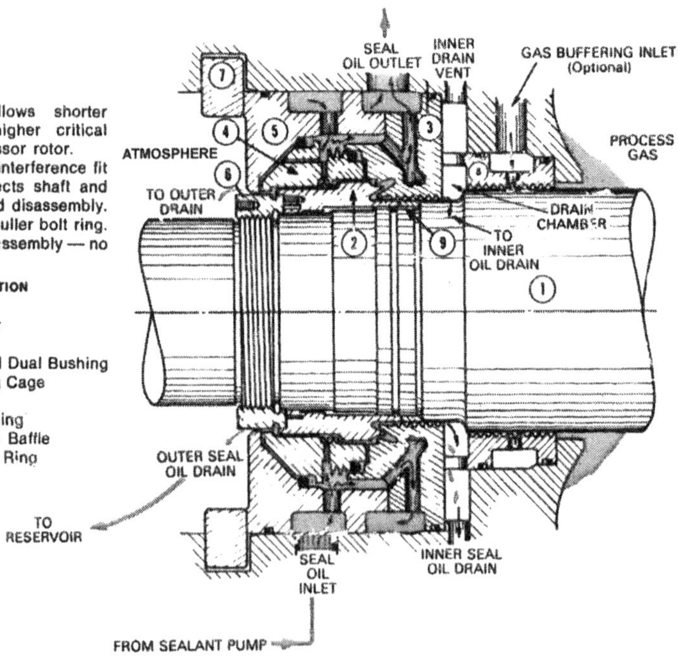

Figure 21.8 Turbo-compressor 'trapped bushing seal' (Courtesy of A.C. Compressor Corp.)

We will now investigate various seal system designs using various seal combinations employing the types of seals that have been discussed in this section.

Seal supply systems

As can be seen from the previous discussion, the type of seal system will depend on the type of seal utilized. We will now examine five different types of seal systems, each utilizing a different type of main compressor shaft seal system. As we proceed through each type, the function of each system will become clear.

Example 1: Contact type gas side seal – bushing type atmospheric side seal with cooling flow

This system incorporates a contact seal on the gas side and a bushing seal on the atmospheric side of each end of the compressor. The inlet pressure of the seal fluid on each end is referenced to the suction pressure of the compressor. It should be noted that some applications

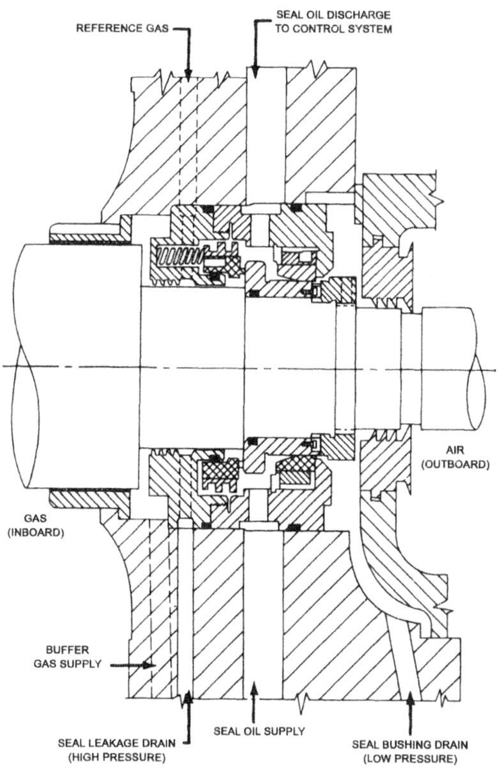

Figure 21.9 Compressor shaft seal (Courtesy of IMO Industries)

employ different reference pressures on each end of the compressor. The reference pressure should be taken off the balance drum end, or high pressure end of the compressor, to assure that the oil to gas differential pressure is always at a minimum acceptable value. Therefore, the low pressure end may experience a slightly higher oil to gas differential then the reference end of the seal. Refer to Figure 21.9.

Proceeding through the seal, the seal oil supply, which is referenced to the gas reference pressure, enters the seal chamber. The differential across the gas side contact seal is maintained by a differential pressure control valve located downstream of the seal. Seal oil flows in three separate directions:

- Through the seal chamber (cooling flow)
- Through the gas side contact seal (10–20 gallons/day)
- Through the atmospheric seal

Let's examine the variants of flows across the equivalent orifice of each portion of this configuration.

The gas side contact seal will experience a constant flow, that for purposes of discussion, can be assumed to be zero gallons per minute (since the maximum flow rate will usually be on the order of ten gallons per day).

The atmospheric side bushing seal flow will vary based upon the referenced gas pressure. At low suction pressure conditions, this flow will be significantly less than it will be under high pressure conditions. The seal system design must consider the maximum reference pressure to be experienced in the compressor case to assure that sufficient seal oil flow is available at maximum pressure conditions.

The seal chamber through flow in this seal design is used to remove any excess frictional heat of the seals and is regulated by the downstream control valve. As an example, let us assume the following values were calculated for this specific seal application.

1. *Gas Side Seal Flow* = 0 GPM
2. *Atmospheric Side Seal Flow*
   Reference Pressure = 0 PSIG
   Seal Flow = 5 GPM
   Reference Pressure = 200 PSIG
   Seal flow = 12 GPM
3. *Flow Through Flow*
   Minimum = 3 GPM (occurring at high ATM bushing flow = 12 GPM)
   Maximum = 12 GPM (occurring at low ATM bushing flow = 3 GPM)
4. *Seal Oil Supply Flow* in both cases = 15 GPM

As shown in the previous example, the required seal oil supply at maximum operating speed required to remove frictional heat is 15 gallons a minute. At start-up, low suction pressure conditions, the control valve must open to allow an additional ten gallons a minute flow through to the seal chamber. At maximum operating pressure, however, the valve only passes a flow of three gallons a minute since 12 gallons a minute exit through the atmospheric bushing. This type of system is less sensitive to low suction pressure operation since flow through oil will remove frictional heat around the atmospheric bushing.

Example 2: Contact type gas side seal – bushing type atmospheric seal with orificed through flow

The only difference between this type of system and the previous system is that the back pressure is maintained constant by a permanently installed through flow orifice. As a result, the differential pressure control valve is installed on the inlet side of the system. The process gas reference is still the same as before, that is, to the highest pressure side of the compressor. Figure 21.10 shows this type of system.

Let us examine the previous example case for this system and observe the differences.

1. *Gas Side Seal Flow = 0 GPM*
2. *Atmospheric Side Seal Flow*
   Reference Pressure = 0 PSIG
   Seal Flow = 5 GPM
   Reference Pressure = 200 PSIG
   Seal Flow = 12 GPM
3. *Flow Through Flow (Orifice)*
   Minimum = 0.5 GPM
   Maximum = 3 GPM

As can be seen, this system is more susceptible to high temperature atmospheric bushing conditions at low suction pressures and must be observed during such operation to assure integrity of the atmospheric bushing. In this system, the control valve will sense supply oil pressure to the seal chamber and control a constant set differential, approximately 35 psid, between the reference gas pressure and the supply pressure. If continued low pressure operation is anticipated with such a system, consideration should be given to a means of changing the minimum flow and maximum flow orifice for various operation points. Externally piped bypass orifices could be arranged such that a bypass line with a large orifice for minimum suction pressure conditions could be installed and opened during this operation. It is important to note, however, that the entire supply system must be designed for this flow condition and control valve must be sized properly to assure

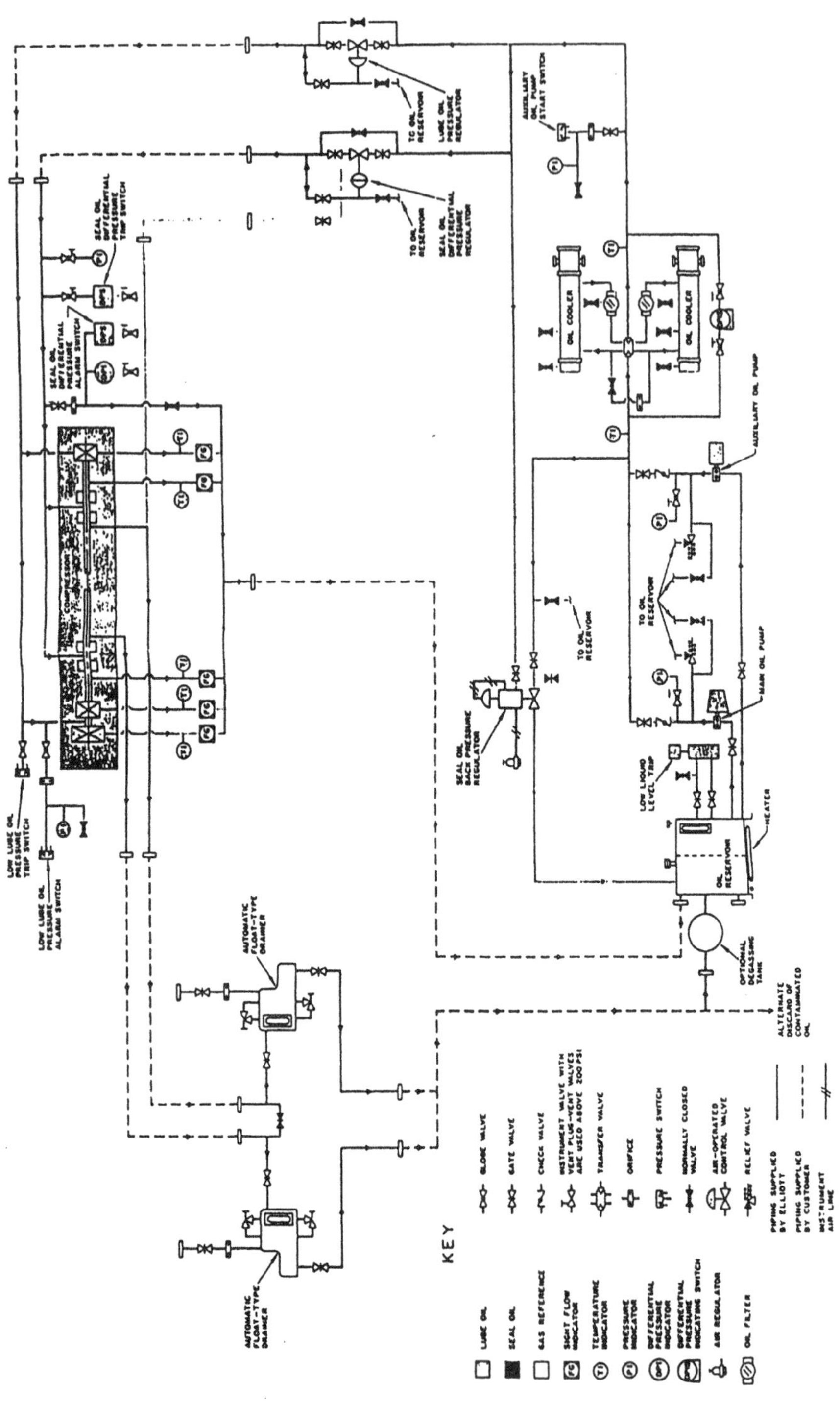

Figure 21.10 API 614 Lube-seal oil system for ISO-carbon seals (Courtesy of Elliott Company)

proper flow at this condition. In addition, the low pressure bypass line must be completely closed during normal high pressure operation.

Example 3: Bushing gas side seal – bushing atmospheric side seal with no flow through provision

Figure 21.1 shows this type of seal system. In this type of system, the differential control valve becomes a level control valve sensing differential from the level in an overhead tank and is positioned upstream of the unit. Both bushings can be easily conceived as equivalent orifices. The gas side bushing flow will remain constant regardless of differential. The atmospheric bushing flow will vary according to seal chamber to atmospheric pressure differential.

Therefore, the atmospheric bushing must be designed to pass a minimum flow at minimum pressure conditions that will remove frictional heat and thus prevent overheating and damage to the seal. Since a gas side bushing seal is utilized, a minimum differential across this orifice must be continuously maintained.

Utilizing the concept of head, the control of differential pressure across the inner seal is maintained by a column of liquid.

As an example, if the required gas side seal differential of oil to gas is 5 psid, by the liquid head equation:

$$\text{Head} = \frac{2.311 \times 5 \text{ psid}}{.85} = 13.6 \text{ ft.}$$

Therefore maintaining a liquid level of 13.6 ft. above the seal while referencing process gas pressure will assure a continuous 5 psid gas side bushing differential. In this configuration, the control valve which senses its signal from the level transmitter, will be sized to continuously supply the required flow to maintain a constant level in the overhead tack. As an example, consider the following system changes from start up to normal operation.

Seal System Flow

Item	Start-up condition	Normal operation
Compressor suction pressure	0 PSIG	200 PSIG
Overhead tank reference pressure	0 PSIG	200 PSIG
Gas side seal bushing flow	0 GPM	0 GPM
Atmospheric side seal bushing flow	5 GPM	12 GPM
Seal pump flow	20 GPM	20 GPM
Bypass valve flow	15 GPM	8 GPM

In this example a change from the start-up to operating condition will increase gas reference pressure on the liquid level in the overhead tank and would tend to push the level downward. Any movement of the level in the tank will result in an increasing signal to the level control valve to open, thus increasing the pressure (assuming a positive displacement pump) to the overhead tank and reestablishing the pre-set level.

In the above example, at 200 psi reference pressure, the bypass valve would close considerably. To increase the pressure supply of the seal oil from 5 psi to 205 psi, the difference of bypass flow through the valve (7 gpm) is equal to the increased flow through the atmospheric bushing at this higher differential pressure condition. Utilizing the concept of equivalent orifices, it can be seen that the additional differential pressure across the atmospheric bushing orifice is compensated for by reducing the effective orifice area of the bypass control valve. This is accomplished by sensing the level in the head tank and maintaining it at a constant value by opening the seal oil supply valve.

As in the case of the orificed through flow example above, this configuration is susceptible to high atmospheric bushing temperatures at low suction pressures and must be monitored during this condition. Repeated high temperatures during low suction pressure conditions should give consideration to re-sizing of atmospheric bushing clearances during the next available turnaround. The original equipment manufacturer should be consulted to assure correct bushing sizing and supply system capability.

Example 4: Gas side bushing seal – atmospheric side bushing seal with through flow design

Refer to Figure 21.11. The only difference between this system and the previous system is that a through flow option is added to allow sufficient flow through the system during changing pressure conditions. The bypass valve in the previous system is replaced in this system by a level control valve referenced from a head tank level transmitter, and is installed downstream of the seal chamber. This system functions in exactly the same way as the system in Example 1. The only difference being that a level control valve in this example replaces the differential control valve in the previous example. Both valves have the same function, that is, to control the differential in the seal chamber between the seal oil supply and the referenced gas pressure. A level control valve is utilized in this example, however, since a bushing seal requires a significantly lower differential between the seal oil supply pressure and the gas reference pressure.

Consider the following example. Assume that a differential control valve would be used as opposed to a level control valve for the system in Figure 21.11. For the start-up case, the differential control valve would have to maintain a differential of 5 psi over the reference gas. When the reference gas pressure were 0 psi, the oil upstream pressure to the valve would be approximately 5 psi. For the operating case, maintaining the same 5 psi differential, the upstream pressure across the valve would be approximately 205 lbs instead of 5 psi. Consequently, the valve position would change significantly, but still would have to control the differential accurately to maintain 5 psi. Reduction of this pressure in any amount below 5 psi could result in instantaneous bushing failure. However, if a level control valve were installed, the accuracy of the valve would be measured in inches of oil instead of psi. Any level control system could control the level within two inches, which would be only a .06 psid variation in pressure differential!

This example shows that the accurate means of controlling differential pressure for systems requiring control of small differential values, is to use level instead of differential control. This system would be designed such that the combination of the atmospheric flow and the through flow through the seal would be equal to the flow from the pump.

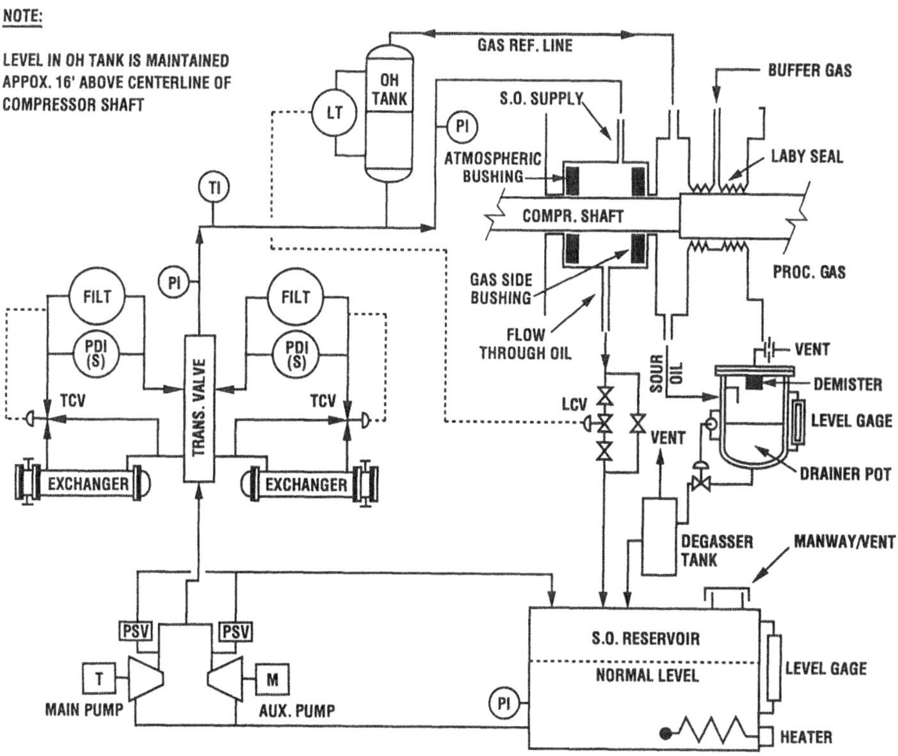

Typical Seal Oil System
(For Clearance Bushing Seal)

Figure 21.11 Typical seal oil system (Courtesy of M.E. Crane Consultant)

Example 5: Trapped bushing gas side seal: atmospheric side bushing seal with flowthrough design

This system would follow exactly the same design as the system described in Example 4. The only difference would be in the amount of flow registered in the seal oil drainer. A trapped bushing system is designed to minimize seal oil drainer pot leakage. Typical values can be less than five gallons/day.

Seal supply system summary

All of the above examples have dealt with a system incorporating one seal assembly. It must be understood that most systems utilize two or more seal system assemblies. Typical multi-stage compressors contain two seal assemblies per compressor body and many applications contain upwards of three compressor bodies in series, or six seal assemblies. Usually each compressor body is maintained at the suction pressure to that body, therefore three discreet seal pressure levels would be required and three differential pressure systems would be utilized. The concepts discussed in this section follow through regardless of the amount of seals in the system. Sometimes, the entire train, that is, all the seals referenced to the same pressure. In this case, one differential seal system could be used across all seals.

In conclusion, remembering the concept of an orifice will help in understanding the operation of these systems. Remember, the gas side bushing is essentially zero flow, the atmospheric side bushing flow varies with changing differential across the seal and any seal chamber through flow will change either as a result of differential across a fixed orifice or the repositioning of the control valve.

Seal liquid leakage system

This seal system sub-system's function is to collect all of the leakage from the gas side seal and return it to the seal reservoir at specified seal fluid conditions. Depending upon the gas condition in the case, this objective may or may not be possible. If the gas being compressed has a tendency to change the specification of the seal oil to off specification conditions, one of two possibilities remain:

- *Introduce a clean buffer gas* between the seal to assure proper oil conditions
- *Dispose of the seal oil leakage*

In most cases, the first alternative is utilized. Once the seal oil is in the drainer, a combination of oil and gas are present. A vent may be installed in the drainer pot to remove some of the gas, or a degassing tank can be incorporated.

This concludes the overview section of seal oil systems. As can be seen from the above discussion, it is evident that the design of a seal oil system follows closely to that of a lube system. The major difference is that the downstream reference pressure of the components (seal) varies, whereas in the case of a lube system it does not. In addition, the collection of the expensive seal oil is required in most cases and a downstream collector, or drainer system, must be utilized. Other than these two exceptions, the design of the seal system is very similar to that of a lube system and the same concepts apply in both cases.

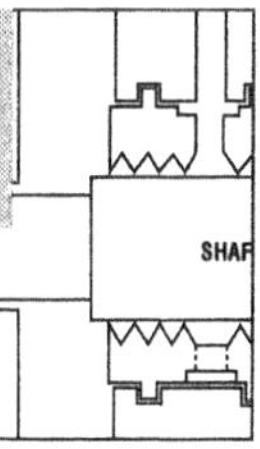

22 Dry gas seal systems

- Introduction
- Dry gas seal design
- Gas seal system types
- Summary

Introduction

The reliability of critical equipment is dependent on the reliability of each component in every auxiliary system connected with the critical equipment unit. How do we maximize critical equipment reliability? The easiest way is to eliminate the auxiliary systems. Imagine the opportunity to eliminate all of the components; pumps, filters, reservoirs, etc. and thereby increase reliability and hopefully, the safety of the equipment. The gas seal as used in compressor applications affords the opportunity to achieve these objectives. However, the gas seal is still part of a system and the entire gas seal system must be properly specified, designed, maintained and operated to achieve the objectives of optimum safety and reliability of the critical equipment.

In this chapter, the principles of gas seal design will be discussed and applied to various gas seal system types. In addition, best practices will be discussed for saturated gas systems as well as shutdown philosophies.

System function

The function of a gas seal system is naturally the same as a liquid seal system. The function of a fluid seal system, remembering that a fluid can be a liquid or a gas, is to continuously supply clean fluid to each specified seal interface point at the required differential pressure,

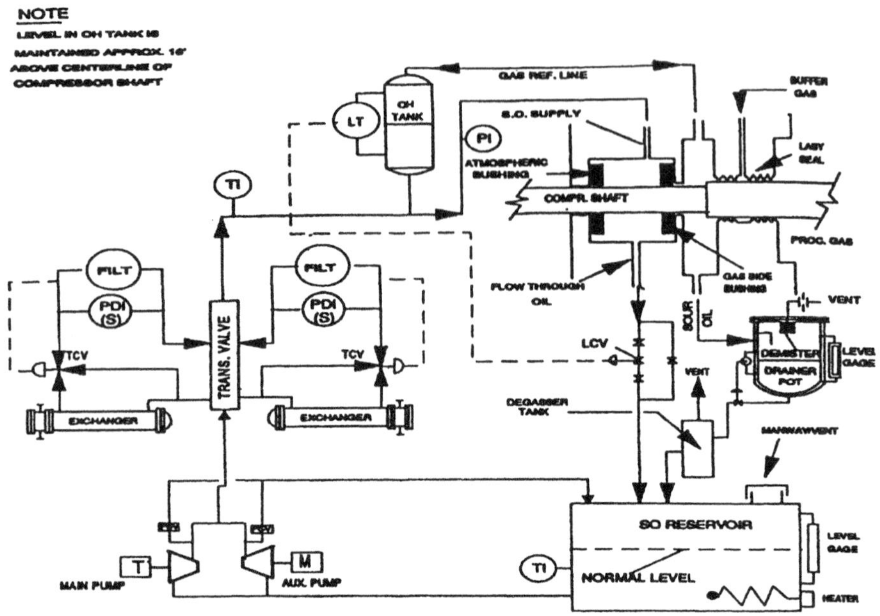

Figure 22.1 Typical seal oil system for clearance bushing seal (Courtesy of M.E. Crane Consultant)

temperature, and flow rate. Therefore, one would expect the design of a gas seal and a liquid seal to be very similar, which, in fact, they are. Then why are their systems so different?

Comparison of a liquid and gas sealing system

Figure 22.1 shows a liquid sealing system.

Compare this system to Figure 22.2 which shows a gas seal system, if the same compressor were retrofitted for a gas seal. WOW!! What a difference. Why are there such a small amount of components for the gas seal system? As an aid, refer to Figure 22.3 which shows a typical pump liquid flush system as specified by the American Petroleum Institute. This system incorporates a liquid mechanical seal and utilizes pump discharge liquid as a flush for the seal. Refer now to Figure 22.2 and observe the similarities. It should be evident that a gas seal system is simplified in compressor applications over a liquid seal system merely because the gas seal utilizes the process fluid. This is exactly the same case for a pump. By using the process fluid, and not a liquid, one can eliminate the need to separate liquid from a gas, thereby totally eliminating the need for a liquid supply system and the need for a contaminated liquid (sour oil), drain system.

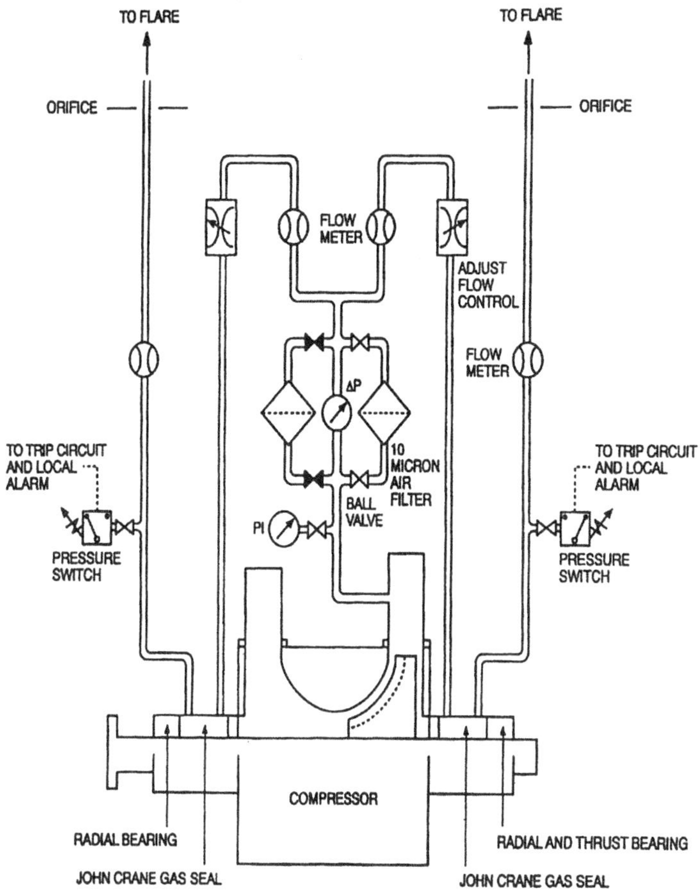

Figure 22.2 Typical gas seal system for dry air or inert gas (Courtesy of John Crane Co.)

Referring back to Figure 22.1, therefore, we can see that the following major components are eliminated:

1. The seal oil reservoir
2. The pumping units
3. The exchangers
4. The temperature control valves
5. The overhead tank
6. The drain pot
7. The degassing tank
8. All control valves
9. A significant amount of instrumentation

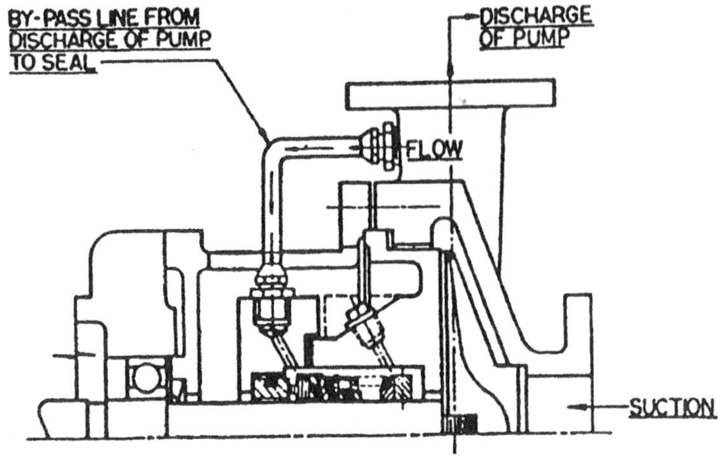

Figure 22.3 Liquid seal flush (Courtesy of John Crane Co.)

Referring back to the function definition of the gas seal system, all requirements are met. 'Continuously supplying fluid' is met by utilizing the discharge pressure of the compressor. The requirements for 'specified differential pressure, temperature and flow rate' are met by the design of the seal itself which can accommodate high differential pressures, high temperatures, and is sized to maintain a flow rate that will remove frictional heat necessary to maintain seal reliability. The only requirement not met is that of supplying a clean dry fluid, and this can be seen in Figure 22.2. This requirement is met by using a dual system coalescing filter.

When one considers all the advantages, the next question to ask is, okay, what are the disadvantages? Naturally, there are disadvantages. However, proper design of the gas seal system can minimize and eliminate many of the disadvantages. Do not forget that the requirements for any system mandate proper specification, design, manufacture, operation and maintenance. One can never eliminate these requirements in any critical equipment system.

Considerations for system design

As mentioned above, there are disadvantages to a gas seal system which are not insurmountable but must be considered in the design of such a system. These considerations are as follows:

Sensitivity to dirt – since clearances between seal faces are usually less than 0.0005 inch and seal design is essential to proper operation, the fluid passing between the faces must be clean (5–10 microns maximum particle size). If it is not, the small grooves (indentations) necessary for seal face separation will become plugged thus causing face contact and seal failure.

Sensitivity to saturated gas – saturated fluids increase the probability of groove (indentation) blockage.

Lift-off speed – as will be explained below, a minimum speed is required for operation. Care must be taken in variable speed operation to assure that operation is always above this speed. It is recommended that the seal test be conducted for a period at turning gear speed to confirm proper 'lift off' followed by seal face inspection.

Positive prevention of toxic gas leaks to atmosphere – since all seals leak, the system must be designed to preclude the possibility of toxic of flammable gas leaks out of the system. This will be discussed in detail below.

Possible oil ingestion from the lube system – a suitable separation seal must be provided to eliminate the possibility of oil ingestion from the bearings. Whenever a gas seal system is utilized, the design of the critical equipment by definition incorporates a separate lube oil and seal system. Consideration must be given during the design or retrofit phases to the separation between the liquid (lube) and gas seal system.

'O' ring (secondary seal components) design and maintenance – most seal vendors state that 'O' ring life is limited and should be changed every five years for operating seals as well as spare seals. The writer's experience has shown that dry gas 'O' ring seals can exceed this limit. It is recommended that seal vendors be required to provide references for similar applications prior to making a decision to change out the seals after five years.

If all of the above considerations are incorporated in the design of a gas seal system, its reliability has the potential to exceed that of a liquid seal system and the operating costs can be reduced.

Before moving to the next section, however, one must consider that relative reliability between gas and liquid seal systems are a function of proper specification, design, etc. as mentioned previously. A properly designed liquid seal system that is operated and maintained can achieve reliabilities of a gas seal system. Also, when one considers operating costs of the two systems, various factors must be considered. While the loss of costly seal oil is positively eliminated, with a gas seal system (assuming oil ingestion from the lube system does not occur) the loss of process gas, while minimal, can be expensive. It is argued that the loss of process gas from a liquid seal system through drainer vents and degassing tank vents, is also significant. While this may be true in many cases, a properly specified, designed and operated liquid seal system can minimize process gas leakage such that it is equal or even less than that of a gas seal.

There is no question that gas seal systems contain far fewer components and are easier to maintain than liquid seal systems. These systems will be used extensively in the years ahead. The intention of this discussion is to point out that existing liquid seal systems that cannot be justified for retrofit or cannot be retrofitted easily, can be modified to minimize outward gas leakage and optimize safety and reliability.

Dry gas seal design

Principles of operation

The intention of this sub-section is to present a brief detail of the principles of operation of a dry gas seal in a conceptual form. The reader is directed to any of the good literature available on this subject for a detailed review of gas seal design.

Refer to Figure 22.4. Figure 22.4 shows a mechanical seal utilized for pump applications, while Figure 22.5 shows a dry gas mechanical seal

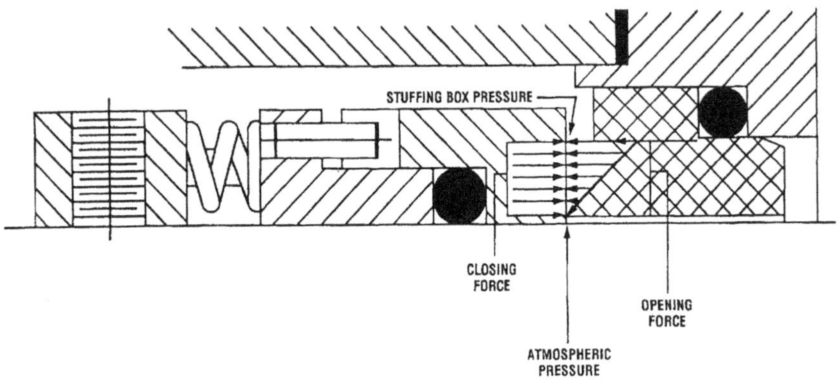

Figure 22.4 Typical pump single mechanical seal

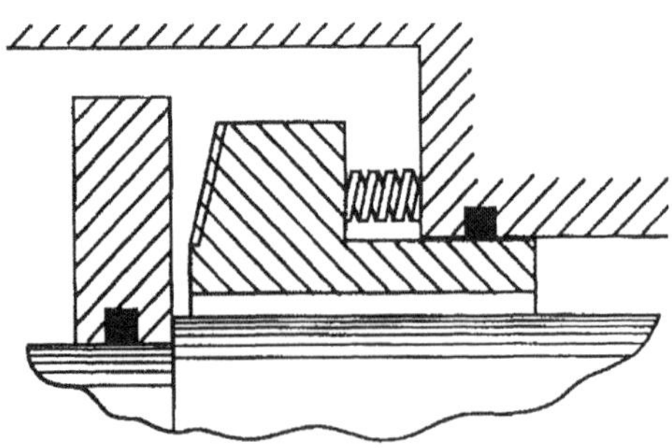

Figure 22.5 Typical design for curved face – spiral groove non-contact seal; curvature may alternately be on rotor (Courtesy of John Crane Co.)

utilized for compressor application. The seal designs appear to be almost identical. Close attention to Figure 22.5, however, will show reliefs of the rotating face of the seal. Considering that both seals operate on a fluid may give some hint as to why the designs are very similar. The objective of seal designs is to positively minimize leakage while removing frictional heat to obtain reliable continuous operation of the seal. In a liquid application, the heat is removed by the fluid which passes between the rotating and stationary faces and the seal flush and changes from a liquid to gaseous state (heat of vaporization). This is precisely why all seals are said to leak and explains the recent movement in the industry to sealless pumps in toxic or flammable service. If the fluid between the rotating faces now becomes a gas, its capacity to absorb frictional heat is significantly less than that of a liquid. Therefore an 'equivalent orifice' must continuously exist between the faces to reduce friction and allow a sufficient amount of fluid to pass and thus take away the heat. The problem obviously is how to obtain this 'equivalent orifice'. There are many different designs of gas seals. However, regardless of the design, the dynamic action of the rotating face must create a dynamic opening force that will overcome the static closing forces acting on the seal to create an opening and hence 'equivalent orifice'.

Refer to Figure 22.6 which shows a typical gas dry seal face. Notice the

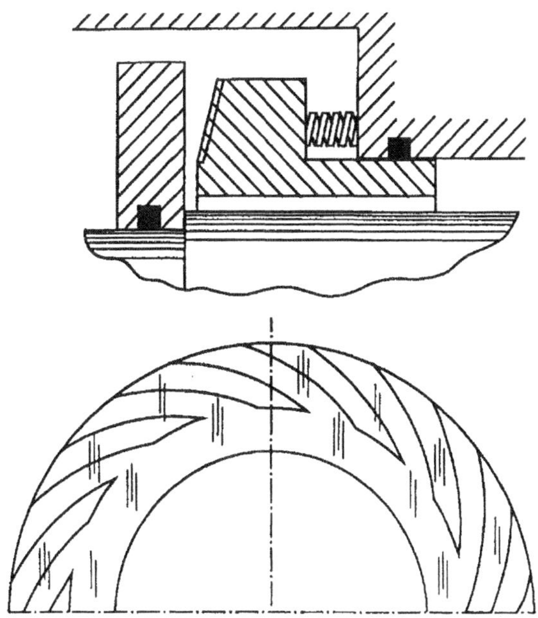

Figure 22.6 Dry gas seal. Top: typical design for curved face – spiral groove non-contact seal; curvature may alternately be on rotor; Bottom: Typical spiral groove pattern on face of seal typical non-contact gas seal (Courtesy of John Crane Co.)

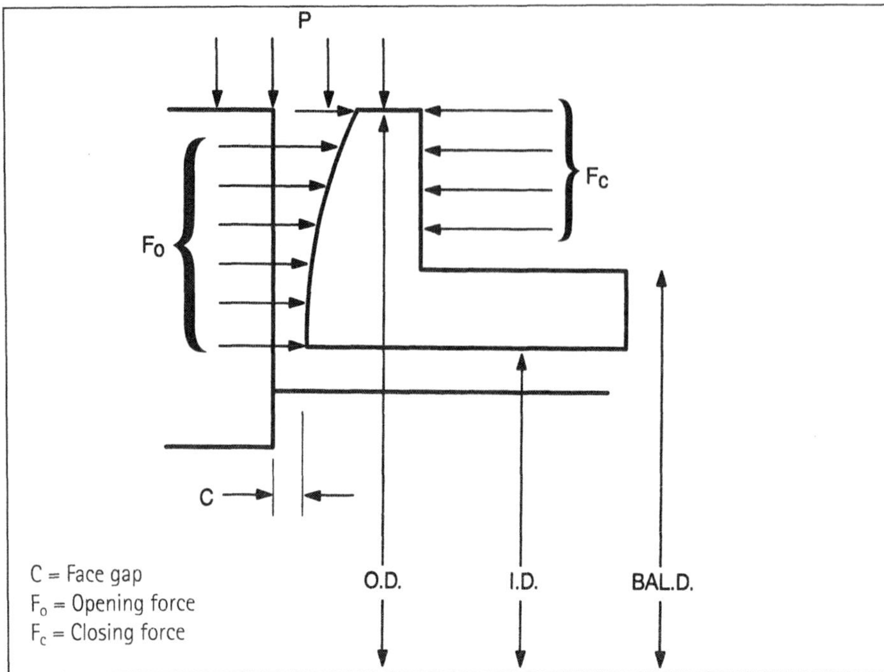

Figure 22.7 Hydrostatic force balance on seal stator ($F_C = F_O$) (Courtesy of John Crane Co.)

spiral grooves in this picture, they are typically machined at a depth of 100–400 micro inches. When rotating, these vanes create a high head low flow impeller that pumps gas into the area between the stationary and the rotating face, thereby increasing the pressure between the faces. When this pressure is greater than the static pressure holding the faces together, the faces will separate thus forming an equivalent orifice. In this specific seal design, the annulus below the vanes forms a tight face such that under static (stationary) conditions, zero leakage can be obtained if the seal is properly pressure balanced. Refer to Figure 22.7 for a force diagram that shows how this operation occurs.

In Figure 22.6, the rotation of the face must be counter-clockwise to force the gas into the passages and create an opening (F_o) force. This design is known as a 'uni-directional' design and requires that the faces always operate in this direction. Alternative face designs are available that all rotate in either direction and they are known as 'bi-directional' designs.

Ranges of operation

Essentially, gas seals can be designed to operate at speeds and pressure differentials equal to or greater than those of liquid seals. Present state-of-the-art limits seal face differentials to approximately 1,000–2,000 psi

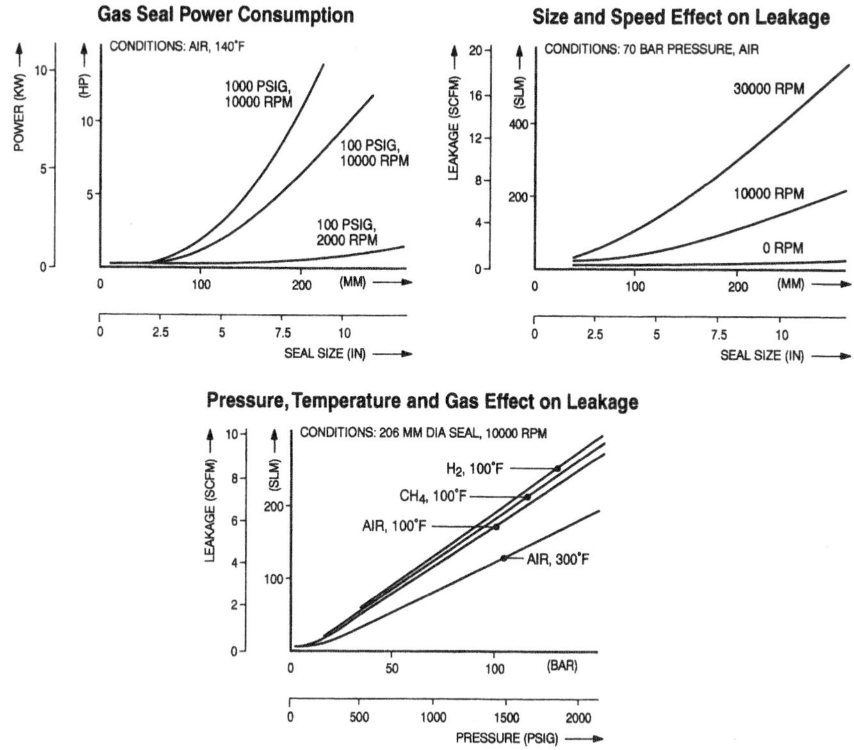

Figure 22.8 Dry gas seal leakage rates (Courtesy of John Crane Co.)

and rubbing speeds to approximately 400 feet per second. Temperatures of operation can reach as high as 1,000°F. Where seal face differential exceeds these values, seals can be used in series (tandem) to meet specifications provided sufficient axial space is available in the seal housing.

Leakage rates

Since the gas seal when operating forms an equivalent orifice, whose differential is equal to the supply pressure minus the seal reference pressure, there will always be a certain amount of leakage. Refer to Figure 22.8 for leakage graphs.

It can be stated in general that for most compressor applications with suction pressures on the order of 500 psi and below, leakage can be maintained on the order of one standard cubic foot per minute per seal. For a high pressure application (1,000–2,000 psi), differential leakage values can be as high as five standard cubic feet per minute per seal. As in any seal design, the total leakage is equal to the leakage across the seal faces and any leakage across secondary seals (O-rings, etc.). There have been reported incidence of explosive O-ring failure on rapid decompression of systems incorporating gas seals, thus resulting in excessive leakage. Consideration must be given to the system in order to tailor system decompression times in order to meet the requirements of the secondary seals. As previously mentioned, all gas seals will leak, but not until the face 'lifts off'. This speed known, oddly enough, as 'lift off speed' is usually less than 500 rpm. Caution must be exercised in variable speed applications to assure the system prevents the operation of the variable speed driver below this minimum lift off value. One recommendation concerning instrumentation is to provide one or two thermocouples in the stationary face of each seal to measure seal face temperature. This information is very valuable in determining lift off speed and condition of the grooves in the rotating seal face. Any clogging of these grooves will result in a higher face temperature and will be a good indication of requirement for seal maintenance.

Gas seal system types

As mentioned in this section, in order to assure the safety and reliability of gas seals, the system must be properly specified and designed. Listed below are typical gas seal system applications in use today.

Low/medium pressure applications – dry air or inert gas

Figure 22.2 shows such a system. This system incorporating a single dry gas seal is identical to that of a liquid pump flush system incorporating relatively clean fluid that meets the requirements of the seal in terms of temperature and pressure. This system takes the motive fluid from the discharge of the compressor through dual filters (ten microns or less) incorporating a differential pressure gage and proportions equal flow through flow meters to each seal on the compressor. Compressors are usually pressure balanced such that the pressure on each end is approximately equal to the suction pressure of the compressor. The clean gas then enters the seal chamber and has two main paths:

A Through the internal labyrinth back to the compressor. Note that the majority of supplied gas takes this path for cooling purposes (99%).

B Across the seal face and back to either the suction of the compressor or to vent.

Since the gas in this application is inert, it can be vented directly to the atmosphere or can be put back to the compressor suction. It must be noted, however, that this port is next to the journal bearing. Therefore a means of positively preventing entry of lube oil into this port must be provided in order to prevent the loss of lube oil or prevent the ingestion of lube oil into the compressor if this line is referenced back to the compressor suction. A suitable design must be incorporated for this bushing. Typically called a disaster bushing, it serves a dual purpose of isolating the lube system from the seal system and providing a means to minimize leakage of process fluid into the lube system in the event of a gas seal failure. In this system, a pressure switch upstream of an orifice in a vent line is used as an alarm and a shutdown to monitor flow. This switch uses the concept of an equivalent vessel in that increased seal leakage will increase the rate of supply versus demand flow in the equivalent vessel (pipe) and result in a higher pressure. When a high flow is reached, the orifice and pressure switch setting are thus sized and selected to alarm and shut down the unit if necessary. As in any system, close attention to changes in operating parameters are required. Flow meters must be properly sized and maintained clean such that relative changes in the flows can be detected in order to adequately plan for seal maintenance.

High pressure applications

In this application, for pressures in excess of 1,000 psi, a tandem seal arrangement or series seal arrangement is usually used. Since failure of the inner seal would cause significant upset of the seal system, and large amounts of gas escaping to the atmosphere, a backup seal is employed. Refer to Figure 22.9 which shows a triple dry gas tandem seal. For present designs up to 2500 psi, double tandem seals are proven and used.

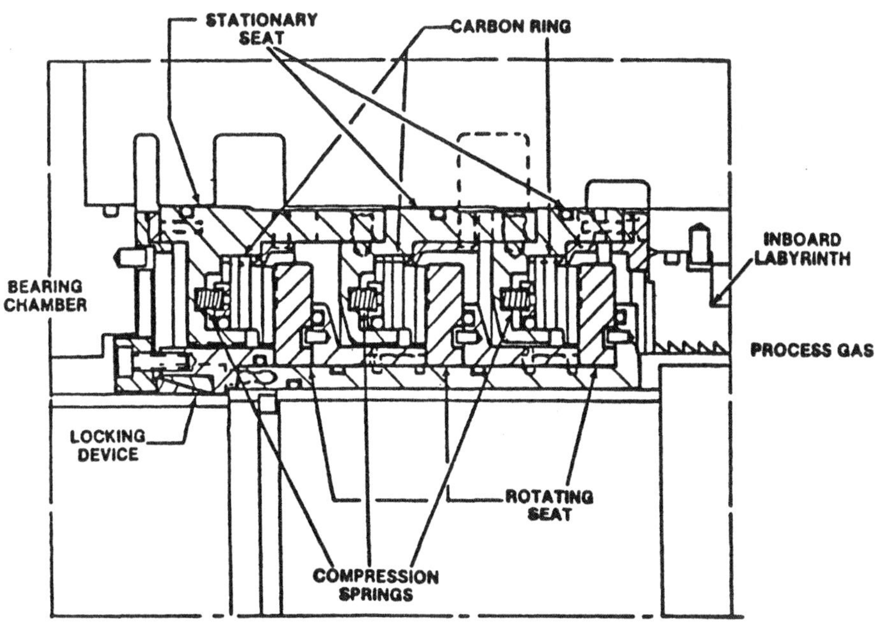

Figure 22.9 Dry gas seal: a triple tandem dry gas seal arrangement (Courtesy of Dresser-Rand Corp.)

The arrangement is essentially the same as low/medium pressure applications except that a backup seal is used in place of the disaster bushing. Most designs still incorporate a disaster bushing between the backup seal and the bearing cavity known these days as the barrier seal. Attention in this design must be given to control of the inter-stage pressure between the primary and backup seal. Experience has shown that low differentials across the backup seal can significantly decrease its life. As in the case of liquid seals, a minimum pressure in the cavity between the seals of 25–30 psi is usually specified. This is achieved by properly sizing the orifice in the vent or reference line back to the suction to assure this pressure is maintained. All instrumentation and filtration are identical to that of the previous system.

Dual seal and system options for toxic and/or flammable gas applications

There are many field proven options available today for use in toxic and/or flammable gas applications. In this section we will discuss the following systems:

- Tandem seals for dry gas applications
- Tandem seals for saturated gas applications
- Tandem seals with interstage labyrinth and nitrogen separation gas
- Double seal system for dry gas or saturated gas applications

Tandem seals for dry gas applications

The tandem seal arrangement for this application is shown in Figure 22.10 and a schematic of this seal in the compressor seal housing is shown in Figure 22.11. Gas from the compressor discharge enters the

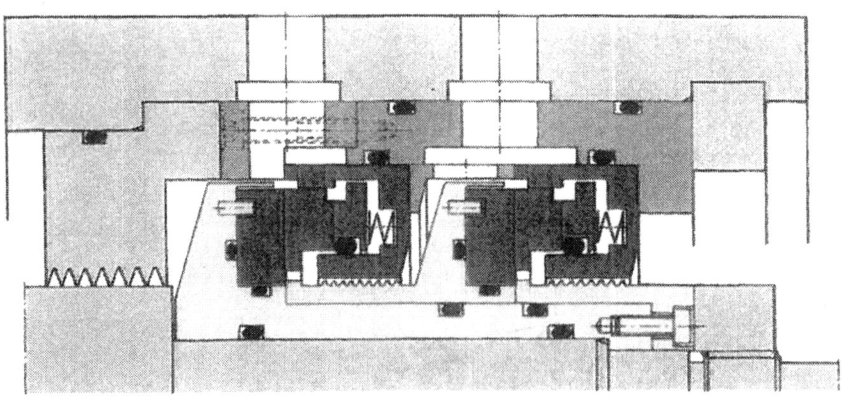

Figure 22.10 Tandem seal (Courtesy of Flowserve Corp.)

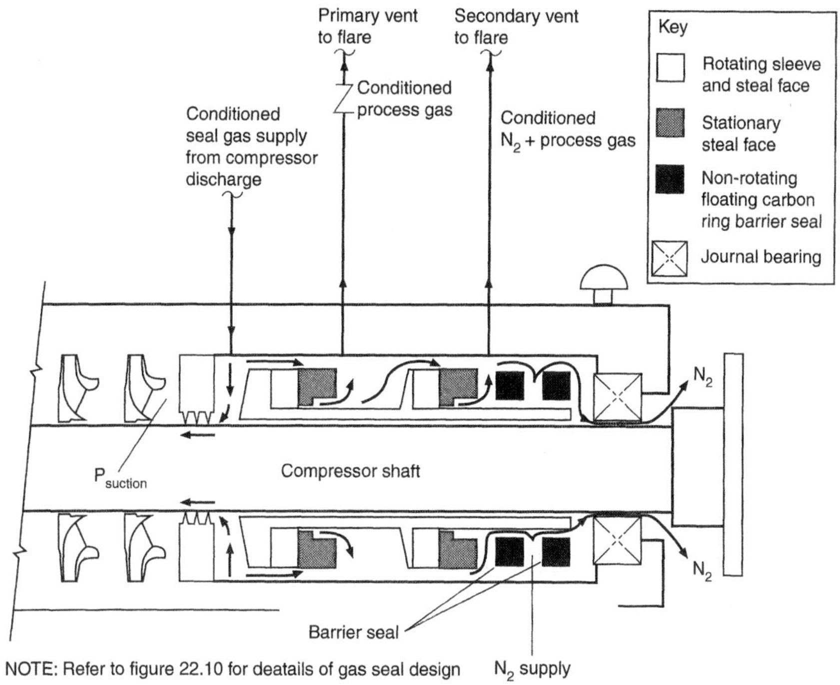

Figure 22.11 Tandem seal and barrier seal typical housing arrangement

port closest to the compressor (labyrinth end) and the majority of the gas enters the compressor thru this labyrinth. To assure that process gas, which is not treated by the dry gas system, does not enter the seal chamber, velocities across the labyrinth should be maintained between 20–50 ft/sec. It is the writer's experience that considering labyrinth wear, the design should be closer to 50 ft/sec. Approximately 1-2 SCFM flow (standard cubic feet per minute) leak across the first tandem seal faces (primary seal) and exit through the primary vent. Based on the backpressure of the primary vent system, 1 SCFM or less will pass through the second tandem seal faces (secondary seal) and exit through the secondary vent. To assure that oil mist from the bearing housing does not enter the dry gas seal chamber and that seal gas does not escape to atmosphere, an additional barrier seal is used and provided with pressurized nitrogen at approximately 5 psi.

A typical seal system for this arrangement is shown in Figure 22.12. As previously mentioned, dry gas seal reliability depends on the condition of the gas entering the seal faces. The function of the seal gas supply system for any dry gas seal option is to continuously supply clean, dry gas to the seal faces. During start-up, when the compressor is not operating with sufficient pressure to supply the seals, an alternate source of gas or a gas pressure booster system should be provided. These items are shown in Figure 22.12 and are typical for any type of dry gas seal application. Note that the following options exist regarding the primary, secondary vent and barrier seal instrumentation and components:

- Primary seal vent triple redundant (2 of 3 voting) flow or differential pressure alarm and shutdown
- Primary seal vent rupture discs in parallel with vent line to rupture at a set pressure and prevent excessive pressure to the secondary seal on primary seal failure
- Spring loaded exercise valves in the primary vent line to exert a backpressure on the primary seal to close the faces in the event of dynamic 'O' ring hang-up
- Secondary vent line flow or differential pressure alarms and trips
- Barrier seal supply pressure alarm and permissive not to start the lube oil system until barrier seal minimum pressure is established.

Tandem seals for saturated gas applications

The tandem seal arrangement for this application can be exactly the same as that shown in Figures 22.10 and 22.11 for the dry gas application. The changes required for this application are solely in the seal system. A typical system is shown in Figure 22.13 and incorporates a cooler, separator and heater in addition to the normal components used for a dry gas application to assure that saturated gas does not enter the seal chamber. Typical values for the cooler are to reduce the gas temperature to 30°F below the saturation temperature of the gas. The typical dimensions for the separator vessel, complete with a demister, are 18 inches diameter and 6 feet high. The typical requirements for the heater are to reheat the gas to 30°F above the saturation temperature. Temperature transmitters are provided upstream and downstream of the cooler and downstream of the heater. As a precaution, in the event of cooler or heater malfunction, a dual filter/coalescer, complete with a drain back to the suction is provided.

Tandem seals with interstage labyrinth

The present industry 'best practice' tandem seal arrangement for dry or saturated gas applications is shown in Figure 22.14. This arrangement features a labyrinth between the primary and secondary seals. This action assures that gas vented from the secondary seal will always be nitrogen since the nitrogen supplied between the primary and secondary seals is differential pressure controlled to always be at a higher pressure than the primary seal vent thus assuring that only nitrogen will be in the chamber between the primary and secondary

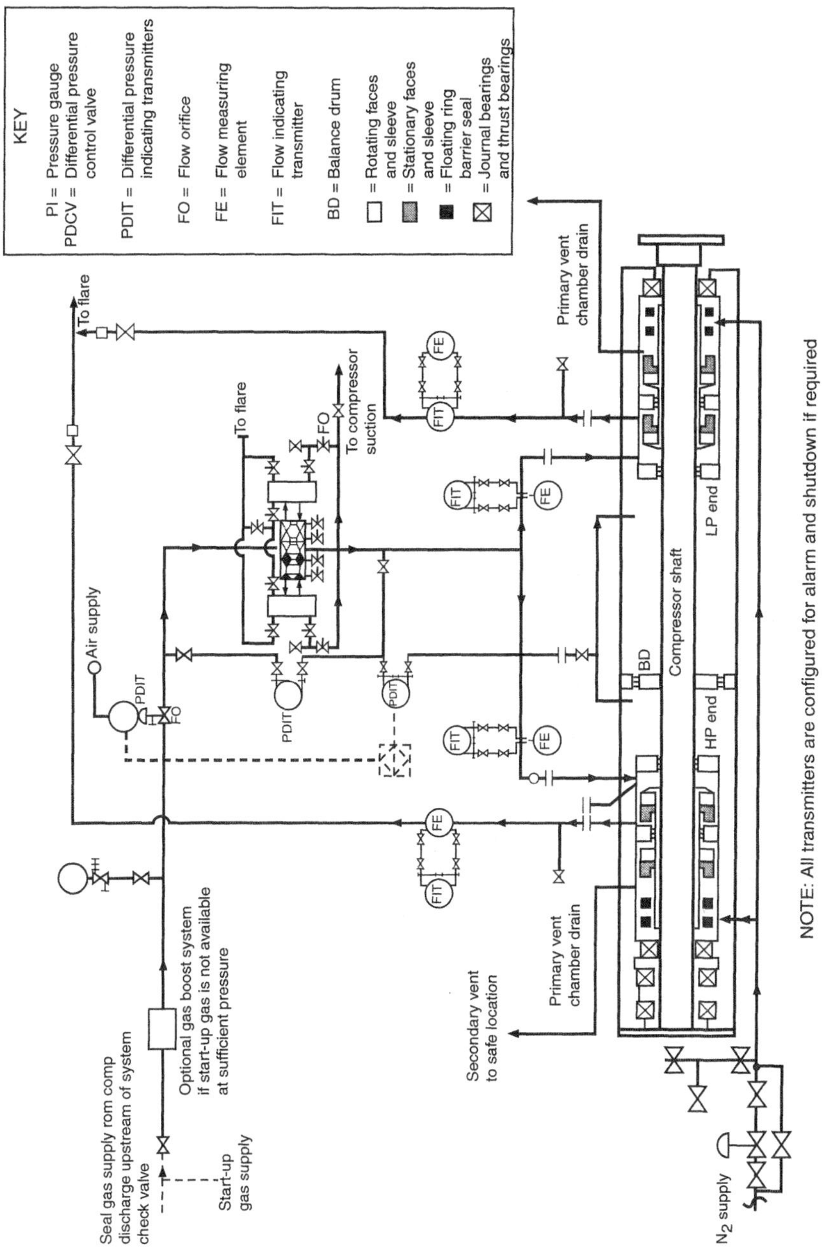

Figure 22.12 Typical tandem seal system for dry process gas

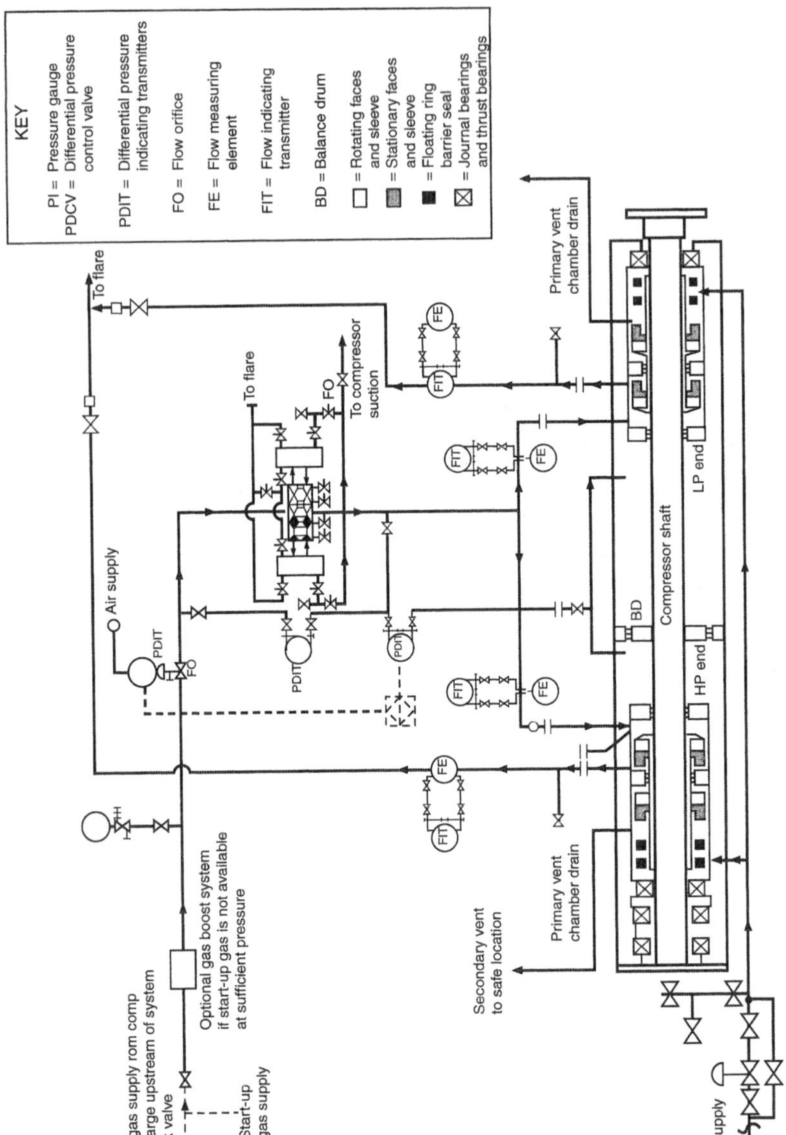

Figure 22.13 Typical tandem seal system for dry process gas

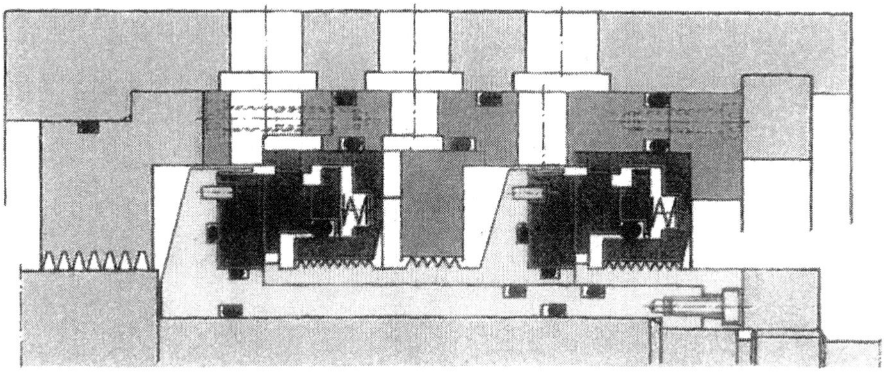

Figure 22.14 Tandem seal with interstage labyrinth (Courtesy of Flowserve Corp.)

seals. Figure 22.15 shows a typical nitrogen supply system used with this tandem seal configuration.

Double seal system for dry gas or saturated gas application

Figure 22.16 depicts a double seal used in either dry gas or saturated gas applications where the process gas is not permitted to exit the compressor case. For this application process gas can be used, after it is conditioned, or an external source can be used if it is compatible with the process gas. If the gas used between the seals is toxic or flammable, a suitable barrier seal, provided with nitrogen, as shown in Figure 22.11 must be used. The seal systems previously shown will be used for the supply of conditioned gas to the seals as required by the condition of the seal gas (dry or saturated).

Summary

Since there are significant advantages to the use of dry gas seals, many units are being retrofitted in the field which incorporates this system. In many cases, significant payouts can be realized.

If a unit is to be retrofitted, it is strongly recommended that the design of the gas seal be thoroughly audited to assure safety and reliability. As mentioned in this section, retrofitting from a liquid to a gas seal system renders the unit a separate system type unit, that is, a separate lube and gas seal system. Naturally, loss of lube oil into the seal system will result in significant costs and could result in seal damage or failure by accumulating debris between the seal rotating and the stationary faces. The adequate design of the separation barriers between the lube and seal face must be thoroughly examined and audited to assure reliable and safe operation of this system. Many unscheduled field shutdowns

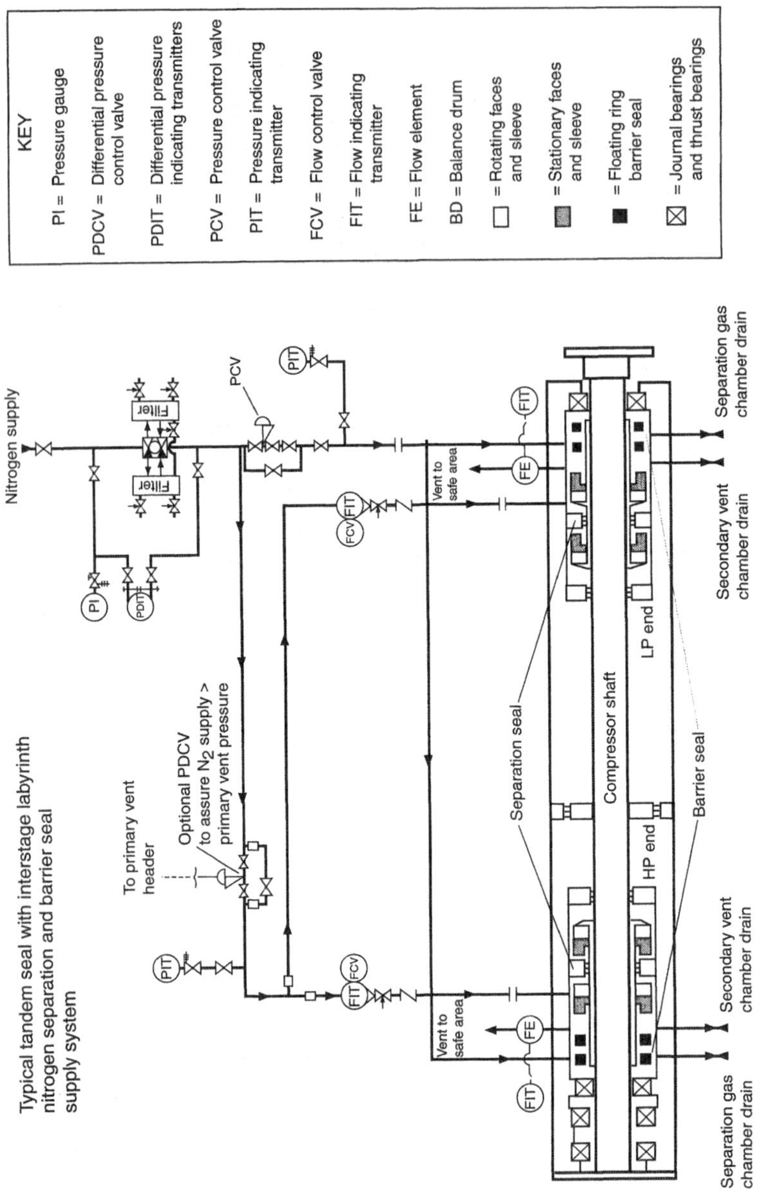

Figure 22.15 Typical tandem seal system with an interstage labyrinth-nitrogen supply

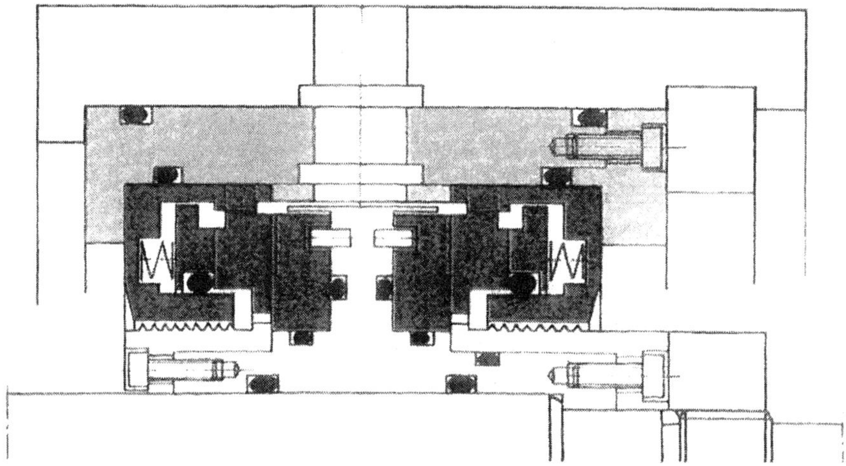

Figure 22.16 Double seal (Courtesy of Flowserve Corp.)

and safety situations have resulted from the improper design of the lube system, seal system separation labyrinth. In addition to the above considerations, a critical speed analysis, rotor response and stability analysis (if the operating discharge pressure is above 500 psi) should always be conducted when retrofitting from liquid to dry gas seals.

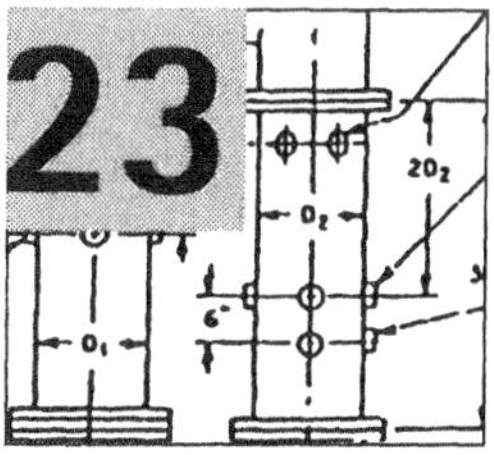

23 Turbo-compressor performance condition monitoring

- The objective
- The parameters
- Accuracy
- The field is not a laboratory!
- Establish a baseline
- Trending

The objective

Now that the principles of turbo-compressor performance have been explained and hopefully understood, they can be implemented to observe internal turbo-compressor condition changes. Always remember that ... we want to know the internal, not the external condition. Figure 23.1 presents the outline of a case history that will show the value of performance condition monitoring.

'The long, long, long, turnaround'

- The first plan
- The second plan
- The third plan

Figure 23.1 'The long, long, long, turnaround'

The first plan

I visited a refinery a few years ago to troubleshoot an existing turbo-compressor problem. While I was on site, another process unit, a reformer, was scheduled for a turnaround. Since I was already on site, I was invited to the pre-turnaround meeting and became involved with turnaround activities.

During this meeting I learned that the recycle compressor was scheduled for a bearing inspection only (radial and thrust). I asked why. The answer was that it was normal practice. I asked if I could see the bearing condition monitoring data (vibration, bearing displacement, bearing temperature, oil flow – valve position and oil sample). I was shown to a room and told – 'It's in there'. In desperation I concluded – 'Oh, what the hey, we usually inspect bearings anyway'.

The second plan

Well, as you might expect, the bearings were removed and they were pretty badly damaged. Wipes on both journal and thrust bearing pads and indications of particle rubs showed that we really needed to both inspect and flush the oil system. It was later learned that an oil system accumulator had a continuous nitrogen purge that was at a higher pressure than the oil system and, oh yes … the bladder had a hole in it! Now back to the bearings … we didn't know the condition or trends of parameters. Were there upsets? High vibrations, temperatures, etc. Were there a lot of surges? Nobody could seem to remember! … sound familiar? Well, as you might expect, we prepared a second plan – better inspect the oil seals. After all, we didn't have to get into the compressor so we had sufficient time.

The third plan

No doubt about it, the seals were really bad. Bushing seals, both the atmospheric bushings and gas side bushings were wiped and the atmospheric side showed typical evidence of high ΔT … probably due to start-up on low pressure N_2 (to save N_2 costs). Since the gas was sweet, seal leakage was returned to the reservoir via a degassing tank and there was no seal condition monitoring trend data available. (Seal oil leakage – gas side, seal oil leakage – atmospheric side, which could have been obtained by trending seal oil valve position). Am I making my point?? Well, it was crunch time (decision time). Were the damaged seals the root cause of bearing failure? Was the dirty lube and seal system or was there another deeper cause inside of the compressor? I could go on and on but in the interest of time let it suffice to say:

- We decided to open the barrel and compressor – we had time, based on schedule
- We could not find tools and when we did, they did not work
- Oh yes, a piece of a suction strainer (supposedly only for start-up) had been lodged in the case/barrel interface for years.
- When we finally got the compressor apart – 2 days after the turnaround was complete, we got lucky – the internals including the balance drum were perfect!

If only we had established a performance condition monitoring program, as well as seal, bearing, balance drum and lube/seal system condition monitoring.

Incidentally, this was a major refinery that placed high value on reliability, maintainability, root cause analysis, etc. They had all the books and had sent people to the right workshops … even one of mine!

The objectives of turbo-compressor condition monitoring are presented in Figure 23.2.

The objectives

- Know turbo-compressor internal condition to determine:
  - Loss of daily revenue
  - Justification for turnaround activities
  - Root cause

Figure 23.2 The objectives

The parameters

What parameters must be measured? Readers should be able to answer this question readily at this point. Figure 23.3 presents the reduced parameters – the answers and Figure 23.4 contains all the factors necessary for calculation.

The turbo-compressor performance condition parameters

- Polytropic (or isentropic) head
- Actual inlet flow rate
- Polytropic (or isentropic) efficiency
- Polytropic exponent
- Horsepower
- Speed (or guide vane position)

Figure 23.3 The turbo-compressor performance condition parameters

The data (factors) required

- From gas analysis and equation of state
  - M.W.
  - K average
  - Z inlet
- P_1, P_2
- T_1, T_2
- Flow rate
- Speed (if applicable)
- Guide vane position (if applicable)

Figure 23.4 The data (factors) required

Accuracy

Accuracy of data involves both calibration and location of instruments. Before proceeding, we need to present some important facts concerning condition monitoring. These facts are presented in Figure 23.5.

Turbo-compressor performance condition monitoring facts

- Specific data (to confirm field guarantees) requires pre-planning to assure accuracy
- Trends produce relative change in values

Figure 23.5 Turbo-compressor performance condition monitoring facts

Refer back to the 'Life cycle of equipment' in Chapter 1, Figure 1.12. If specific data is required to confirm field guarantees, the project team must pre-invest and plan for the proper instruments, number of and location in the design phase of the project!

One, (especially the rotating equipment engineer) must remember the words in Figure 23.6.

'The field is a place to produce product, not conduct laboratory experiments'

- 1981 A.D. ... a production supervisor

Figure 23.6

Figure 23.7 presents important facts concerning instrument calibration.

Instrument calibration facts

All performance condition monitoring instruments must have known, accurate calibration values!

If:
- ΔT as measured is off 5%, efficiency is affected 20%!
- Inaccurate gas analysis of 5% can affect efficiency 20%
- Pressure gauge inaccuracy has *much less* of an affect on efficiency!

Figure 23.7 Instrument calibration facts

Inaccurate gas analysis procedures can produce some pretty wild results! Efficiencies that exceed 100%! Figure 23.8 presents guidelines for accurate gas analysis.

Gas analysis guidelines

- Take samples from the top of pipe
- Measure gas temperature
- Be sure to analyze gas at temperatures equal or greater than field conditions
- Confirm laboratory experience and methods

Figure 23.8 Gas analysis guidelines

Location and number of field instruments are just as important as instrument calibration. I cannot overemphasize the importance of this fact. Convince management (plant management first, then project management) to pre-invest in turbo-compressor performance and instrumentation. If you don't accurately know what's happening inside the patient, you can't effectively *prevent* problems! ... That old medical analogy to a turbo-compressor again! Figure 23.9 presents some guidelines from the ASME Turbo-compressor Test Code (PTC-10) regarding the number and location of instruments.

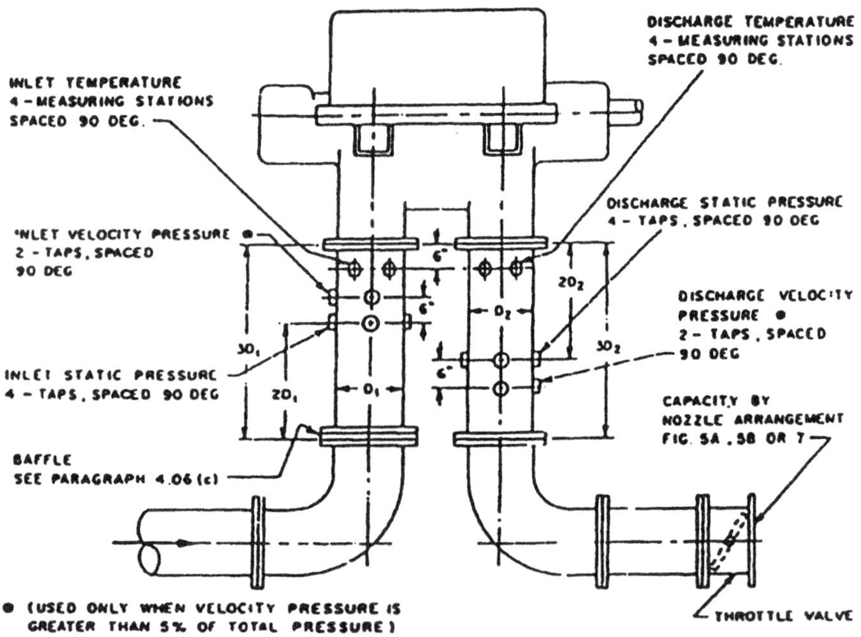

Figure 23.9 Performance instrumentation location guidelines (Courtesy of ASME PTC-10) (typical for multi-stage compressor)

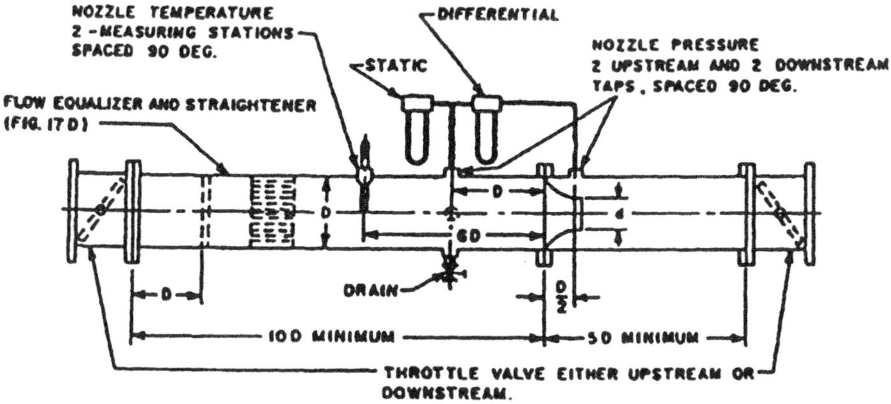

Figure 23.9 Continued

The field is not a laboratory!

Field turbo-compressor condition monitoring must be planned such that it does not impact production rates.

Establish a baseline

Before we discuss trending or fully understand turbo-compressor performance condition monitoring, the baseline condition must be defined. Since performance condition monitoring, as well as any type of condition monitoring, is concerned with relative change, the starting point must be established. Figure 23.10 defines the baseline condition.

The baseline condition

- The baseline condition establishes the initial ('new' or 'rebuilt') condition from which all changes (trends) are measured.

Figure 23.10 The baseline condition

I can't remember how many times in my career I have said ... 'I sure wish we had established the 'baseline condition'.

The baseline turbo-compressor performance condition should be established once the process unit is on spec, immediately, not one day, week, month or year later. The sooner the baseline is established the more data in your machine historical file. Many problems occur in the initial period of operation, be sure you have recorded the data. The baseline condition then represents the first point on the trend graph of any measured parameter. Figure 23.11 recommends when baseline conditions should be taken.

Baseline performance conditions – when?

Baseline performance conditions should be established when the process is on spec. immediately after:

- New compressor start-up
- Turnaround (if internals, or thrust bearing was changed)

Figure 23.11 Baseline performance conditions – when?

Trending

Trending requires that a parameter be monitored over a period of time to determine if significant change occurs. These factors are defined in Figure 23.12.

Trending guidelines

- Establish baseline condition
- Trend parameter (Y axis) vs time (X axis)
- Compare related trends with same time scale (X axis)
- A significant change in a parameter is a 25% – 50% change.

Figure 23.12 Trending guidelines

It should be noted that significant change is not synonymous with alarm point. If an initial value is small, a small change can be significant and still be far away from the set alarm point!

Details concerning field performance testing will be discussed in the next chapter. We have presented suggested turbo-compressor performance trend parameters in the final figure of this chapter … Figure 23.13.

Useful trend parameters

- Polytropic head ÷ speed^2 (Hd_{POLY}/N^2)
- Actual flow ÷ speed (Q/N)
- Polytropic efficiency (η_{POLY})
- In certain cases,
  - Molecular weight
  - Inlet temperature

Figure 23.13 Useful trend parameters

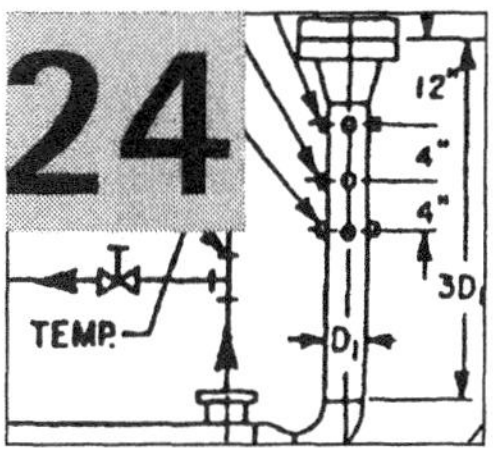

24 Turbo-compressor performance testing

- Introduction
- Shop performance test overview
- Equivalent speed determination
- Field testing

Introduction

The objective of this chapter is to present practical guidelines for turbo-compressor field performance determination. However, it is my opinion that successful field performance determination begins with the shop performance test. After all, it is the baseline for any field performance assessment.

Not all end users require a shop performance test. It is my opinion that a shop performance test should always be required for the following reasons noted in Figure 24.1.

Reasons to shop performance test

- Low performance will have a long term (3 year minimum) impact on profit
- Vendors may build in extra margin to minimize re-test.
- Field performance is difficult to accurately determine

Figure 24.1 Reasons to shop performance test

Therefore, this chapter will first focus on the basics of a shop performance test before discussing field performance testing.

As mentioned in Figure 24.1, field performance is difficult to accurately determine. In fact, it has been my personal experience that the accuracy of a field test is within 10% at best and depends heavily on flow meter accuracy. This is primarily due to the lack of pre-investment in test grade instrumentation and location of instrumentation used for performance testing.

Since turbo-compressor field performance directly influences unit profit and since the life of a process unit is typically 30–40 years, one can easily justify a pre-investment in 'test quality' instrumentation field set ups. Refer to the previous chapter, Turbo-Compressor Performance Condition Monitoring for details concerning field performance test instrumentation, its location and the number of instruments to use.

Finally, before we begin, a question!

... After the field acceptance test of a new or revamped compressor, should we continue to test or should we trend performance?

Let's discuss this important point later in this chapter.

Shop performance test overview

The industry accepted guidelines for shop performance testing is the ASME Power Test Code for compressors and exhausters. It is better known as 'PTC-10'. The objective of a shop test performed in accordance with PTC-10 is described in Figure 24.2.

PTC-10 objective

To perform a shop test that will accurately define compressor performance by assuring an accurate match of a compressor section between the 'field' and 'shop' conditions for the following parameters:

- Volume ratio
- Capacity/speed ratio
- Machine mach number
- Machine Reynolds number

Figure 24.2 PTC-10 objective

By assuring the 'field' and 'shop test' parameters in Figure 24.2 match within defined limits, dynamic similarity is assured. Dynamic similarity is defined in Figure 24.3.

The principles of dynamic similarity

Assures that the aerodynamic flow pattern of test gas compressed at an equivalent test speed will be the same as that operating on the field gas at specified conditions (MW, T_1, speed, P_1, P_2, K and Z)

Figure 24.3 The principles of dynamic similarity

Equivalent speed determination

The test equivalent speed is determined by the relationship noted in Figure 24.4.

Shop test equivalent speed determination

$$N_{EQUIV.SPEED} = N_t = N_{sp} \sqrt{\frac{T_{it}}{T_{isp}} \; \frac{Z_t}{Z_{sp}} \; \frac{M_{sp}}{M_t} \; \frac{\left(\frac{n}{n-1}\right)_t}{\left(\frac{n}{n-1}\right)_{sp}} \; \frac{\left(rp^{\frac{n-1}{n}} - 1\right)_t}{\left(rp^{\frac{n-1}{n}} - 1\right)_{sp}}}$$

Where: ■ t = Test conditions ■ M = mol. weight ■ Rp = Pressure ration
■ sp = Specified conditions ■ n = polytropic exponent

Figure 24.4 Shop test equivalent speed determination (Courtesy of ASME PTC-10)

Once the equivalent test speed is calculated, a compressor calculation is performed using the proposed test gas at the test gas equivalent speed. The field to test value ratio of the parameter noted in Figure 24.2 must fall within the limits of Table 3 of the PTC-10 to constitute an acceptable equivalent speed. These facts are presented in Figure 24.5.

Equivalent speed validation procedure

- Calculate compressor section equivalent shop test speed
- Perform stage by stage section calculation on test gas at equivalent speed
- Calculate volume ratio, capacity/speed ratio mach number and Reynolds number test values
- The values must fall within the limits of design values as noted in PTC-10 Table 3

Figure 24.5 Equivalent speed validation procedure

PTC-10 Table 3 is presented in Figure 24.6.

PTC-10 Table

Variable	Symbol	Range of test values Limits – % of design value Min	Max
Volume ratio	q_i/q_d	95	105
Capacity-speed ratio	$q_i/N\sim$	96	104
Machine mach number	M_m		
0 to 0.8		50	105
Above 0.8		95	105
Machine Reynolds number where design value is:	Re		
Below 200,000 centrifugal		90	105
Above 200,000 centrifugal		10*	200
Below 100,000 axial comp.		90	105
Above 100,000 axial comp.		10**	200

Mechanical allowable losses shall not exceed 10 percent of the total shaft power input at test conditions.

*Minimum allowable test Machine Reynolds number is 180,000

**Minimum allowable test Machine Reynolds number is 90,000

Figure 24.6 PTC-10 table (Courtesy of ASME)

Typical shop test gases are usually inert and are noted in Figure 24.7.

Typical shop test gases

- Nitrogen
- Carbon dioxide
- Helium
- Freon substitute
- Air

Figure 24.7 Typical shop test gases

The appropriate test gas should be the gas that satisfies PTC-10 Table 3 at the equivalent test speed closest to the rated speed. Mixtures of the gases noted in Figure 24.7 have also been used successfully. In applications where a section consists of three (3) or more stages, it is advisable to have the vendor calculate the required PTC-10 parameters for each stage at the test equivalent speed and compare these values to the design parameters. This action assures that the calculated 'section' equivalent speed and 'stage' equivalent speed are approximately equal. In other words, it assures there will be no stage mismatching during the shop test. If the section and stage equivalent speeds are not equal, shop test values may not match field test values.

It should also be noted that some compressor vendors have the capability to performance test the compressor on field gases (usually natural gas). Since the test gas is not at the exact field condition, the test equivalent speed will be slightly different than the rated field speed.

Test procedure

The shop performance test procedure should be issued for review by the contractor and end user a minimum of two months before the shop test. Figure 24.8 contains the minimum performance test agenda requirements.

PTC-10 test agenda minimum requirements

- Equivalent test speed calculation
- Verification of compliance with PTC-10 Table 3 (allowable departure from specified parameters)
- Description of test procedure including test gas to be used
- Drawing and bill of material of test loop and instrumentation
- Calibration details of instrumentation
- Test data reduction procedure
- Listing of specified limits for each measured parameter

Figure 24.8 PTC-10 test agenda minimum requirements

Pre-test meeting

When the project requires more than one compressor train from the same vendor (ethylene plant, methanol plant, etc.), it is advisable to have a pre-test meeting after the test procedure has been reviewed. This meeting is usually held approximately one month before the scheduled shop test and will take 1–2 days depending on project scope. Figure 24.9 presents a suggested pre-test meeting agenda.

Pre-test meeting agenda

- Confirm test requirements
- Confirm all test parameters have established limits mutually agreed to
- Review all instrument calibration procedures
- Review test set-up drawing
- Review data calculation (data reduction) methods

Figure 24.9 Pre-test meeting agenda

Shop test activity

We have provided a shop test activity check list in Figure 24.10. This list will aid in optimizing performance shop test results.

An example of a PTC-10 inert gas performance test loop is shown in Figure 24.11.

Shop test activity checklist

1. Scope
   - Appropriate industry specs included ☐ (ANSI, API, NEMA, etc.)
   - In-house and/or E & C specs included ☐
   - Project specific requirements
     - ☐ Performance test ☐ All rotors ☐ One rotor
       - ☐ Test (equivalent) conditions
       - ☐ Field (actual) conditions
     - ☐ Mechanical test ☐ All rotors ☐ One rotor
       - ☐ Test (equivalent) conditions
       - ☐ Field (actual) conditions
       - ☐ Unit test of all equipment (string test)
         - ☐ No load ☐ Includes auxiliary systems
         - ☐ Full load ☐ Does not include auxiliary systems
     - ☐ Use of job couplings and coupling guards

Figure 24.10 Shop test activity checklist

- ☐ Testing of instrumentation, control and protection devices
- ☐ Auxiliary system test
  - ☐ Lube oil ☐ Test press ☐ Full press
  - ☐ Control oil ☐ Test press ☐ Full press
  - ☐ Seal oil ☐ Test press ☐ Full press
  - ☐ Seal gas ☐ Test press ☐ Full press
  - ☐ Fuel ☐ Test press ☐ Full press
  - ☐ Flow measurement required
  - ☐ Time base recording of transient events required
- ☐ Use of all special tools during test (rotor removal, coupling, etc.)
- ☐ Shop test attendance (includes assembly and disassembly)
  - ☐ Site reliability ☐ Site maintenance ☐ Site operations
  - ☐ Review of instruction book during shop test visit
  - ☐ Test agenda requirements
    - ☐ Mutually agreed limits for each measured parameter
      - ☐ Issue for approval 2 months prior to contract test date

2. Pre-test meeting agenda

- ■ Meet with test department prior to test to:
  - ☐ Confirm test agenda requirements
  - ☐ Confirm all test parameters have established limits mutually agreed to
  - ☐ Review all instrument calibration procedures
  - ☐ Review test set-up drawing
  - ☐ Review data calculation (data reduction) methods
  - ☐ Define work scopes for site personnel (Assembly and disassembly witness, video or still frame pictures, etc.)
  - ☐ Confirm assigned vendor service engineers will be in attendance for:
    - ☐ Assembly
    - ☐ Disassembly
    - ☐ Test

3. Shop test activity

- ■ Review and understand test agenda prior to test
- ■ Immediately prior to test meet with assigned test engineer to:
  - ☐ Review schedule of events
  - ☐ Designate a team leader
  - ☐ Confirm test team leader will be notified prior to each event
  - ☐ 'Walk' test set-up to identify each instrumented point
  - ☐ Confirm calibration of each test instrument
  - ☐ Obtain documents for data reduction check – if applicable (flow meter equations, gas data, etc.)
- ■ During test
  Note: coordinate with test personnel to avoid interference

Figure 24.10 Continued – Shop test activity checklist

- ☐ Review 'as measured' raw data for consistency
- ☐ 'Walk' equipment – look for leaks, contract instrument, piping and baseplate vibration, etc.
- ☐ Use test team effectively – assign a station to each individual
- ☐ Ask all questions now, not later, while an opportunity exists to correct the problem
- ☐ Check vendor's data reduction for rated point – if applicable

- ■ After test
  - ☐ Review data reduction of performance data corrected to guarantee conditions
  - ☐ Generate list of action (if applicable) prior to acceptance of test
  - ☐ Approve or reject test

Figure 24.10 Continued – Shop test activity checklist

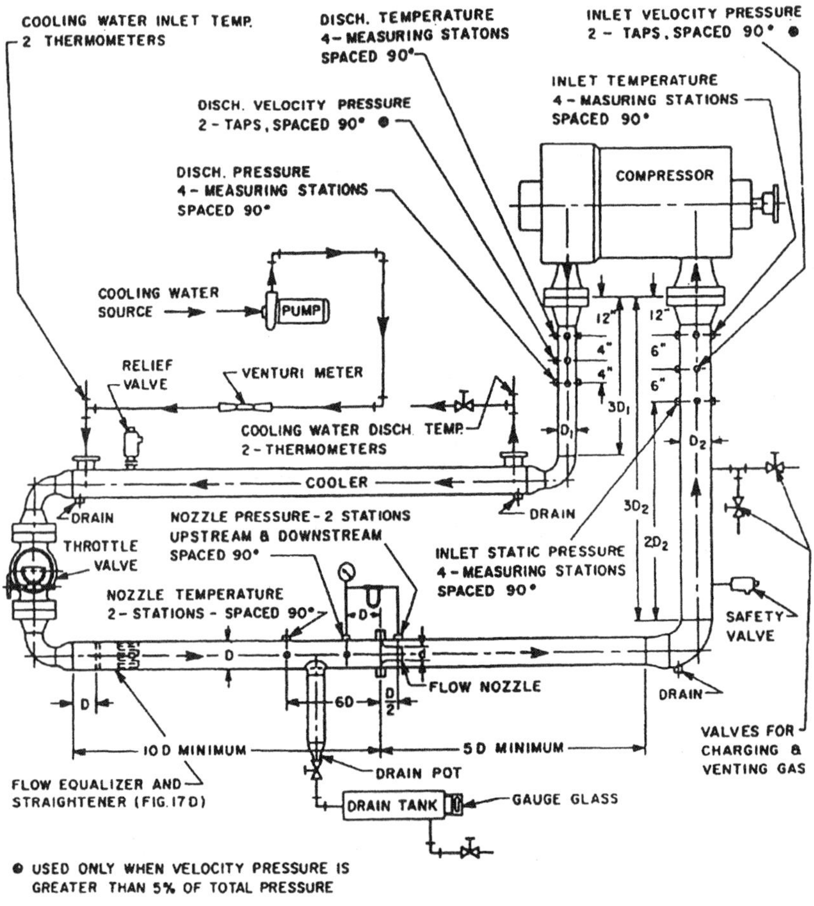

Figure 24.11 PTC-10 Performance test closed loop arrangement (Courtesy of ASME)

Field testing

Field testing is usually performed when new turbo-compressors are installed. The purpose of the test is to confirm that the contractor's guarantees concerning process unit production rates are met. In addition, field tests are conducted when turbo-compressors are upgraded (re-wheeled). In both cases above, these tests constitute acceptance of the work performed if the guaranteed conditions are met.

These tests are different than performance trending. Performance trending is concerned with relative change whereas acceptance testing is concerned with specific values. These facts are presented in Figure 24.12.

Acceptance testing and performance trending

- Acceptance testing is concerned with specific values
- Performance trending is concerned with the change of relative values over a specific time period.

Figure 24.12 Acceptance testing and performance trending

Acceptance testing

Since acceptance tests must confirm that guaranteed values are met, instrument accuracy, location and procedural accuracy are of the utmost importance. An acceptance test procedure outline is presented in Figure 24.13.

Field acceptance test procedure outline

1. Confirm test instrumentation is installed in accordance with pre-established agreements (see previous chapter, Figure 23.9 for guidelines)
2. Confirm all test instrumentation has been properly calibrated – especially flow meters.
3. A minimum of five points, excluding the surge point, at a constant speed should be obtained. Each point should settle out prior to data being recorded. ΔT should remain constant within $\pm$ 2°F for 10 minutes before data is recorded.
4. Gas analysis is to be recorded in accordance with gas analysis guidelines in Figure 23.8.

Figure 24.13 Field acceptance test procedure outline

An acceptance test data reduction outline for a section of compression is presented in Figure 24.14. Figure 24.15 presents an acceptance test data reduction outline for a refrigeration compressor.

Data reduction outline

1. Calculate actual flow to section 1
   Calculate inlet ACFM to section 1, using appropriate flow meter equation.
2. Use measured section P_1, P_2, T_1, T_2 to calculate the polytropic exponent.

$$\frac{n-1}{n} = \frac{LN\left(\frac{T_2}{T_1}\right)}{LN\left(\frac{P_2}{P_1}\right)}$$

3. Use K from gas analysis to determine the polytropic efficiency.

$$\eta_{poly} = \frac{\frac{k-1}{k_{avg.}}}{\frac{n-1}{n}}$$

4. Calculate section polytropic head.

$$HD_{poly} = \frac{1545}{\text{M.W.}} \times T_1 \times \frac{n}{n-1} \times Z_{avg} \left[\left(\frac{P_2}{P_1}\right)^{\frac{n-1}{n}} - 1\right]$$

Figure 24.14 Data reduction outline

Data reduction outline (refrigeration compressor)

1. Calculate actual flow to section 1

   Calculate inlet ACFM to section 1, using appropriate flow meter equation.
2. Determine trial section 1 discharge temperature

   Options available:

   A. Assume polytropic efficiency from section performance curve at calculated ACFM. Then solve for polytropic exponent from:

$$\frac{n-1}{n} = \frac{\frac{k-1}{k_{avg.}}}{\eta_{poly\,(\text{assumed})}}$$

Figure 24.15 Data reduction outline (refrigeration compressor)

Solve for T_2 from:

$$T_{2\,(assumed)} = T_1 \times \left(\frac{P_2}{P_1}\right)^{\left(\frac{n-1}{n}\right)}$$

Note: Section P_2 is slightly less than induction pressure.

B. Use measured section discharge temperature

This value can be obtained from a thermocouple in a case drain.

Note: the measure value will vary from the actual section discharge temperature but can be used for the *assumed* discharge temperature.

The polytropic exponent can be obtained from:

$$\frac{n-1}{n} = \frac{LN\left(\frac{T_2}{T_1}\right)}{LN\left(\frac{P_2}{P_1}\right)}$$

The polytropic efficiency can be obtained from:

$$\eta_{poly} = \frac{\frac{k-1}{k_{avg.}}}{\frac{n-1}{n}}$$

3. Calculate section 1 polytropic head

Calculate $head_{poly}$ from:

$$HD_{poly} = \frac{1545}{M.W.} \times T_1 \times \frac{n}{n-1} \times Z_{avg}\left[\left(\frac{P_2}{P_1}\right)^{\frac{n-1}{n}} - 1\right]$$

4. Determine mass flow rate to downstream compressor section N

The mass flow to the downstream section is simply the addition of total upstream mass flow and the induction mass flow.

$$M_{SECTION\ N} = M_{SECTION\ N-1} + M_{SECTION\ N}$$

5. Calculate the assumed section inlet temperature

The assumed section inlet temperature is obtained from:

$$T_{INLET\ N} = \frac{[(M_{N-1})(T_{2\ N-1\ (ASSUMED)})] + [(M_N)(T_{INDUCTION1})]}{M_{N-1} + M_N}$$

Where M = mass flow

Figure 24.15 Continued – Data reduction outline (refrigeration compressor)

6. Determine trial section N discharge temperature

   Proceed as in Step 2 based on measured section pressures (induction pressures) assumed efficiency, and assumed section inlet temperature determined in Step 5 above.

7. Calculate individual section performance and calculate final compressor discharge temperature

Final step: Compare calculated discharge temperature to measured temperature and iterate entire calculation procedure until calculated discharge temperature equals measured discharge temperature.

Figure 24.15 Continued – Data reduction outline (refrigeration compressor)

In addition to procedure outlines, we have provided some helpful field acceptance test hints in Figure 24.16.

Helpful hints

- Be sure to locate all pressure and temperature instruments close to equipment flanges downstream of suction screens and silencers and upstream of discharge silencers.
- The lower the compressor ΔT, the greater the required accuracy, i.e.
  - Calibration
  - Attention to heat loss or addition to case
- Conduct gas analysis carefully. Pay attention to details.

Figure 24.16 Helpful hints

Performance trending

Performance trending can range from conducting performance checks at regular intervals to continuous trending of head, efficiency and speed. A suggested practice of conducting performance checks at regular intervals is presented in Figure 24.17.

'Regular performance checks' guidelines

1. Check $head_{POLY}$, η_{POLY} at or as close as possible to normal operating flow and speed (or IGVs) after turnaround (T & I).
2. Recheck yearly or after turbo-compressor upset is noted.
3. Check before turnaround planning meeting.

Figure 24.17 'Regular performance checks' guidelines

If performance trending is selected in lieu of periodic performance checks, the parameters to monitor will depend on the control methods used and the compressor application. Figure 24.18 presents suggested parameters for various applications, control configurations and machinery types.

Performance trending parameter

Application	Control scheme	Compressor type	Trending parameters
Instrument air	Constant speed, suction throttle on flow or pressure control	Integral gear	Stage head, stage η, inlet valve position, inlet ACFM, intercooler(s) ΔT
Main air blower	Adjustable guide vanes on flow or pressure control	Axial	Inlet ACFM, head, η, IGV position
Main air blower	Pressure or flow control to turbine speed	Axial	Inlet, ACFM, speed, head, η
Wet gas	Suction throttle valve on overhead accumulator pressure control	Centrifugal	Section head, section η, section flow, valve position, M.W., intercooler ΔT, phase angle
Wet gas	Suction pressure or flow control to steam turbine speed	Centrifugal	Section head, section η, section flow, speed, M.W. intercooler ΔT, phase angle
Cracked gas	Suction pressure or flow control to steam turbine speed	Centrifugal	Section head, section η, section flow, speed, intercooler ΔT, phase angle

Figure 24.18 Performance trending parameter

Application	Control scheme	Compressor type	Trending parameters
Cracked gas	Constant speed suction throttle on flow or pressure control	Centrifugal	Section head, section η, section flow, valve position, phase angle
Refrigeration (ethylene, propane, propylene, freon)	Suction pressure or flow control to steam turbine speed	Centrifugal	Section head, section η, * section flow, speed, inlet and side load temperatures.
Refrigeration (ethylene, propane, propylene, freon)	Suction pressure or flow control to suction throttle valve	Centrifugal	Section head, section η, section flow valve position, inlet and side load temperatures
Pipeline	Suction pressure to gas turbine speed	Centrifugal	Head, η, flow, speed
Pipeline	Suction pressure to IGV	Centrifugal	Head, η, flow, IGV position
H_2 recycle	Suction pressure to steam turbine speed	Centrifugal	Head, η, flow speed, M.W., phase angle
H_2 recycle	Suction pressure to suction throttle valve	Centrifugal	Head, η, flow, valve position, M.W. phase angle
LLDPE recycle	Suction pressure to IGV position	Centrifugal	Head, η, flow, IGV position, phase angle
LLDPE recycle	Suction pressure to suction throttle valve	Centrifugal	Head, η, flow, IGV position, phase angle
Methanol syn-gas	Suction pressure to steam turbine speed	Centrifugal	Section head, section η, section flow, speed
Ammonia refrigeration	Suction pressure to steam turbine speed	Centrifugal	Section head, section η, section flow, speed
Reinjection	Suction pressure to steam turbine speed	Centrifugal	Section head, section η, section flow, speed, M.W.

*iteration procedure – see Figure 24.15

Figure 24.18 Continued – Performance trending parameter

'Real time' operating point

'Real time' trending of the operating point (flow vs head, speed and efficiency) is becoming more and more popular with the use of DCS systems and dedicated surge protection systems. In both cases, the actual operating point is plotted on the compressor curve. It is recommended that the 'As tested in shop' performance grid be programmed to remain constant. Plotting of the actual operating point head, η and flow (also speed if applicable) will show up any performance deterioration.

When to take action

Based on the fact that turbo-compressor performance is determined by many parameters, it has been my personal experience that maintenance should be planned when head and/or η deterioration approaches 10% change from original values. If the compressor indicates fouling (phase angle change and performance change) washing should be planned. Figure 24.19 presents these facts.

Taking action

- If performance parameters reduce 10% from baseline values,
  - Maintenance should be planned if fouling is not indicated (phase angle changes)
  - If fouling is indicated, the compressor should be cleaned at the first opportunity.

Figure 24.19 Taking action

Reciprocating compressors major component functions

- Introduction
- Frame and running gear
- Cylinder distance piece
- Piston rod packing
- Cylinder and liner
- Reciprocating compressor cylinder valves
- Piston assembly
- Pulsation dampeners
- Cylinder and packing lubricators
- Cooling system

Introduction

In this chapter we will define the functions of each major component of a reciprocating compressor. That is, what the purpose of each component is or "What It Does". By understanding what each component is supposed to do, you will be in a better position to know if it is performing its duty correctly. We will present each major component starting with the crankcase, state its function, operating limits and what to look for. After presenting each component's general information, we will present specific information concerning site compressors.

Frame and running gear

Figure 25.1 presents a picture of a seven throw crank shaft arrangement along with a sectional view of two throws. The crankcase supports the crank shaft bearings, provides a sump for the bearing and crosshead lube oil and provides support for the crosshead assembly.

Typical crank case condition monitoring and safety devices are:

- Relief device – To prevent crank case breakage in the event of explosion (caused by entrance of process gas into the crank case).
- Breather vent – To allow removal of entrained air from the lube oil.
- Crank case oil level gauge – Allows continuous monitoring of crank case lube oil level.
- Crank case oil temperature gauge – Allows continuous monitoring of crank case lube oil temperature.
- Crank case vibration detector (optional) – Provides information concerning crank case vibration useful in detecting dynamic changes in running gear.
- Crank case low oil level switch (optional) – Provides alarm signal on low crank case oil level.
- Main lube oil pump – shaft driven (optional) – Directly connected to crank shaft and usually discharges oil directly to crank shaft

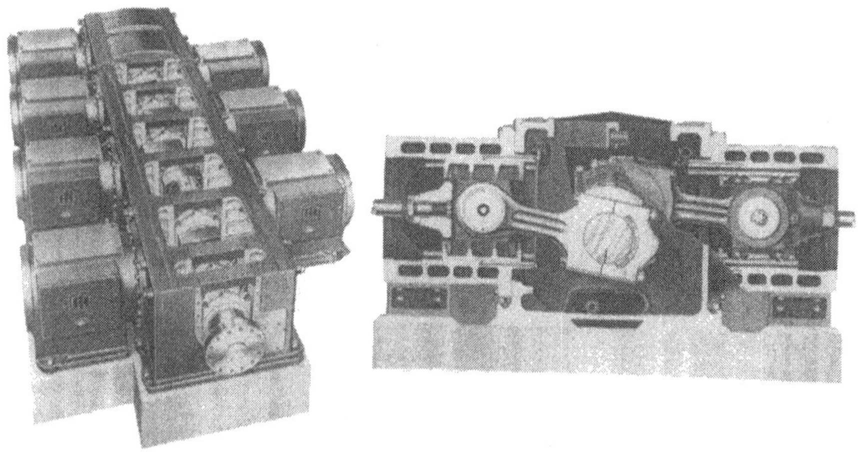

FUNCTIONS:
- TRANSMITS POWER
- CONTAINS LUBE OIL
- CONVERTS ROTARY TO RECIPROCATING MOTION

Figure 25.1 Frame and running gear (Crank case and crosshead) (Courtesy of Dresser Rand)

bearings, connecting rod bearing, crosshead shoes and crosshead pin bushing via precision bore in crank shaft and connecting rod bearing.

- Main lube oil pump discharge pressure gauge (when supplied) – Allows continuous monitoring of main lube oil pump discharge pressure.

An important reliability consideration is to assure that the crank case is securely mounted and level. This requires proper grouting and maintaining a crack free (continuous) crank case base support. Since the dynamic forces on the crank case and crosshead mounting feet can be very large, it is usually common to use an epoxy grout. Epoxy grouts provide high bond strengths and are oil resistant. All reciprocating baseplates should be continuously checked for any evidence of grout foundation cracks (discontinuities) and repaired at the first opportunity.

Figures 25.2 and 25.3 show plan, elevation and side views of a two throw balanced opposed crank case assembly.

The crosshead assembly shown has the function of continuously assuring vibration free reciprocating motion of the piston and piston rod. The crosshead pads (or shoes) and supports are usually made from

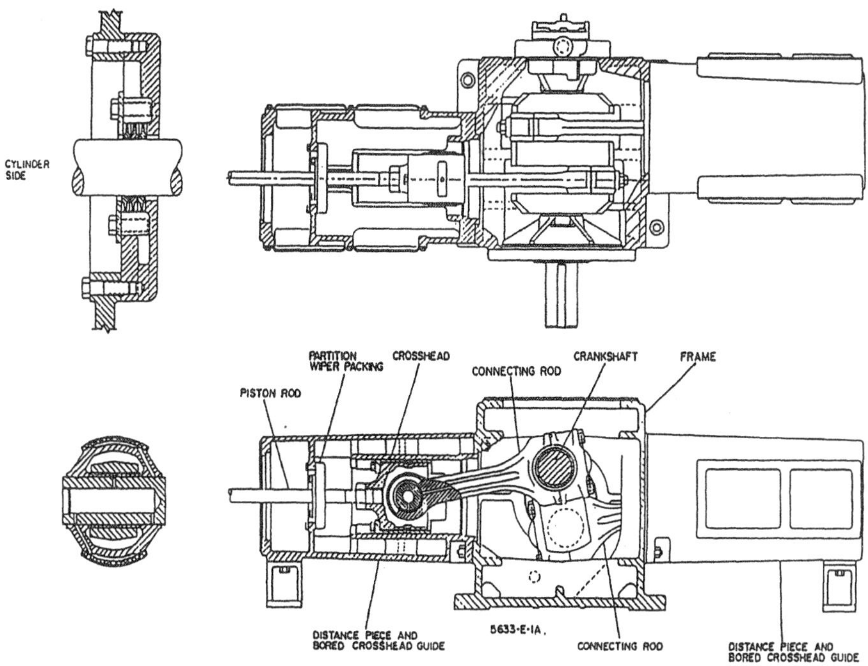

Figure 25.2 HDS off-gas running gear (Courtesy of Dresser Rand)

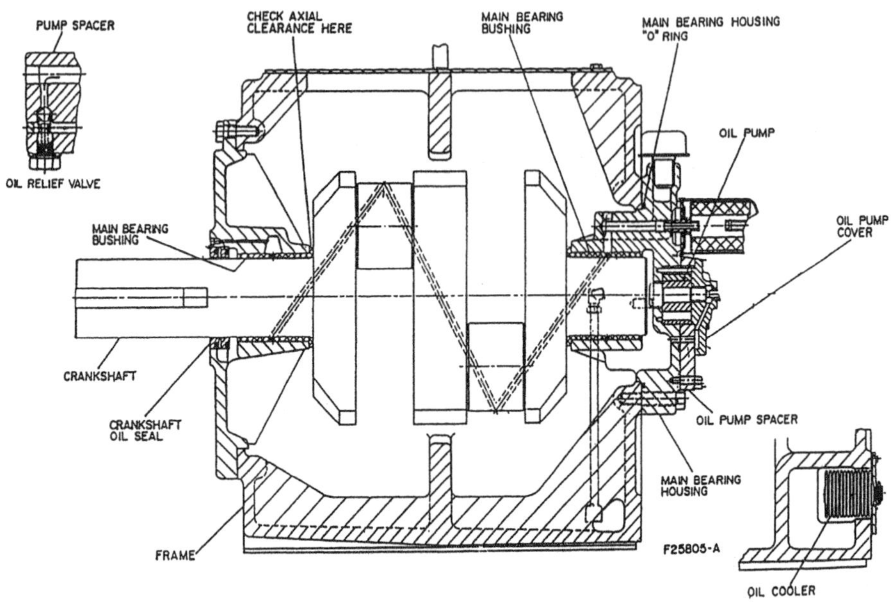

Figure 25.3 HDS off-gas running gear (Courtesy of Dresser Rand)

Babbitt or aluminum (smaller size units). Crosshead assembly lubrication is supplied via a pressure drilled hole (rifle drilled) in the connecting rod which in turn lubricates the crosshead pin bushing and crosshead shoes (see Figure 25.3).

Cylinder distance piece

Figure 25.4 presents the functions of the cylinder distance piece.

The proper operation of the distance piece baffles and seals is essential to maintaining reciprocating compressor safety and reliability in process gas applications. In most refinery process gas applications, a double compartment distance piece is used to assure contamination of the crank case or cylinders does not occur. Usually, the cylinder end compartment contains a partial N_2 atmosphere since the packing rings is usually N_2 purged.

Reliability considerations concerning this assembly are assuring proper packing, partition packing and wiper ring clearances.

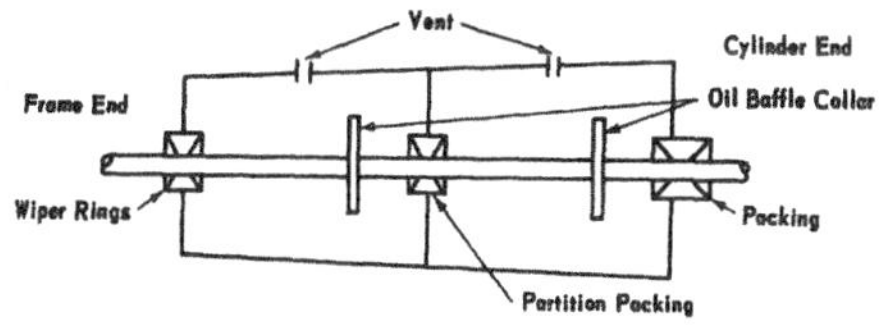

DOUBLE COMPARTMENT DISTANCE PIECE

DOUBLE COMPARTMENT SCHEMATIC

FUNCTIONS:
- PREVENTS CONTAMINATION OF PROCESS GAS
- PREVENTS CONTAMINATION OF CRANKCASE OIL

Figure 25.4 Cylinder distance piece

Piston rod packing

Figure 25.5 depicts a sectional and exterior view of a cartridge packing assembly.

The number of packing rings and type of arrangement is varied according to the cylinder maximum operating pressures. It is important to note that the packing does not provide an absolute seal, but only minimizes the leakage from the cylinder. Shown in the upper portion of the section drawing in Figure 25.5 is the vent port which carries the leakage gas either to a safe vent location (atmosphere, flare or fuel gas system) or back to the cylinder suction.

A means should be available to provide easy detection of excessive packing clearances. Alternatives are:

- Packing line flow switch
- Packing line orifice and pressure switch (only if compressor pressures are controlled to be constant)

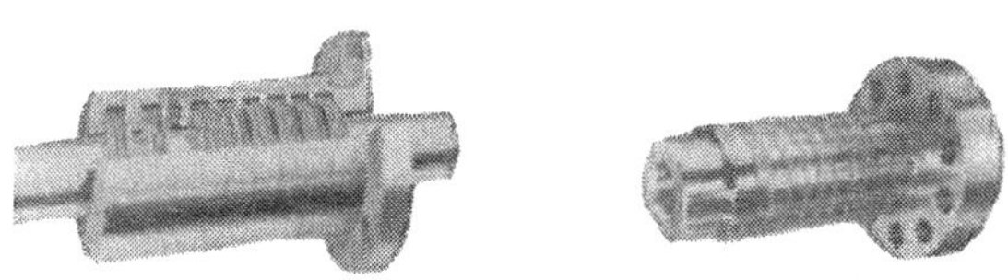

FUNCTION: MINIMIZES GAS LEAKAGE FROM CYLINDER

Figure 25.5 Cylinder packing

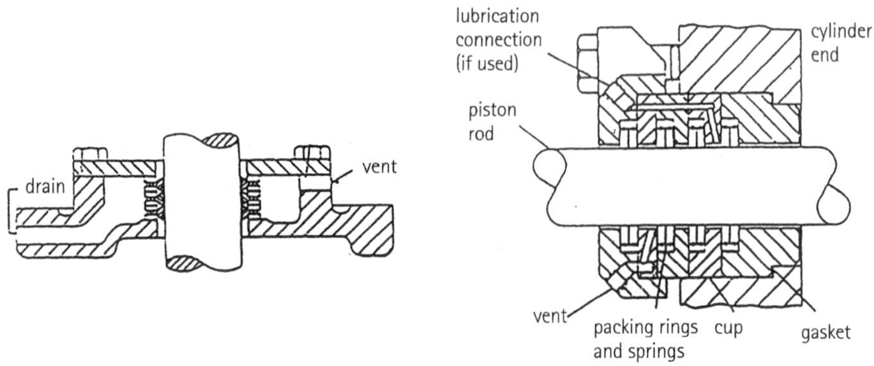

Figure 25.6 Left: Oil scraper ring arrangement for 9" & 11" ESH/V, HSE units; Right: Piston rod packing (Courtesy of Dresser Rand)

- Visual detection of gas flow (vent). Note: Flammable or toxic process gas must be purged with N_2 to attain a non-flammable mixture if the gas is to be vented to atmosphere.

Figure 25.6 shows additional packing assembly details.

The figure on the left side of the drawing is typical of a packing arrangement used between sections of a distance piece. Mounted horizontally, the assembly is equipped with a gravity drain and top vent.

The figure on the right side of Figure 25.6 shows a four (4) ring piston rod packing assembly. The upper part of the drawing shows the lubrication connections that are used when lubricated packing is required. Lube packing is normally used if the lubricant is compatible with the process stream.

If dry packing is used, piston rod speeds are usually slower and PTFE materials are usually employed. In the lower half of the drawing, the vent connections are shown and perform as previously discussed. The cup supports and positions an individual packing ring. Not shown is a purge connection which is inserted between the last and next to last packing ring (rings closest to the distance piece).

Cylinder and liner

Shown in Figure 25.7 are the two most common cylinder arrangements; double acting and single acting.

Most process reciprocating compressors are supplied with a replaceable cylinder liner. All cylinders are either jacked for cooling H_2O or finned for air cooling. Some older design cylinders use gaskets to isolate

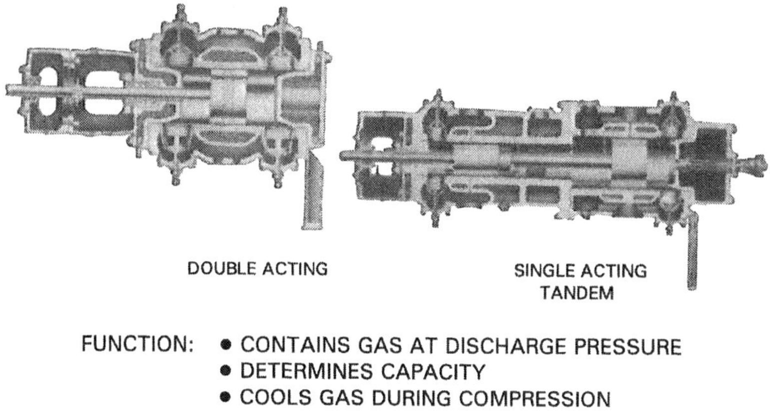

Figure 25.7 Cylinder and liner (Courtesy of Dresser Rand)

cooling water jackets from the cylinder. This design exposes the user to breakage from excessive cylinder H_2O entrainment if the gasket fails. Most reciprocating specifications today do not allow gaskets to be used in the cylinder. A double acting cylinder is designed to compress gas on both ends of the cylinder. (crank end and cylinder head end) while a single acting cylinder is designed for compression only on one end of the cylinder.

Reciprocating compressor cylinder valves

There are many different types of reciprocating compressor valves. Regardless of their design, all valves perform the same function … they allow gas to enter the cylinder, prevent recirculation flow back to the suction piping and allow gas to pass into the discharge system when the process discharge pressure at the compressor flange is exceeded. Valves are the highest maintenance item in reciprocating compressors. Their life is dependent on gas composition and condition, gas temperature and piston speed. Typical valve lives are:

- Process gas service in excess of one (1) year
- H_2 gas service – 8 – 12 months

In hydrogen service, particular attention should be paid to cylinder discharge temperature in order to obtain maximum valve life, cylinder discharge temperature for service with > 60% H_2 should be limited to 250°F. Recently, light weight, non metallic valves (PEEK) have been used successfully to increase the valve life in H_2 service above one (1) year.

Figure 25.8 shows a typical channel valve assembly.

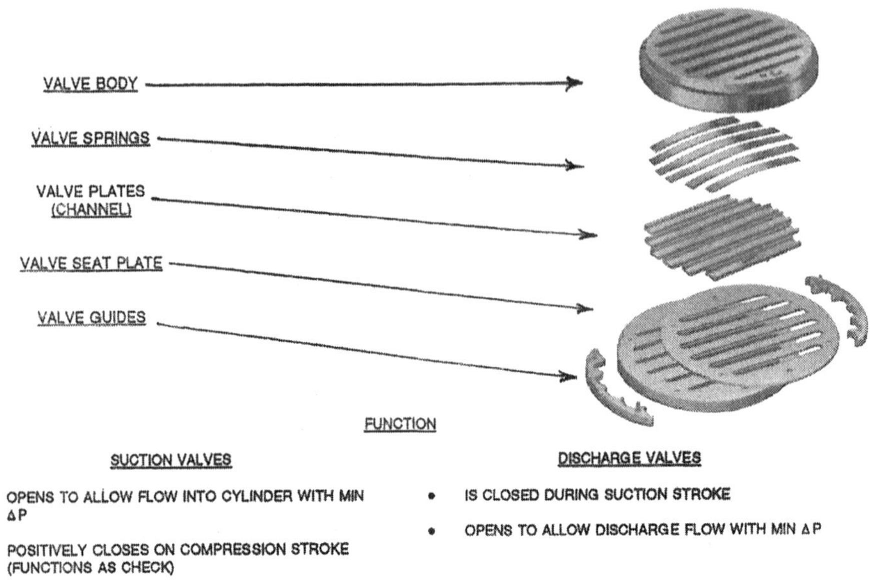

Figure 25.8 Reciprocating compressor valves – (Channel type, Suction and Discharge) (Courtesy of Dresser Rand)

The life of the channel valves shown is controlled by the spring force of the valve springs. The channel arrangement reduces the forces on the valve seal and usually results in increased valve life.

Figure 25.9 depicts a ring or plate valve assembly. This type of valve is most widely used.

Regardless of the type of valve, condition monitoring of valves is important to the profitability of any operation. The following parameter should be monitored.

Type valve

Suction
- Valve body temperature
- Compressor volume flow rate

Discharge
- Interstage process gas temperature
- Compressor volume flow rate

Changes in these parameters in excess of 10% from original (baseline) values should be cause for component inspection and replacement.

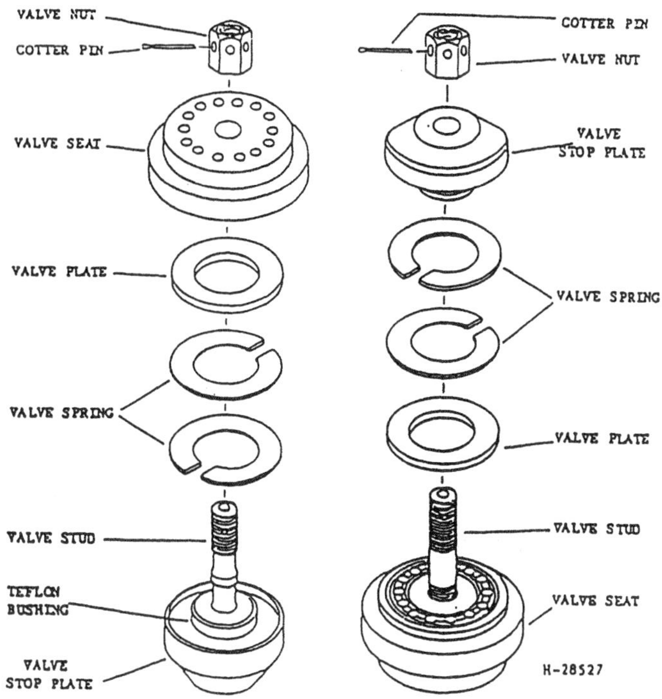

Figure 25.9 HDS suction and discharge valves (Courtesy of Dresser Rand)

Piston assembly

Figure 25.10 presents a typical piston assembly consisting of the piston rod nut, piston rod, piston and piston nut.

Piston rod materials are hardened steel and can include metal spray in packing areas to extend rod life. Piston materials can be steel, cast nodular iron or aluminum. The most common being cast iron due to its durability. Aluminum pistons are used in large cylinder applications (usually 1st stage) to minimize piston rod assembly weight.

Figure 25.11 shows piston rider bands (2) items 111 and piston rings (3) items 113.

Rider band and ring material is dependent on cylinder lubrication. If the cylinder is lubricated, carbon materials or compounds are used. If non-lubricated service is required, PTFE materials or other Teflon derivatives are used. Note also the piston hollowed area for piston weight control. Rider band and ring life is a function of piston speed, cylinder gas temperature and cleanliness of the process gas. In many process applications, a strainer is required upstream of the compressor to prevent excessive ring wear.

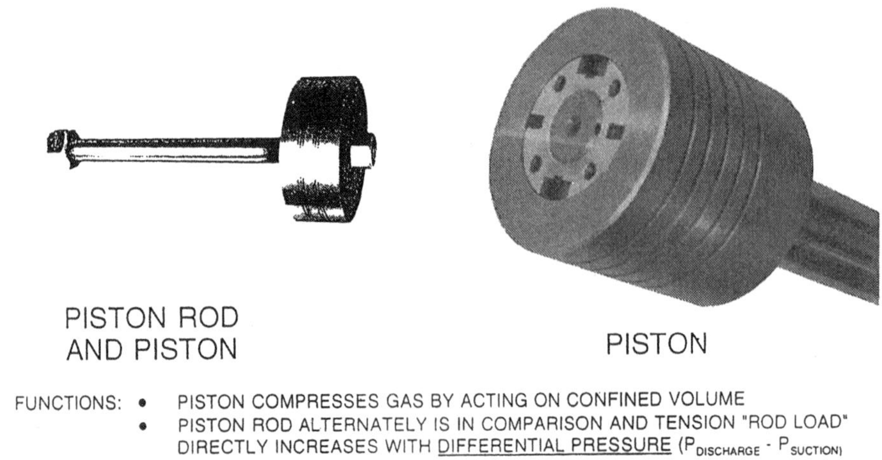

Figure 25.10 Rod and piston

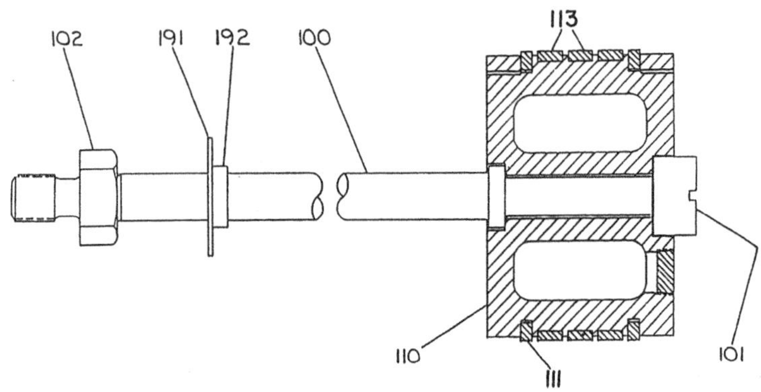

Figure 25.11 HDS off-gas-piston rod and piston (Courtesy of Dresser Rand)

Condition monitoring of rider band and piston ring wear can be accomplished by measuring and trending the vertical distance between each piston rod and a fixed point (known as rod drop). This can be accomplished either by mechanical or electrical (Bentley Nevada proximity probe) means. Of importance in piston assembly design is rod loading and rod reversal.

Rod loading is the stress (tension or compression) in the piston road and crosshead assembly caused by the ΔP across the piston. Rod load limits the maximum compression ratio that a cylinder can tolerate. This is the reason that many first stage cylinders are supplied with a suction pressure switch. Rod reversal is necessary so that the piston rod reaction forces on the crosshead pin will change allowing oil to enter the pin bushing. If the position of the pin in the bushing did not change (reverse) with each stroke, the bushing could not be sufficiently lubricated and would prematurely fail.

Pulsation dampeners

Since the action of the piston is non continuous, pressure pulsations will be generated. Depending upon the piping arrangement, these pulsations can be magnified to destructive levels. The use of pulsation dampeners, as shown in Figure 25.12 can reduce pulsations to 2% or lower.

There are methods available to evaluate and simulate the effect of pulsation dampeners prior to field operation. However, the correlation between predicted and actual results can be large and field modifications (installation of orifices or pipe modifications) may be necessary.

FUNCTION: TO REDUCE GAS PULSATION (EXCITATION FORCE THAT CAUSES PIPE MOVEMENT)

NOTE: EXCITATION FORCES VARY WITH PISTON SPEED AND CYLINDER LOADING

Figure 25.12 Pulsation dampeners (Courtesy of Dresser Rand)

Cylinder and packing lubricators

Whenever mineral oil is compatible with the process, lubricators will be used. Lubricators can be either positive displacement or dynamic type. Attendees are asked to review lubrication details in the appropriate instruction book. Lubricators will increase piston ring and packing life by reducing friction.

Figure 25.13 presents a typical lube oil system and its function.

ALL instruments in the lube oil system should be continuously monitored (baseline and current conditions). Remember, component (bearing) failure will occur if any major component in the system fails to function.

Figure 25.14 shows a lube oil system containing a shaft driven main lube oil pump with an internal relief valve.

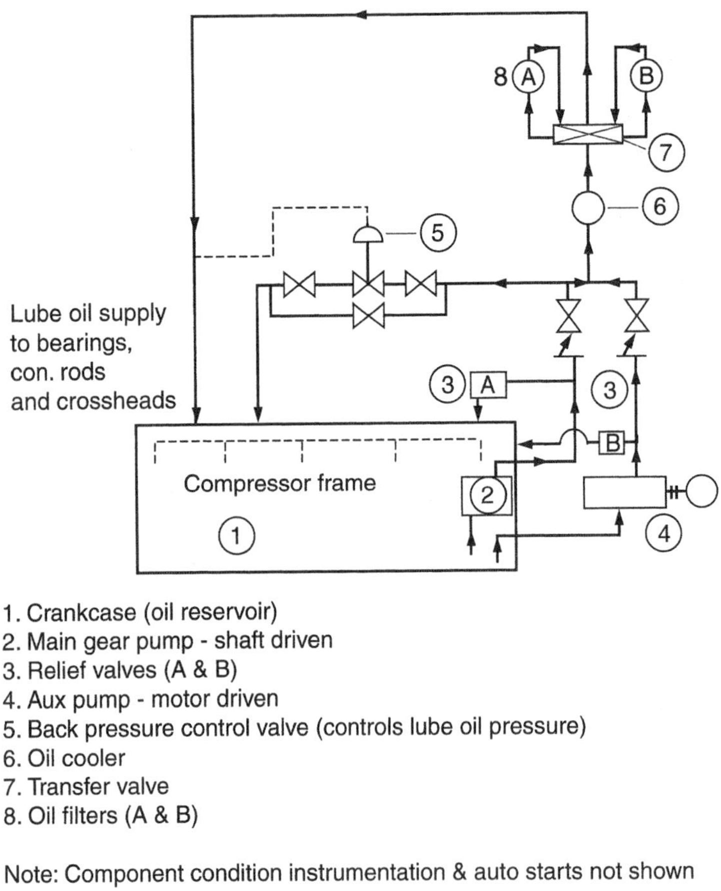

Figure 25.13 Lube oil system

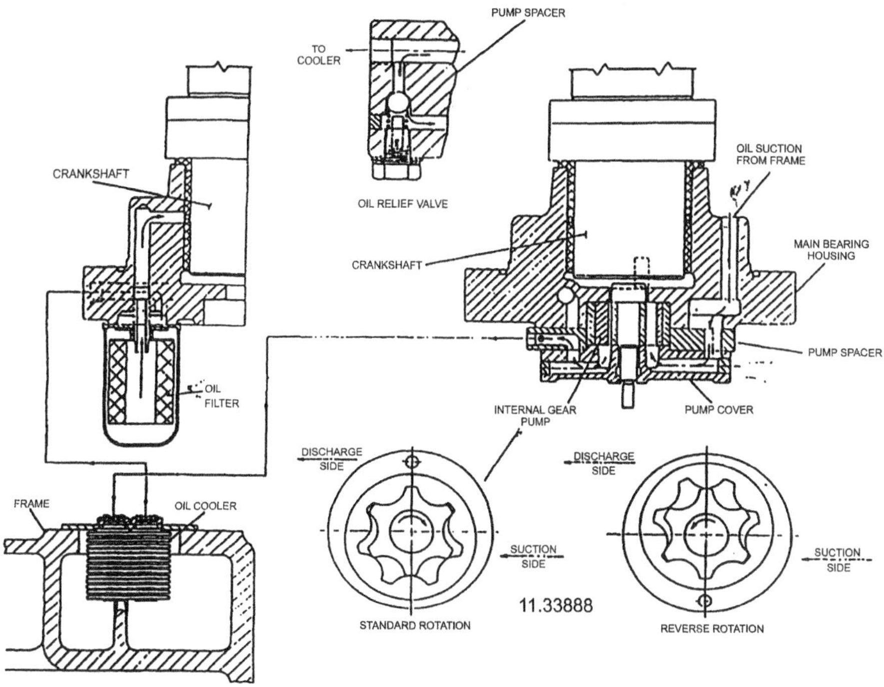

Figure 25.14 HDS off-gas-lube oil system (Courtesy of Dresser Rand)

This arrangement is a common one. Failure of the relief valve to seat can cause a low lube oil pressure trip.

Cooling system

The final topic to be covered is the cooling system. The cylinders, packing and process gas must be cooled to extend run time and minimize maintenance. Figure 25.15 presents a typical water cooled circuit.

In addition to cooling, the temperature of the cooling water must be regulated so that moisture (condensate) will not form in the cylinder in wet gas applications. It is recommended that the tempered water system temperature in the cylinder be maintained a minimum of 10–15°F above the cylinder inlet gas temperature.

Careful monitoring of the cooling circuit is essential in determining cooler, jacket, cylinder maintenance (cleaning) requirements.

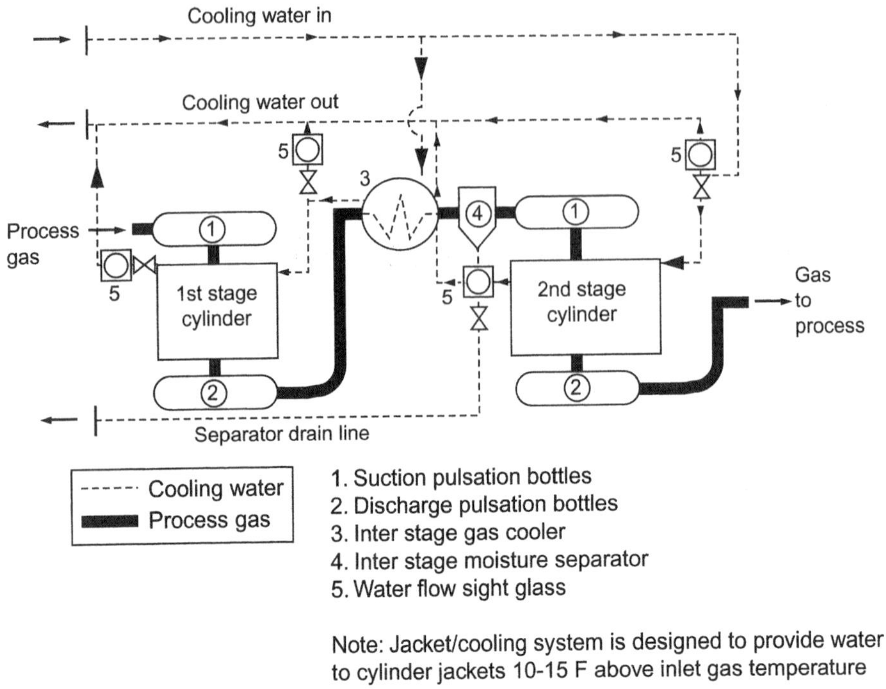

Figure 25.15 Cylinder, packing and intercooler cooling water system

26 Reciprocating compressor inspection guidelines

- Introduction
- Component function
- The vendor instruction book
- Safety requirements
- Documentation of results
- Frame running gear components
- Crankshaft main bearings
- Corrective measures
- Disassembly inspection
- Crankshaft web deflection
- Crankpin bearing
- Corrective measures and disassembly inspection
- Crosshead assembly (pin bushing, crosshead shoes and piston rod attachment)
- Crosshead pin bushing
- Corrective measures
- Inspection requirements
- Crosshead assembly, corrective action and inspection
- Piston rod alignment
- The piston assembly
- Cylinders
- Compressor valves
- Valve unloaders
- Cylinder packing

Introduction

This chapter presents assembly, disassembly and inspection guidelines for each major reciprocating compressor component. The material is arranged in the order of power transmission, from the running gear (frame) to the cylinder assemblies.

Component function

The function of each major component is presented first followed by the assembly, disassembly and inspection guidelines. Understanding the component function or 'what it is supposed to do' is key in performing complete and effective maintenance procedures.

The vendor instruction book

Use of the instruction book is a must for any procedure, regardless of the experience level of the Millwright. Clearances, dimensional tolerances and procedures can and do change. 'Rules of thumb' are helpful but must be confirmed by specific instruction book requirements.

Even if you have worked on the same compressor model before, there is no guarantee that clearances, tolerances or procedures will be the same. Different manufacturing locations, design changes and/or cost cutting measures can often result in a change in maintenance requirements. Figure 26.1 presents a summary of the guidelines to follow regarding the instruction book.

Always use the instruction book to confirm:

- Tolerances
- Clearances
- Procedures

And to avoid:

- Assembly errors

Figure 26.1 Always use the instruction book to confirm

Safety requirements

Heavy components, gas pressures, toxic or flammable gases, sharp edges etc. can cause serious accidents. Never assume a safety procedure has been followed – confirm it! If it cannot be confirmed, assume that the procedure was not performed. Figure 26.2 presents safety guidelines for reciprocating compressors.

Reciprocating compressor safety guidelines

- Obtain proper work permit from operations 'hot work permit' – if any welding or cutting is to be performed, otherwise 'normal work permit'
- Block out, vent, purge and tag compressor cylinders, pressure packing and vent lines
- Disconnect and tag motor power source
- Confirm crankcase is gas free and tag
- Use proper component lifting points and procedures
- Look out for sharp edges, if you can't see, use your mirror
- Make sure parts are dry and oil free before lifting

Figure 26.2 Reciprocating compressor safety guidelines

Documentation of results

Don't be 'doctor like'. By that we mean – put it in writing, not in code! All required dimensions and observations should be neatly and accurately detailed. Don't worry about spelling. If it's not in writing, it does not exist! Don't ever be accused of not putting it in writing. Your information is very important. Figure 26.3 presents documentation guidelines.

Documentation guidelines

- Obtain all required forms (if not available, make them)
- Be complete, accurate and neat
- Date and sign all forms
- Carefully list any observations for each component
- Put yourself in the position of the person who will read your documents

Figure 26.3 Documentation guidelines

Frame running gear components

The frame running gear consists of the following components:

- The crankcase and associated components (relief device, vibration detectors etc.)
- Crankshaft, flywheel (if supplied) and coupling
- Main bearings
- Crankshaft
- Connecting rods
- Crossheads
- Crosshead pins and bushings

The overall function of the frame running gear is to convert rotary (continuous) motion to reciprocating (alternating linear) motion while transmitting power to the piston with a minimum of external forces. See Figure 26.4.

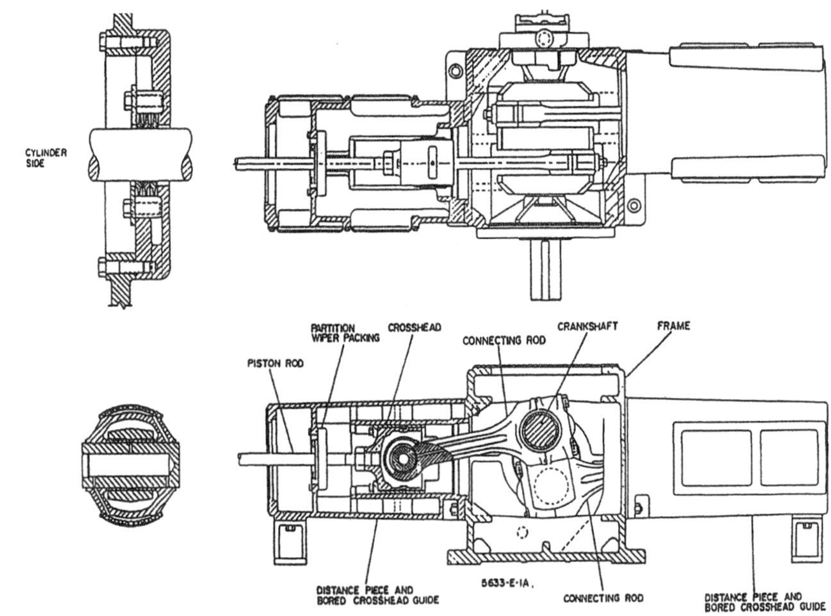

Function: transmits power and converts rotary motion to reciprocating motion with a minimum of external forces also acts as the main support for the machine and serves as the oil reservoir.

Figure 26.4 Frame running gear function (Courtesy of Dresser-Rand)

Crankshaft main bearings

The main bearings support the crankshaft and restrict its axial motion. If more than one crank is provided, an additional main bearing is supplied for each additional throw. Figure 26.5 presents a picture of a typical main bearing and its function.

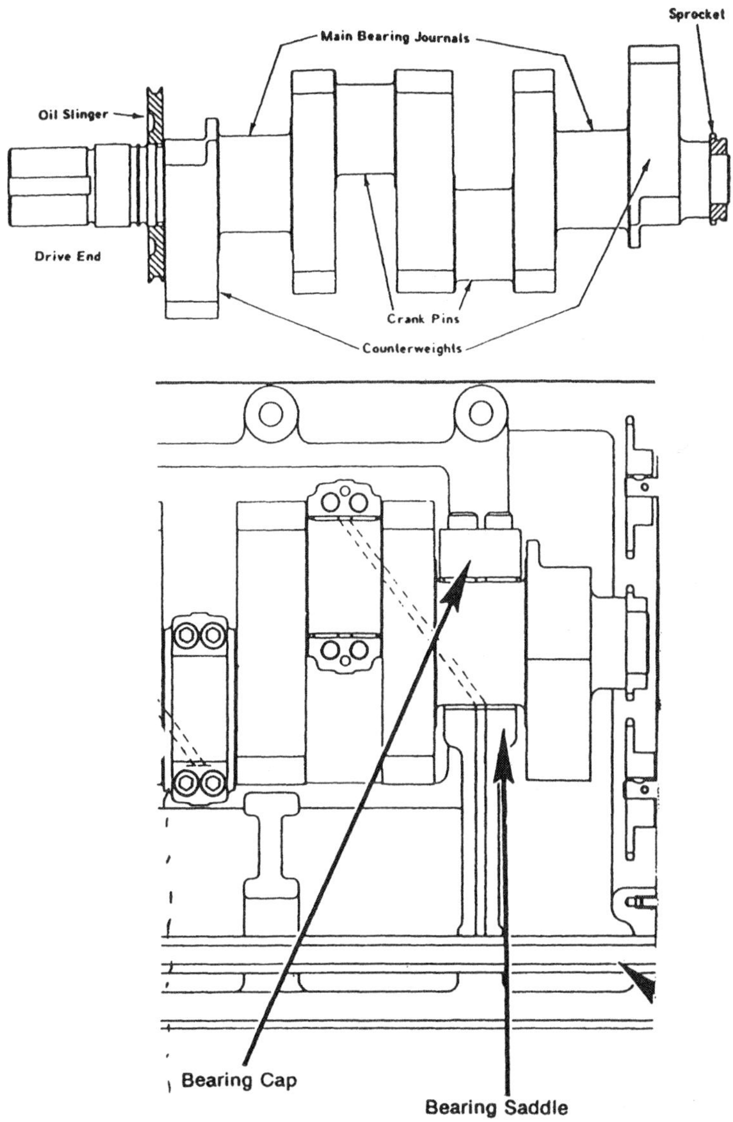

Function: support the crankshaft in the radial direction, absorb crosshead/connrod forces and locate the crankshaft axially.

Figure 26.5 Main bearing function (Courtesy of Ariel)

Main bearings are frequently constructed of aluminum rather than the traditional Babbitt bearings to absorb additional loads. Regardless of the bearing shell material, proper assembly preparation is essential to long bearing life and reliability. Figure 26.6 presents guidelines for bearing assembly preparation.

Bearing assembly preparation guidelines

- Set up a clean work area with all the necessary solvents, tools and measurement devices
- Thoroughly clean all mating surfaces with appropriate solvent and lint free rags
- Provide trays to hold parts to prevent loss, damage and contamination
- Confirm all oil ports are clean and free (open without restriction)
- Aluminum surface bearings require special care as follows:
  - Because aluminum is not as forgiving as Babbitt, a finer filtration is usually used in the oil supply system
  - Clearances are slightly more critical in aluminum bearing machines

Figure 26.6 Bearing assembly preparation guidelines

Main bearing crush measurements should be confirmed and corrected prior to obtaining bearing clearances as noted in Figure 26.7.

Main bearing crush check guidelines

- Assure all surfaces of the bearing shell and housing are clean
- Position bearing shell(s) in the lower and upper half of the bearing retainer
- Confirm that bearing shell fits properly in retainers (dye check if necessary)
- Assemble top bearing retainer and measure retainer gap on each side using a feeler gauge – record this value as crush
- Refer to instruction book for required values of crush

Figure 26.7 Main bearing crush check guidelines

Bearing radial clearances should be checked after proper bearing crush is confirmed using Plasti Gage (or soft wire) as noted in Figure 26.8.

Main bearing clearance guidelines

- Confirm crankcase soft foot values are within acceptable limits (0.002" max.)
- With all main bearing lower half shells in place, lay crankshaft (lubricate bearing shells with STP or equal before installing shaft)
- Using feeler gauges, check side clearance at each bearing. This clearance should be very close to the same on each side. (If not, check for bowed shaft or crankcase misalignment)
- Confirm proper alignment of all main bearing locating surfaces
- Dye check for proper lower bearing contact (Check with manufacturer for minimum contact)
- Position proper size Plasti gage or wire across top of shaft surface (in axial direction)
- Carefully position bearing top retainer and tighten bolts to the torque specified
- Remove top retainer and measure clearance by calibrated width of Plasti gage
- Also use micrometer to confirm Plasti gage clearance values (values should agree within 10%)
- Move shaft axially and measure axial clearance using a dial indicator
- Refer to the instruction book for the acceptable values of radial and axial clearance

Figure 26.8 Main bearing clearance guidelines

Corrective measures

If clearance and/or crush values are out of spec., the following corrective measures are recommended. Refer to Figure 26.9.

Main bearing clearance and crush corrective measures

1 Bearing clearances
- Excessive – increase crush to maximum allowable value
- Too tight – hand work
  - Babbitt – use proper scraping tools
  - Aluminum – corrective action is more complicated for aluminum bearings since they are usually a precision shell. Advice should be obtained from the manufacturer before making corrections. It is possible that larger bearings are available. It could be necessary to re-machine the bearing bore.

2 Bearing crush
- Excessive – shim bearing retainer at split line using full S.S.Shims. (Shims should not extend into the shell split line)
- Insufficient – surface grind bearing retainer at split line to obtain proper crush

Figure 26.9 Main bearing clearance and crush corrective measures

Disassembly inspection

Observing and recording all disassembly measurements and abnormal conditions is an essential ingredient of 'good maintenance practices'. Figure 26.10 presents guidelines for main bearing disassembly. These guidelines can generally be followed for the disassembly of any component. Always remember to 'think system' for bearings, this requires close examination of bearing lubrication passages.

Main bearing disassembly and inspection guidelines

- Obtain a digital camera for inspection documentation
- Thoroughly inspect and record bearing surface conditions. Look for:
  - Wipes
  - Debris and scratches
  - Evidence of high temperature
  - Electrostatic discharge
- Inspect O.D. of bearing shell for debris and evidence of 'non contact'
- Inspect all oil passages for evidence of debris and restriction
- Assure all oil passages properly line up
- Record crush, radial and axial clearances
- Be concise and accurate in recording all measurements and observations

Figure 26.10 Main bearing disassembly and inspection guidelines

Crankshaft web deflection

Any significant amount of crankshaft web deflection (in excess of 0.0005 ") will reduce the life of the components noted in Figure 26.11.

Crankshaft web deflection produces dynamic running forces that effect the life of:

- Crankshaft
- Main bearings
- Frame

Figure 26.11 Crankshaft web deflection

As a result, accurate measurement of web deflection is a key contributor to reciprocating compressor reliability. Figure 26.12 outlines web deflection measurement guidelines.

Web deflection measurement guidelines

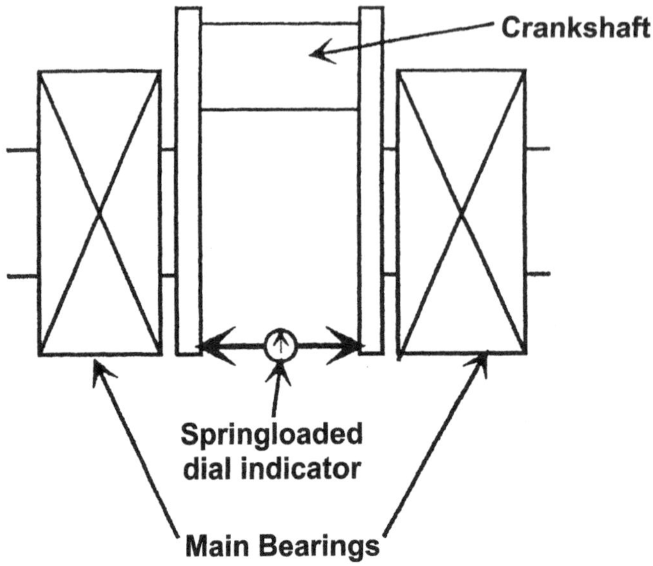

- Always reference throws from the flywheel end
- Indicator set up (conn rod out)
  - Make accurate center punch marks on opposing faces of counter balance flanges
  - Install spring loaded dial indicator in center punch marks as shown above.
  - Zero out indicator at 12 O'clock and check for repeatability.
  - Recorded at 4 positions (12, 3, 6 and 9 O'clock)
  - Refer to vendor instruction book for tolerances.
- Record readings above and below conn rod (Note: It may be necessary to use an I.D. MIC if indicator is not accessible).

Figure 26.12 Web deflection measurement guidelines (Courtesy of M.E. Crane consultant)

If the web deflection exceeds the specified values, the following corrective measures are recommended in Figure 26.13.

Excessive web deflection corrective measures

- Check crankcase for soft foot at all locations.
  (Assuming this is initial installation. If the machine has been grouted, the grout must be removed and the machine must be re-leveled and re-grouted)
- Confirm alignment of main bearings (this could involve the use of wire line measurements)
- Consult vendor – (excessive web deflection is a *very serious problem* and any corrective action should be discussed with the vendor)

Figure 26.13 Excessive web deflection corrective measures

Crankpin bearing

Figure 26.14 shows a typical connecting rod crankpin bearing and it's function.

Connecting rod crankpin bearing function

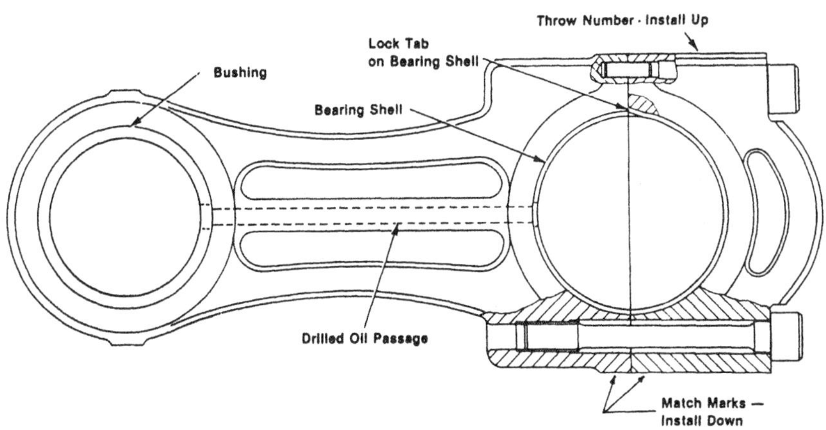

Function: to transmit all loads from crankshaft to connecting rod, crosshead and piston and to support process loads from all of these components including rod load and rod reversal.

Note that, in this illustration, the crankpin bearing is a precision shell insert.

Figure 26.14 Connecting rod crankpin bearing function

Connecting rod crankpin bearing assembly preparation guidelines should follow the bearing assembly guidelines shown in Figure 26.6.

First perform the crankpin bearing crush check as noted in Figure 26.15.

Crankpin bearing crush check

- Assure all bearing shell and conn rod surfaces are clean
- Position bearing shell (s) and confirm proper contact (dye check if necessary)
- Assure that bearing shell lubrication hole is aligned with conn rod oil passage
- Position bearing shell in conn rod and measure gap with conn rod bolts loose using feeler gauge. Record as crush
- Refer to instruction book for required valves of crush

Figure 26.15 Crankpin bearing crush check

After proper crankpin crush is confirmed, bearing radial and side clearance should be checked as outlined in Figure 26.16.

Crankpin bearing clearance guidelines (conn rod not installed)

- Confirm web deflection values are acceptable
- Confirm all conn rod and bearing shell surfaces are clean
- Assure all oil passages are clean and fully open
- Install crankpin bearing shells being sure that all oil holes are completely lined up with conn rod oil passages
- Fully torque conn rod bolts and record bearing i.d. at oil hole location and 90 degrees away
- Mic throw at oil hole location and 90 degrees away
- Subtract values and record 'mic clearance'
- Refer to vendor's instruction book for acceptable values

Figure 26.16 Crankpin bearing clearance guidelines (conn rod not installed)

After crankpin radial mic clearance checks are performed, all connecting rods should be properly installed with proper torque and crush values. It is recommended that STP or equal low friction lubrication be used on crankpin bearing surfaces during final assembly. With the connecting rod crank pin bushings properly torqued, bump check and side clearance must be checked at each crankpin bearing as outlined in Figure 26.17.

Crankpin bearing axial clearance and radial clearance (bump check)

- Record side clearances of crankpin bearing using feeler gauges
- Refer to vendor's instruction book for acceptable clearances
- 'Bump checks' can be performed by setting the crankshaft throw at 90 degrees to the connecting rod and locking it with the jacking gear then mounting a dial indicator on the crankshaft to read longitudinal movement of the connecting rod. The connecting rod can be bumped back and forth at the crosshead and movement recorded between the crankshaft and the connecting rod. Acceptable readings should be obtained from the compressor vendor.

Figure 26.17 Crankpin bearing axial clearance and radial clearance (bump check)

Corrective measures and disassembly inspection

The same comments apply as for the main bearings. Refer to figures:

- Figure 26.9 – Corrective measures for crush and bearing clearance
- Figure 26.10 – Disassembly and inspection guidelines

Crosshead assembly (pin bushing, crosshead shoes and piston rod attachment)

Figure 26.18 shows a typical balanced opposed crosshead assembly, the major components and it's function.

Frame with crosshead assembly in place

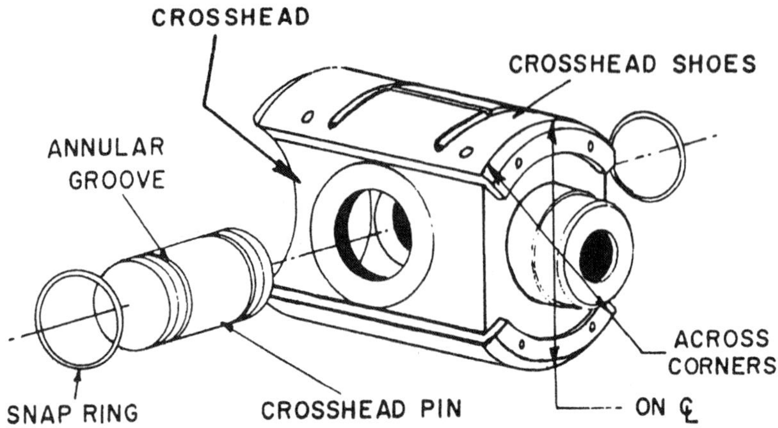

Major components:

- Pin bushing
- Shoes
- Piston rod attachment

Function: smoothly transmit power to piston without excessive rod runout and vibration

Figure 26.18 Frame with crosshead assembly in place (Courtesy of Dresser-Rand)

There are various types of crosshead designs. The most common types are:

- Adjustable shoe (separate shoe) types
- Integral shoe types

Figure 26.19 shows a crosshead for a high speed (1200 rpm+) non-API compressor.

Most compressors (API – 618), use crossheads with adjustable shoes.

Crosshead pin bushing

Before proceeding with the crosshead, we will cover the crosshead pin bushing. This important component must absorb all of the running gear and piston assembly inertia and pressure forces while assuring sufficient oil film is produced between the pin and the bushing.

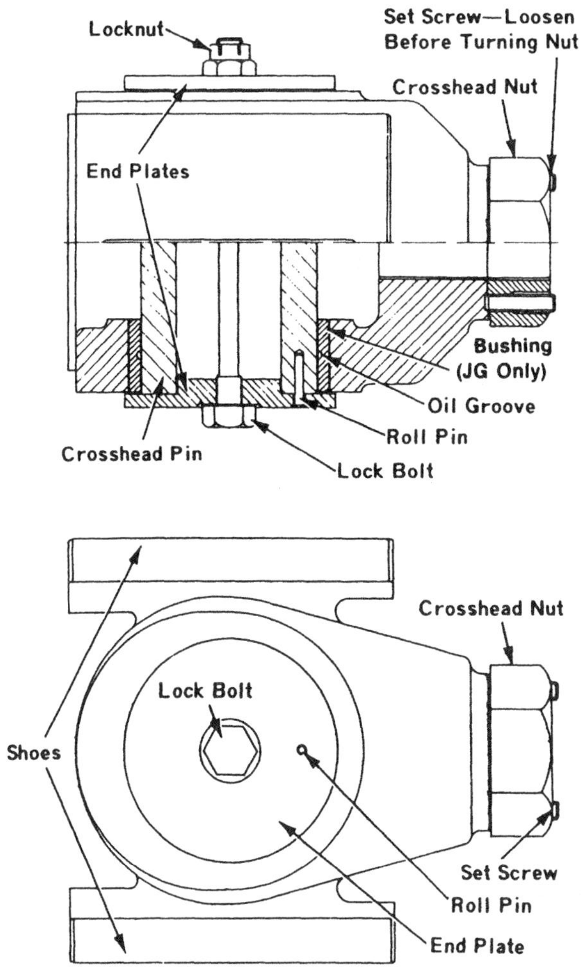

Figure 26.19 Crosshead types (Courtesy of Ariel)

Crosshead pin and bushing are shown in Figure 26.20.

Since the bushing does not experience continuous rotation, sufficient movement of the Conn-Rod is required to produce and oil film at the load point. This is accomplished by designing the connecting rod and piston rod assembly inertia's to assure sufficient movement of the pin in the bushing during each stroke. This is known as 'rod reversal'. Normally compressors are designed to produce a minimum of 25 degrees rod reversal in the pin bushing (degrees of rotation of pin in the bushing).

Rod reversal, can be affected by the manner in which the compressor is operated (degree of unloading) by:

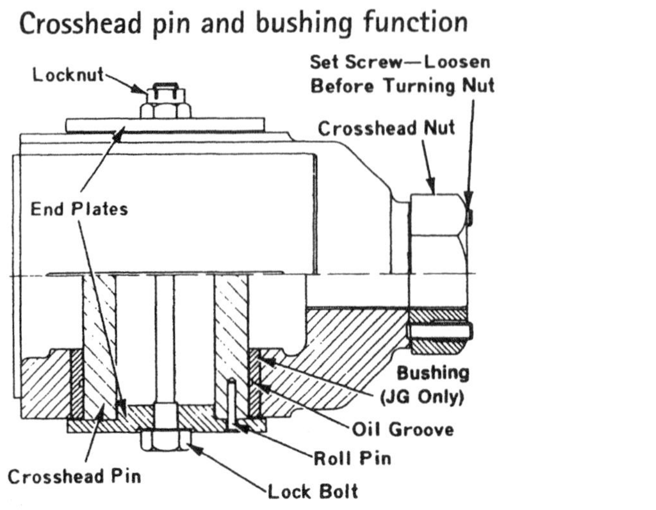

Figure 26.20 Crosshead pin and busing function (Courtesy of Ariel)

- Valve unloaders
- Clearance volume pockets
- Bypass valves

Figure 26.21 contains crosshead bushing assembly guidelines.

Crosshead pin bushing assembly and clearance guidelines

- Confirm proper assembly of conn-rod crosshead end bearing (proper shrink fit) and alignment with conn-rod oil passage
- Mic id of conn-rod crosshead end bearing and confirm correct dimension
- Confirm proper assembly (shrink fit) of bushings in crosshead
- Mic pin bushing diameter on flywheel and outboard ends and confirm correct dimension
- Mic crosshead bushing ID on flywheel and outboard ends
- Subtract values and record flywheel and outboard end clearances
- Consult instruction book for required values
- The side clearance between the conn rod bearing and the inside of the crosshead should be obtained and compared with the manufacturers recommendation.

Figure 26.21 Crosshead pin bushing assembly and clearance guidelines

Corrective measures

Proper fit and oil passage alignment is essential to crosshead pin bushing life. All fits and clearances must be within manufacturers tolerances. Refer to Figure 26.22 for guidelines.

Crosshead pin bushing assembly corrective measures

- Conn-rod crosshead end bearing
  - Excessive shrink fit – grind bearing shell OD as required
  - Insufficient shrink fit – replace bearing
- Conn-rod crosshead end bearing to pin clearance
  - Excessive – replace bearing
  - Insufficient – scrape bearing if allowable (babitt or Bronze)
- Crosshead bushing shrink fit
  - Excessive – grind bushing shell OD as required
  - Insufficient – replace bushing
- Crosshead bushing pin clearance
  - Excessive – replace bushing
  - Insufficient – scrape bushing if allowable
- Crosshead pin OD
  - Excessive – grind as required
  - Insufficient – replace pin

Figure 26.22 Crosshead pin bushing assembly corrective measures

Inspection requirements

Please refer to Figure 26.23 for crosshead pin bushing inspection requirements.

Crosshead clearance corrective measures

- Insufficient shoe contact – confirm proper support adjustment and absence of cocking of the crosshead in the guide.
- For adjustable shoe types – check measurements to ascertain which way the crosshead is cocked (or not properly machined). Analysis of the measurements should tell you which of these problems exists. Note that the problem could be either in the machining of the crosshead or the crosshead guide or in the conn-rod alignment. If the piston rod is installed and the lock nut is tight, loosen the crosshead lock nut to be sure the piston rod is not forcing the crosshead out of line. See Figure 26.27 for comments regarding piston rod lock nuts.
- Excessive clearance
- Adjustable shoe types – adjust as required.
- Non-adjustable, integral type – remove crosshead and machine for adding babitt to increase O.D. remachine to proper O.D.
- Insufficient clearance
- Adjustable shoe type: shim as required in order to provide specified clearance.
- Integral (non-adjustable) shoe type: remove crosshead and machine to correct dimensions. Care must be taken to assure that there is adequate babbitt on the crosshead for the procedure.

Figure 26.26 Crosshead clearance corrective measures

There are many types of crosshead to piston rod attachment methods. Regardless of the method, the attachment must assure that all piston assembly static and inertial forces are absorbed without any relative motion of the piston rod to the crosshead attachment. Figure 26.27 describes the most common type of piston rod locking device and its function.

Crosshead – piston rod locking device

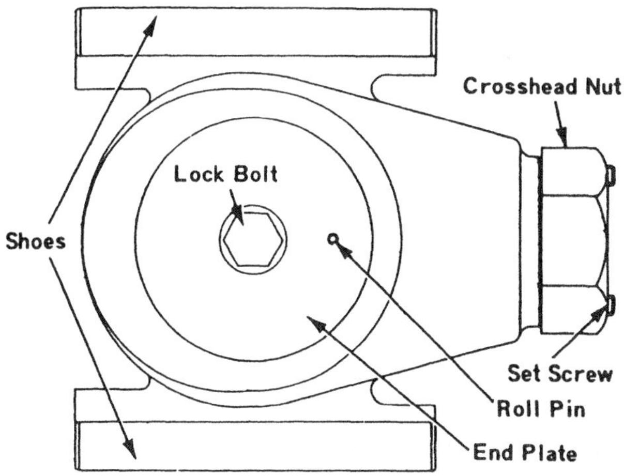

- Function: absorb all static and inertial forces without allowing any relative movement between rod and crosshead attachment area.
  - The normal mode for locking the piston rod to the crosshead is a lock nut
  - There are many variations on these lock nuts
  - One of the primary problems with the lock nuts is that they can cause a deflection in the rod when they are tightened. Machining anomalies make the face of the lock nut 'out-of-parallel' with the contact face on the crosshead. When the lock nut is tightened, it places an angular force on the rod and causes misalignment.
  - There are patented lock nuts on the market that can minimize this problem. Most have axial screws in the nut that are tightened in accordance with nut designer's instructions to avoid placing offset forces on the rod.
  - Another correction is to use spherical washers between the lock nut and the face of the crosshead. This will also avoid offset forces on the rod.
  - Care should be taken to follow the guidelines from the manufacturer of the compressor or the patented lock nut regarding application of proper torque to these locking devices.

Figure 26.27 Crosshead – piston rod locking device (Courtesy of Ariel)

To assure proper rod to crosshead attachment, the torque procedure guidelines outlined in Figure 26.28, should be followed, as a minimum. Consult the instruction book for all requirements for the specific compressor model.

Piston rod crosshead attachment torque guidelines

- Assure threads are clean and burr free
- Use thread lubrication on rod, locknut and crosshead
- Install locknut on piston rod
- Remove a valve from each end of the cylinder
- Turn the crank to the full retracted position
- Screw the piston rod into the crosshead
- Check the float clearance on the inboard side of the piston. This clearance should be at the approximate value for final adjustment
- Tighten the locknut just to the snug position
- Take a total float clearance check (rotate the crankshaft with the jacking gear) using soft wire and determine the total float value.
- Loosen the locknut and adjust as required to obtain the correct inboard and outboard piston float clearances. (These clearances must be in accordance with the vendor recommendations regarding cold settings)
- When these float clearances are correct, set the locknut in accordance with the compressor manufacturer's (or proprietary lock nut manufacturer's) instructions
- Repeat the total and individual float measurements to be sure they have remained the same

Figure 26.28 Piston rod crosshead attachment torque guidelines

Piston rod alignment

Proper piston rod alignment is essential to reciprocating compressor reliability and directly affects the life of the following components:

- Distance piece packing
- Cylinder packing
- Cylinder liners
- Rider bands
- Piston rings

Experience has shown that if cold (installed) piston rod runout is maintained below 0.002" optimum component life and reliability will be achieved.

Figure 26.29 shows the typical set up of piston rod runout checks for a double acting horizontal cylinder.

Rod runout definition and basic practice guidelines

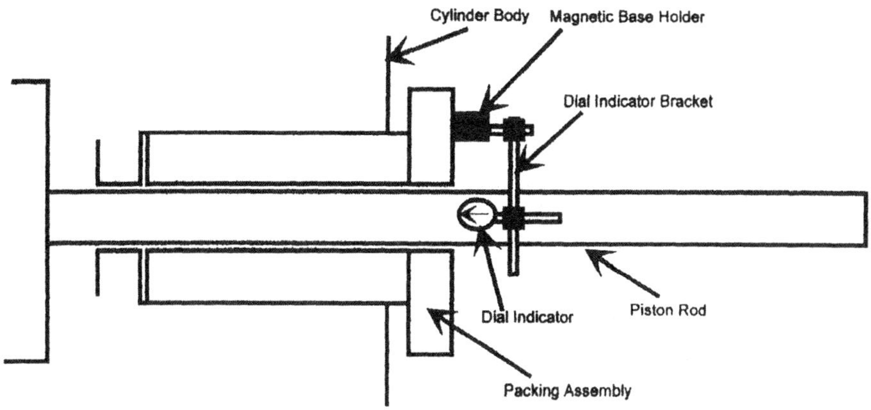

Definition: The measurement variation (millimeters or thousandths of an inch) in dial indicators as the rod is moved slowly through its stroke.

Basic practices

1. Set up one dial indicator as shown to read the vertical rod run-out.
2. Set up a second dial indicator at 90 degrees to the first one to read horizontal rod run-out. (In accordance with previous comments, it is preferable to all horizontal readings at the same 'time' position). This makes it simpler to correlate any misalignment with crank or conn rod problems.
3. Both indicators should be as close to the packing gland as possible without touching or interfering with the visibility of the indicator face.
4. Starting with the piston in the full inboard position, zero both indicators and slowly rotate the crankshaft 180 degrees and record the readings of both indicators. (Take care to note readings as horizontal or vertical and use + or – signs to indicate a positive or negative direction).
5. Continue to rotate the crankshaft another 180 degrees in the same direction to the starting position and record the readings of both indicators.
6. If the last readings are not zero on both indicators, repeat steps 4 and 5 to determine what is causing the lack of repeatability.
7. Steps 4 and 5 should be repeated with the rod to crosshead lock nut loose and again with the lock nut torqued to specifications. This is to prove that the faces of the lock nut and crossheads are true and not causing deflection of the rod in relationship to the crosshead.

Figure 26.29 Rod runout definition and basic practice guidelines (Courtesy of M.E. Crane Consultant)

Figure 26.30 contains limits for rod runout.

Maximum rod runout guidelines

- Horizontal limits – 0.002" runout maximum
- Vertical limits – 0.002" runout maximum

Note: excessive rod runout will result in shortened packing life and could be cause for problems with piston rings and rider bands.

Figure 26.30 Maximum rod runout guidelines

The definition of hot rod runout and acceptable limits are shown in Figure 26.31.

Hot rod runout

- Definition: the actual rod measured runout at specified operating conditions (use method in Figure 26.29)
- Limits: both vertical and horizontal = 0.002" rod runout
- Significant differences between hot and cold rod runout can exist with aluminum pistons and/or fluorocarbon rider bands due to thermal expansion

Figure 26.31 Hot rod runout

If rod runout values are exceeded, Figure 26.32 presents guidelines for corrective action.

Rod runout adjustment guidelines

- Confirm proper cylinder support (acceptable soft foot check)
- Confirm cylinder alignment
- Adjust shims under crosshead shoes (if possible).

  Note: consult with compressor vendor before taking action. (Any adjustment on the shims under the cylinder support will place distortional stress on the cylinder and distance piece. The vendor should give the limits on the amount of change allowable.)
- Do not vary from recommended maximum shim thickness under cylinder support feet
- All shims under cylinder supports must be full contact shims

Figure 26.32 Rod runout adjustment guidelines

Maintaining concise, accurate records of rod runout are good practice and important to the end user. Expected values of vertical rod runout should be obtained from the instruction book (from the vendor if not available in the instruction book).

The piston assembly

The piston assembly consists of the following components:

- Piston rod
- Piston
- Crosshead attachment nut
- Piston nut
- Piston rings
- Rider bands

Refer to Figure 26.33.

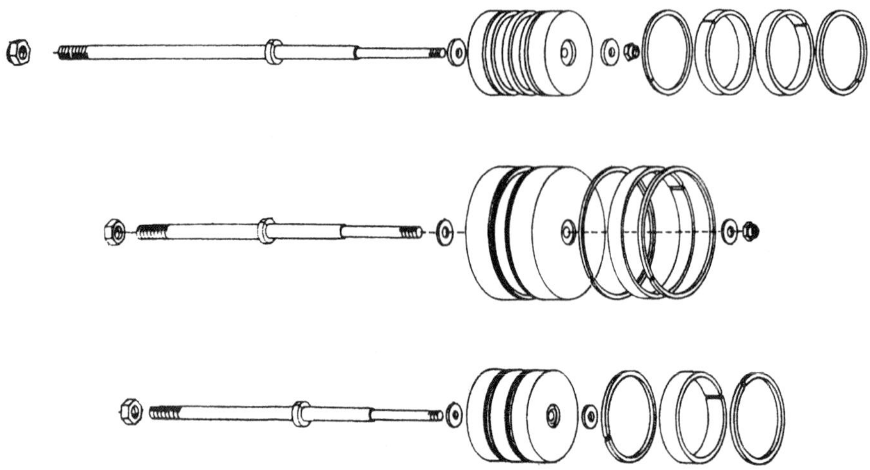

Figure 26.33.1, 2, 3 Typical piston assemblies

Figure 26.34 presents guidelines for piston to rod attachment.

Depending upon the application, piston diameter and stroke speed, pistons may be supplied with rider bands in addition to the piston rings. Figure 26.35 defines the function of the piston rings and the rider bands.

Piston rod attachment guidelines

- Confirm proper thread cleanliness and condition.
- Apply recommended lubrication to all threads.
- Check the instrument manual to determine the accepted method for holding the rod while torqueing the rod nut. Improper procedures here can cause serious damage to the rod. This procedure can require special tools.
- Install piston and measure piston head end rod to piston face dimension.
- Properly torque piston nut. Note that, if the manufacturer's instructions only include torque values, then rod stretch and/or piston crush measurements are not needed.
- Repeat piston head end face to head end measurement and record piston crush or rod stretch.
- Confirm the measured value in the instruction book.

Figure 26.34 Piston rod attachment guidelines

Piston ring and rider band function

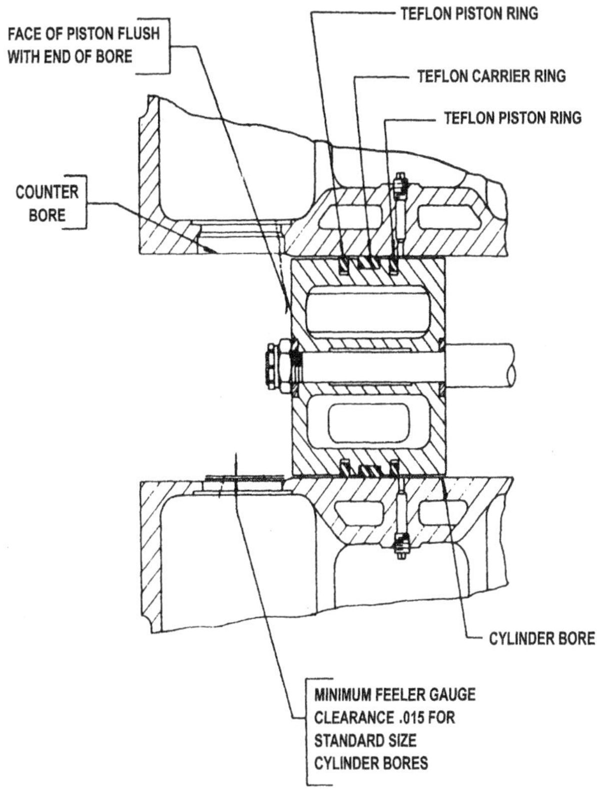

Piston rings: minimize the internal leakage between the cylinder liner and piston to provide maximum cylinder capacity and efficiency

Rider bands: protect the cylinder liner against contact with the piston. (They also allow operation with a lower lubrication feed rate for processes that are damaged by lube oil.)

Figure 26.35 Piston ring and rider band function (Courtesy of Ariel)

Guidelines concerning piston ring and rider band installation are presented in Figure 26.36.

Piston ring and rider bands installation guidelines

- Measure and confirm proper piston ring and rider band dimensions
- Check groove widths and profiles as follows:
  - Groove widths must be in accordance with manufacturer's specifications. If there is too much clearance relative to the rings, it can cause rapid ring wear and/or ring damage. If the rings fit too tightly in the grooves, they will not be able to float and accomplish an acceptable seal to the cylinder bore.
  - Groove profiles must be such that radii in the bottoms will not interfere with ring movement within the groove. Groove faces must be 90 degrees relative to the linear turn of the piston.
- Install piston rings and rider bands in accordance with manufacturers instruction manual. It is critical that proper sequences be used in this step. (Particularly in the case of the rider bands). Each manufacturer has made provision for gas passage around the rider bands and if they are not installed in accordance with their design, the gas passages could be blocked or occluded causing the rider bands to act as piston rings with resulting serious damage.
- Measure piston ring side clearances and end gaps
  - This must be done in accordance with manufacturers guidelines. If side clearances are excessive, rings must be replaced. If they are too tight, they must be fitted. It is best to consult with the vendor before revising ring dimensions. End gaps must be measured with rings installed in the piston grooves (piston outside of machine) and O.D.'s flush with piston diameter. Measurements are to be taken as indicated by vendor manual instructions. Note that the rings are either angle cut or step cut and the manual should tell you how and where to take the end gap readings.

Figure 26.36 Piston ring and rider bands installation guidelines

Cylinders

The cylinders contain the gas and provide a compression chamber in which the piston can compress the gas. They provide cooling (through water cooled cylinder jackets) and contain the suction valves, discharge valves and piston rod packing. The cylinders and their components are designed to allow gas compression at the highest volumetric cylinder flow and efficiency. Examples of single acting, double acting and tandem cylinders are shown in Figure 26.37.

Compressor cylinder types and function

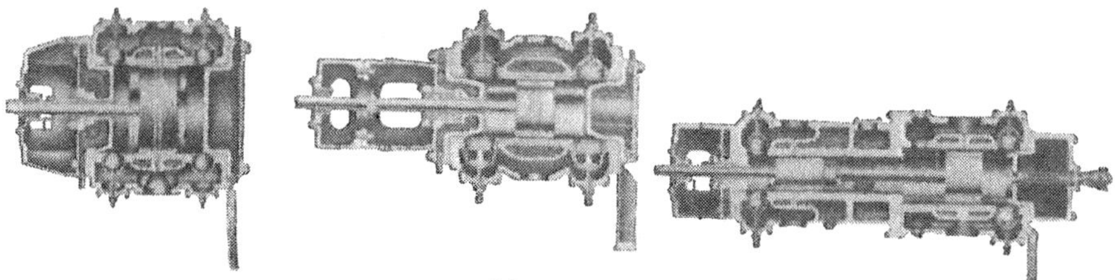

DOUBLE ACTING (LUBRICATED) | DOUBLE ACTING (NON-LUBRICATED) | DOUBLE ACTING TANDEM (LUBRICATED)

Functions: Contain the gas, support all components that determine the cylinder volumetric flow and efficiency and cool the gas during compression

Figure 26.37 Compressor cylinder types and function (Courtesy of Dresser-Rand)

With the piston installed in the cylinder, Figure 26.38 defines the clearances to be measured and suggested guidelines.

Piston to cylinder clearance measurement guidelines (piston installed)

- Measure rider band to bore clearances using a feeler gauge at top, bottom and determine total value
- Measure piston ring to cylinder side clearance using feeler gauges (record largest and smallest values)
- Measure piston to cylinder liner clearances using feeler gauge at top, bottom and determine total value if rider bands are installed.
- Install cylinder heads, remove valves as required and obtain 'float clearance'.
  - Cranked clearance
  - Head end clearance
  - Total actual clearance

Figure 26.38 Piston to cylinder clearance measurement guidelines (piston installed)

All cylinders today are provided with cylinder liners. Figure 26.39 defines the proper methods for cylinder liner installation, removal and inspection.

Cylinder liner installation, removal and inspection guidelines

1. Installation
   - Measure bore and liner rod
   - Confirm proper shrink fit with above measurements
   - Use dry ice for installation or follow manufacturer's maintenance manual
2. Removal
   - Remove using dry ice if possible.
   - In some cases it is necessary to machine the liner out
3. Inspection
   - Use digital camera to record findings
   - Stroke piston and observe condition of liner as follows
   - Location of damage
   - Evidence of debris
   - Lubricator hole condition (if provided)
   - Evidence of lack of lubrication
   - Evidence of excessive moisture
   - Accurately record all findings

Figure 26.39 Cylinder liner installation, removal and inspection guidelines

If any cylinder critical dimension is out of tolerance, Figure 26.40 guidelines for corrective action.

Cylinder assembly corrective action guidelines

1. Head end or crank end clearance
   - Excessive H.E. or C.E. – consult compressor vendor
   - Insufficient H.E. or C.E. – consult compressor vendor.

   Note: consult vendor before performing any work on C.E. or H.E. heads
2. Piston to liner clearance
   - Excessive – consult vendor regarding acceptable piston ring and rider band clearance. Consider non-standard rings and bands
   - Insufficient – consult vendor. Consider non-standard rings and bands

   Note: do not machine liner without vendor approval
3. Rider band and piston ring clearances
   - Excessive – consult vendor for the effect on capacity and efficiency
   - Insufficient

Figure 26.40 Cylinder assembly corrective action guidelines

Compressor valves

In order to fully understand the function of compressor valves it is well to simplify their definition by viewing them as special check valves. They operate on the same principal as any check valve by allowing free flow of the gas or fluid in one direction and blocking flow in the other direction. Because of the frequency of flow pulsations involved in the compressor valve, they must be carefully designed to provide reliable service for long periods. The valve designer must work within the confines of weights of components, resistance of the material to the gas properties, ability to handle contaminants, flow efficiencies and longevity of service. In their efforts to accomplish this, the valve manufacturers and designers have created a very competitive market. The compressor manufacturers each have their own valve designs, which they provide with the machine. In addition to these valves there are many companies who specialize in the design and manufacture of valves only (or packing and valves). They are in direct competition for the replacement part business of the end user.

There are many types of compressor valves. Due to the comments in the above paragraph the installed valves might not be the valves that were originally furnished with the cylinder. In this case, observe the sealing surfaces for evidence of bypass and confirm valve overall dimensions against the original valve dimensions if possible. Figures 26.41 through 26.44 illustrate different types of valves showing the major components.

All the valve types shown in the following illustrations are and have been in use for many years and have provided excellent service. The advantage or disadvantage of any one type over another is dependent on the following:

- Process gas mol weight, density, pressure and flow rate.
- Compressor speed (stroke frequency).
- Corrosive components of gas.
- Amount of solid contaminants and liquid in the gas stream.

Figures 26.42, 26.43, and 26.44 illustrate different types of valves showing the major components.

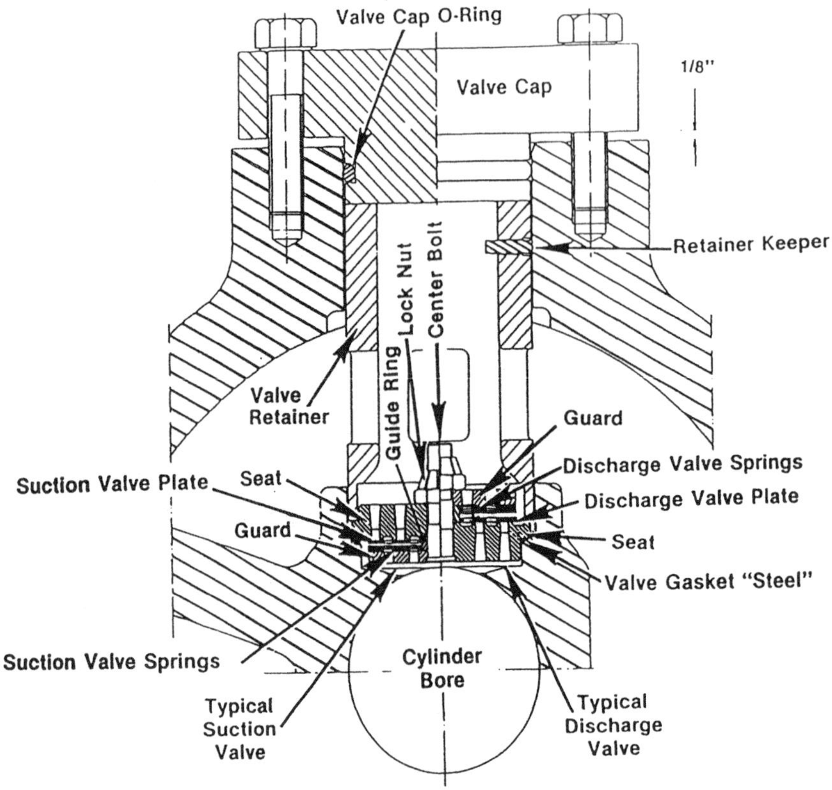

Figure 26.41 Typical valve assembly (Courtesy of Ariel)

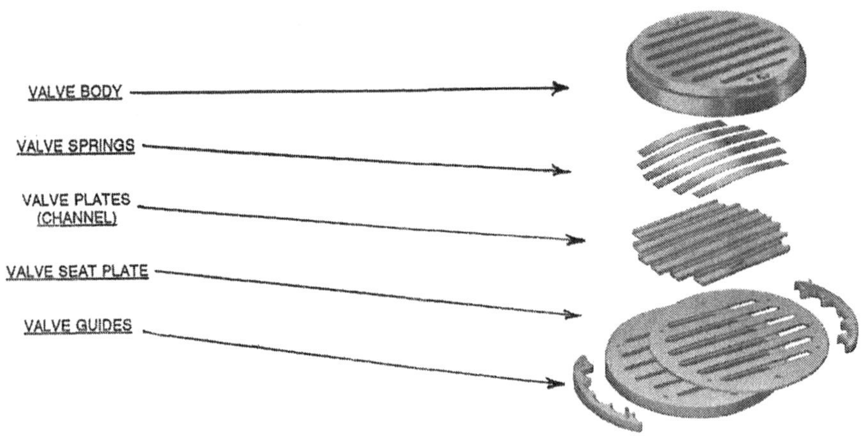

Figure 26.42 Channel valve components (Courtesy of Dresser-Rand)

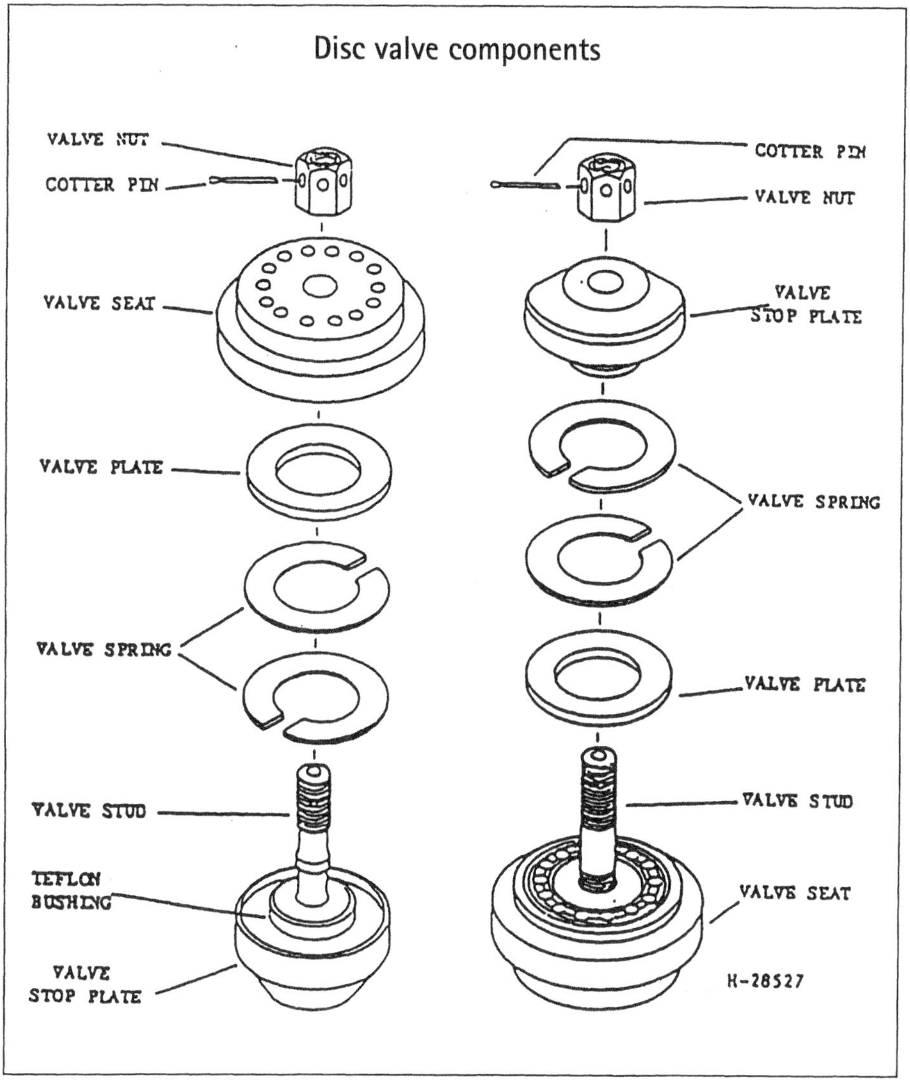

Figure 26.43 Disc valve components

Poppett valve components

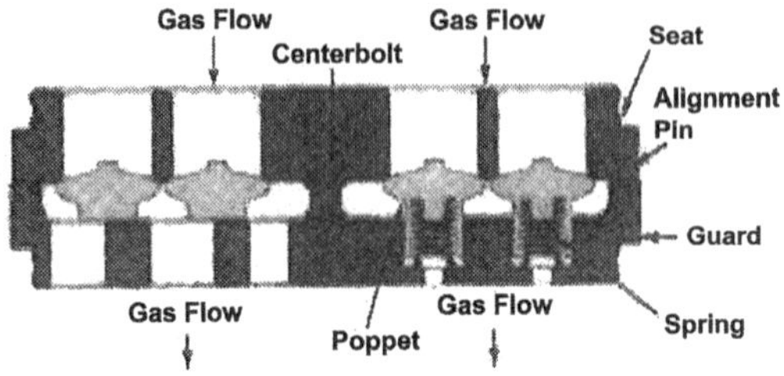

Features	Benefits
Single size poppet for all valves	■ Reduces parts inventory and simplifies assembly
Streamlined gas path; larger flow holes and spring vent holes	■ Minimizes entrapment of impurities and increases valve life ■ Minimizes pressure drop for improved efficiency
Enhanced poppet geometry with smooth radii	■ Provides high strength and streamlined flow paths
Hi-Temp® PEEK poppets	■ Dresser-Rand's proven blend of PEEK has high impact strength as well as excellent resistance to the effects of dirt and corrosion ■ Improves durability to increase valve reliability

Figure 26.44 Poppett valve components (Courtesy of Dresser-Rand)

The functions of the major valve components are defined in Figure 26.45.

Valve parameter and component functions

- Valve seat – provides flat sealing surface for valve to provide minimum leakage and smoothly direct gas to and from cylinder with a minimum pressure drop.
- Valve plate, channel or popett – these are the major moving parts in the valve assembly. They must be able to perform the required movement in concert with the valve springs and valve plate, channel or poppet guides and provide flat sealing surface to assure minimum leakage
- Valve springs – allow valves to open smoothly, with minimum pressure drop and turbulence. These springs also hold the valves closed during start-up and shut-down modes
- Valve guides – maintain proper alignment of valve plates, poppets or channels with valve seats and springs and assure that valves function with a minimum of friction
- Valve seat provides smooth, flat surface for valves to seal on and to stop each stroke without causing excessive valve wear and force on components.
- Valve lift – the total distance that the valve travels in the valve assembly. This value varies with gas properties and flow rates.

Note: valves cannot be substituted from one compressor service to another without checking with the valve vendor!

Figure 26.45 Valve parameter and component functions

Valve assembly and lift check guidelines are presented in Figure 26.46.

Valve assembly and lift check guidelines

- Thoroughly clean all cylinder retainer and valve surfaces
- Install proper gaskets as required
- Check valve for proper lift as required. This can be done by measuring from the outside of the valve seat to the valve plate with a depth mic and depressing the valve plate to full open position and measuring again. The difference in these two measurements is the 'valve lift'.
- Confirm suction and discharge valves are properly installed and not reversed. Note that in newer units that meet API 618 requirements, suction and discharge valves are designed to prevent inadvertent switching of suction valve to a discharge port and vice versa.
- Check all valve assembly seating surfaces for full contact by torquing all valve hold down nuts to specified value
- Once seating surface contact is confirmed, reinstall valves using proper valve hold down torque values

Figure 26.46 Valve assembly and lift check guidelines

Figure 26.47 describes suggested valve inspection guidelines.

Valve removal inspection guidelines

- Use a digital camera for documentation
- Check all seating surfaces for signs of bypass and record findings
- If time permits, disassemble valve and check the following conditions:
  - Valve seat
  - Valve body
  - Valve springs valve guide
  - Valve stop
  - Thoroughly clean valve and record valve lift
  - Check valve lift against valve specifications

Figure 26.47 Valve removal inspection guidelines

Valve unloaders

The function of the valve unloaders is to reduce cylinder volume flow rates. Figure 26.48A defines their function.

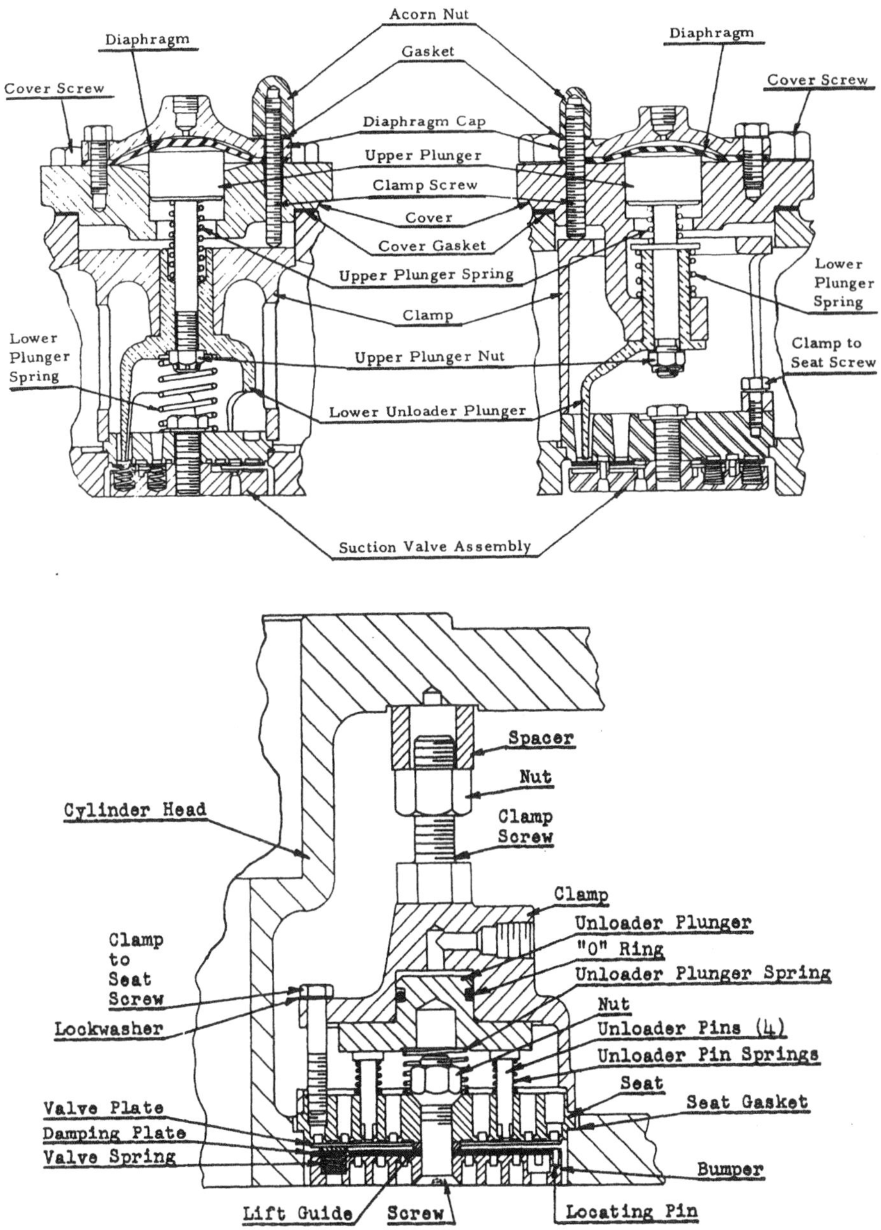

Figure 26.48A Cylinder valve unloader function. Diaphragm type unloader; piston type unloader; piston type unloader (Courtesy of Cooper Industries)

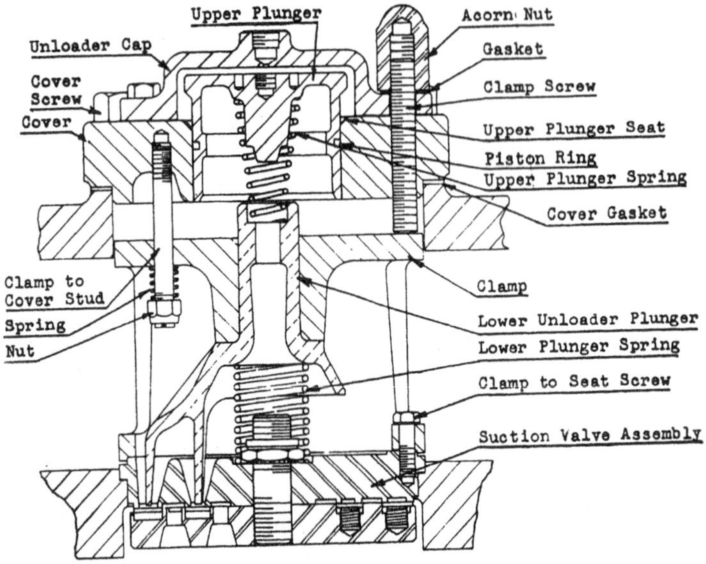

Function: Hold suction valve open to prevent compression and gas flow in the designated cylinder end

Figure 26.48A – Continued – cylinder valve unloader function. Diaphragm type unloader; piston type unloader; piston type unloader (Courtesy of Cooper Industries)

Cylinder valve unloader facts

Function: hold suction valve open to prevent compression and gas flow in the designated cylinder end.

- Valve unloaders are manual or pneumatic operated devices used to hold suction valve plates, popetts or channels in the open position.
- With the suction valves held open, the flow capacity of that end of the cylinder is reduced to zero.
- On multiple stage compressors, the unloading sequence is critical and must be carefully calculated by the design engineer. If, for instance, the second stage of a three stage machine is partially unloaded, the interstage pressure ahead of that cylinder will increase and the interstage pressure downstream of that cylinder will decrease. The system designer must be certain that changing these interstage pressures does not create an unacceptable condition in the piping, coolers or other vessels in that system.
- In all case, it is necessary to unload all suction valves in the end of the cylinder being unloaded. This is to assure adequate flow capacity through each valve to avoid over heating.

Figure 26.48B Cylinder valve unloader facts

- Valve unloading sequences should not be altered without an engineering evaluation of the consequences.

Figure 26.48B Continued - Cylinder valve unloader facts

Depending upon process flow requirements, the types of cylinders (single or double acting) and the number of cylinders, the reduction in capacity can vary from 100% to 25%. Clearance pockets and bypass valves can also be used to allow accurate volume control to any specified value. It should be noted that most designs avoid the use of bypass control except for start-up functions due to the power wasted with this type of control. In these cases, the control system can be very sophisticated and will usually utilize a PLC (programmable logic controller) to control the volume flow rate. The PLC is programmed to avoid unacceptable combinations of cylinder unloading. Figure 26.49 defines proper unloader removal/installation and set-up guidelines.

Valve unloader, removal, installation and set-up guidelines

1. Removal
   - Be certain that the compressor has been properly shut down, locked out electrically, isolated, depressured and purged as required for safety.
   - Be sure you have all necessary tools and supplies on site to complete the work. This must include replacement gaskets, parts cleaning equipment, work benches, special tools, etc. If the machine is very large, proper lifting equipment must be available for handling valves and valve covers.
   - Remove covers and valves in accordance with the manual guidelines.
   - Place any valve covers equipped with unloaders on a solid work bench for disassembly. Review manufacturers drawings before beginning the work and disassemble in the proper sequence.
   - Repair after cleaning, using new diaphragms, pistons, rings and/or springs as required to restore the unloader to a usable condition. Follow vendor guidelines for all assembly procedures.
2. Installation
   - Installing new seat and cover gaskets as required, err on the side of safety and replace if questionable.
   - Be certain all gaskets are properly seated either on the valve or in the port seat before installing valves or valve covers. Remember that a gasket that is cut or damaged will result in a leak and the need to shut the machine down again for replacement.

Figure 26.49 Valve unloader, removal, installation and set-up guidelines

- After covers are set in place, install all cover nuts or bolts and tighten just snug. Check to be sure the cover is not cocked or not started into the port diameter.
- Using a torque wrench, tighten cover bolts or nuts in a logical sequence to the specified torque.

3. Set-up
   - Install all control lines to the valve unloaders.
   - With the help of instrument technicians, apply unloader signal to the valve unloaders and check for leaks.
   - If the unloader stem is visible externally, check for proper movement.

Figure 26.49 Continued – Valve unloader, removal, installation and set-up guidelines

Cylinder packing

Figure 26.50 shows a typical cylinder packing cartridge ring types.

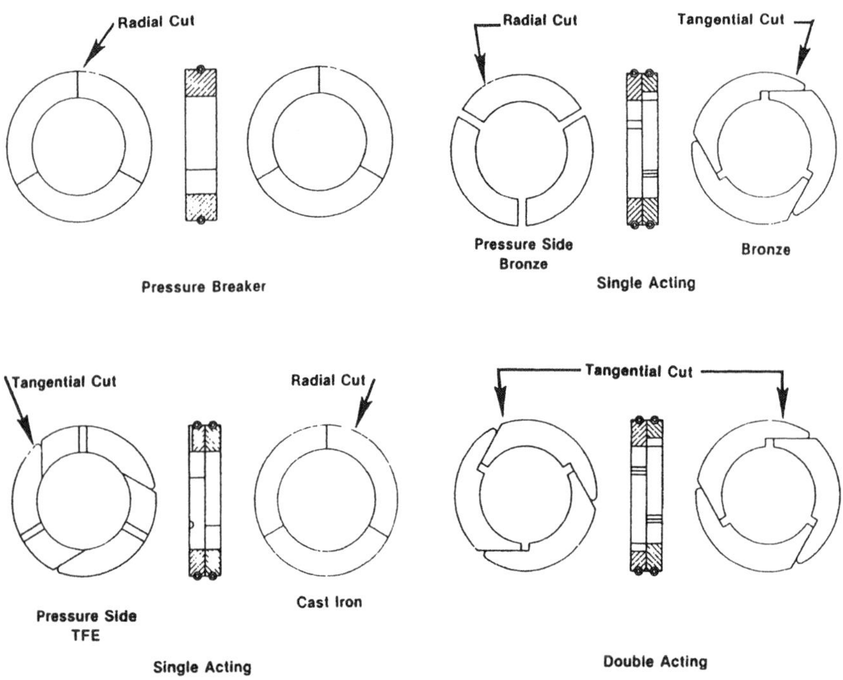

Figure 26.50 Cylinder packing ring types

Figure 26.51 presents an exploded view of a typical packing cartridge assembly and its function.

Typical cylinder packing cartridge function

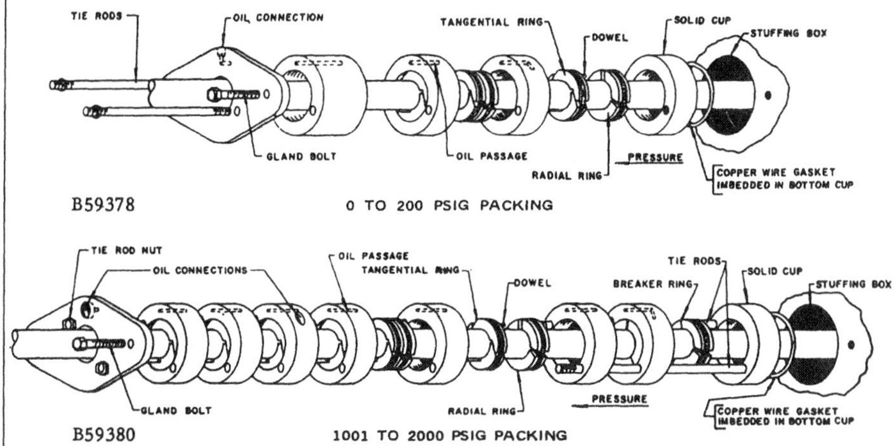

Function: Minimizes gas leakage from the cylinder and directs leakage to a safe location.
Also provides piston rod/packing interface cooling and lubrication when required.

Figure 26.51 Typical cylinder packing cartridge function (Courtesy of Ariel)

Figure 26.52 provides guidelines regarding the packing cartridge assembly.

Packing cartridge assembly guidelines

- Packing cartridges can be assembled in two ways.
  1. Complete assembly outside machine
  2. Assembly with rod in place in machine
- Complete assembly outside machine
  1. This involves bench assembly of all components in accordance with manufacturers drawings and instructions.
  2. Some special tools are required.
  3. All packing cups, plates and rings must be examined carefully for contamination of any kind.
  4. Preliminary assembly should be done in a vertical stack.
  5. Place inner end cup with the tie bolts up on bench and stack all parts in correct orientation and order on the tie bolts.
  6. Be certain that all oil and cooling water holes are properly aligned and that all 'O' rings are firmly set in their grooves.
  7. Apply light oil film to all packing rings as they are installed.
  8. When all parts are correctly stacked on tie bolts, place packing gland cover plate on top of stack and install tie bolt nuts. These nuts should be just snug to start.
  9. Lay the assembly on its side in a piece of angle iron or other vee groove and gently tap all cups and plates to assure they are completely aligned and concentric.
  10. Tighten all tie rod nuts to maintain this concentricity during installation of the packing assembly in the compressor.
  11. Install packing assembly in compressor cylinder and tighten packing gland nuts just enough to hold the assembly in place while installing the piston and rod.
  12. Since the rod has threads on the end which could damage the packing ring bores, it is necessary to use a protective cover on the end of the rod while inserting through the packing assembly. This cover is a special tool and is commonly known as a 'condom'. It is usually made from brass (soft) and is quite thin to avoid over-expanding the packing rings to fit through the packing cups and plates.
  13. Install the piston and rod through the new packing, being careful not to force it and damage the packing rings.
  14. Install the cylinder head and torque to specifications.
  15. Follow steps 13 through 15 under 'assembly with piston and rod in the machine' below.

Figure 26.52 Packing cartridge assembly guidelines

- Assembly with piston and rod in the machine.
  1. This is normally only done if there is a reason to leave the piston and rod in the compressor.
  2. It is still necessary to remove the rod from the crosshead to perform this type of assembly since all packing cups, plates and 'O' rings must be slipped over the end of the rod.
  3. This method is also more time consuming and requires working in a restricted space while installing and assembling the packing ring groups.
  4. Start by loosening the piston rod/crosshead locknut, unscrewing the rod and moving the rod and piston all the way outboard. Remove the locknut from the rod.
  5. Remove the crosshead to the inboard end of its stroke. With the rod and crosshead in this position, it is possible to disassemble and remove the old packing assembly over the end of the rod.
  6. After removing the old packing and thoroughly cleaning the rod and packing gland, insert all of the packing cups, plates and 'O' rings over the rod. Be certain that all components are in the correct order and orientation.
  7. Packing ring groups can now be assembled on the rod and moved into place in the cups and 'O' rings (if any) can be put in place. Assemble each packing group in sequence starting with the pressure end of the assembly. As cups, 'O' rings, packing and plates are assembled, move them on to tie bolts.
  8. When all parts (including the packing cover plate) are assembled on the tie bolts, install the tie bolt nuts to a snug position.
  9. Let the entire assembly rest in the rod and gently tap tap the cups and plates down to be sure they are all concentric.
  10. Fully tighten the tie bolt nuts to hold the parts in this position for installation.
  11. Reconnect the piston rod to the crosshead and set the lock nut snug.
  12. Install the packing assembly in the gland and set the packing nut snug.
  13. Remove one suction valve from each end of the cylinder and perform a total float and inboard/outboard ratio setting.
  14. Fully torque the lock nut using the procedures outlined in Figure 26.28.
  15. Torque the packing gland bolts while using a feeler guage between the rod and the bore of the packing cover plate to be sure the packing assembly is centered relative to the rod.

Figure 26.52 Continued – Packing cartridge assembly guidelines

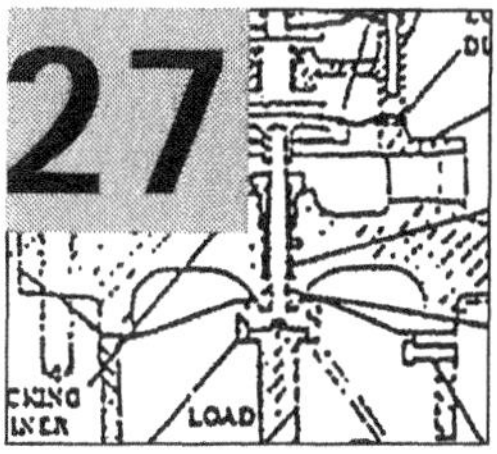

27 Reciprocating compressor control and operation

- Introduction
- Capacity control
- Methods of capacity control
- Start-up methods

Introduction

Regardless of process conditions a reciprocating compressor essentially produces a constant volume flow. In this chapter you will learn what is done to control and operate a reciprocating compressor in processes that require variable volume flows. The subject of actual compressor cylinder volume will be discussed and certain factors that can reduce capacity will be covered. The available means of adjusting compressor throughput during operation will be presented. The items to watch concerning unloading bypassing will be discussed.

After presenting the general facts concerning control and operation, the specific start-up and shutdown control and operation schemes for the make-up compressor will be presented.

Capacity control

As previously discussed, a performance characteristic of all types of positive displacement compressors is constant delivered volume flow rate regardless of process conditions. Since a reciprocating compressor is a positive displacement type, the actual volume delivered to the process will not be significantly affected by process conditions. This fact is very important since any reduction in actual flow rate must be the result of an internal mechanical problem (worn piston rings, rider bands, packing and/or valves). This is not true for centrifugal compressors, since their flow rate is affected by process conditions. Figure 27.1 describes the causes of reduced actual volume flow rate.

Reciprocating compressor volume

- Actual volume is the volume delivered to the process
- It therefore is the displaced volume minus the volume that is lost by:
  - valve losses
  - piston ring leakage
  - packing leakage
  - 'dead' volume

Figure 27.1 Reciprocating compressor volume

Effective performance condition monitoring of reciprocating compressors will therefore require baseline (start-up or after maintenance) and condition monitoring of the actual volume flow rate. Note: Mass flow can be reduced to actual volume flow by multiplying the mass flow by the inlet gas density. If the actual volume flow of a reciprocating compressor decreases more than 10% (assuming constant speed), the compressor should be inspected at the first opportunity. Figure 27.2 presents these facts.

Reciprocating compressor volume (continued)

- For a given cylinder and compressor speed:
- The actual volume per stroke is constant as long as:
  - nothing changes (valves, piston rings, packing, etc)
  - speed remains constant

Figure 27.2 Reciprocating compressor volume (continued)

Methods of capacity control

Since the actual volume flow rate delivered to the process by a positive displacement compressor is constant, some means of compressor control to vary the flow rate must be provided. Figure 27.3 describes the available types of capacity control.

Reciprocating compressor volume (continued)

- Therefore, reciprocating compressor flow is *constant*
- Since process flow requirements change, the reciprocating compressor must be supplied with a *control system* that changes compressor flow
- Three options:
  - Cylinder unloading
  - Volume (clearance) pockets
  - Bypass

Figure 27.3 Reciprocating compressor volume (continued)

The most common form of reciprocating compressor capacity control is cylinder unloading. Figure 27.4 explains the function of cylinder unloaders.

Reciprocating compressor capacity control cylinder unloading

- A reciprocating compressor compresses flow by positively displacing flow out of the cylinder
- To accomplish this action, the suction valve must remain closed and *all flow* must exit the cylinder
- Cylinder unloaders either:
  - Keep suction valves open
  - Open a port in the cylinder

Figure 27.4 Reciprocating compressor capacity control cylinder unloading

Cylinder unloading provides fixed amounts of flow rates since either approximately 50% or 100% of a cylinder's flow rate is available. Refer to Figure 27.4. The most common unloaders are shown in Figure 27.5.

Cylinder unloaders

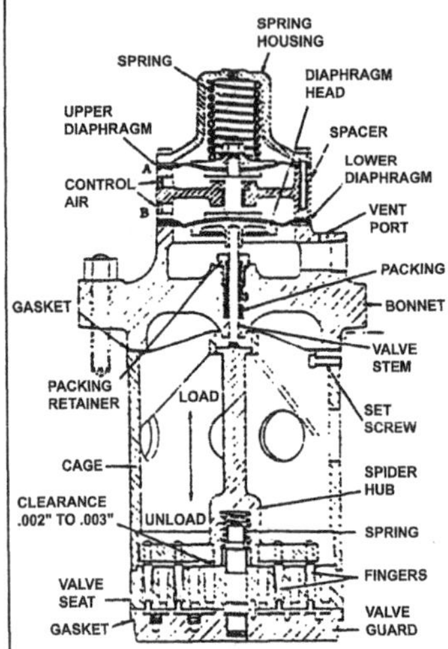

Valve unloader
Function: Suction valve is held open to prevent compression in cylinder end

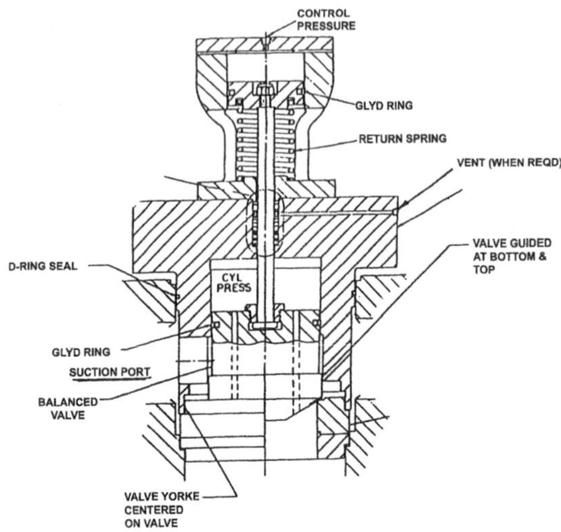

Port unloader
Function: Closes off port to valve thereby reducing capacity

Figure 27.5 Cylinder unloaders (Courtesy of Cooper Industries)

Figure 27.6 presents the most common type of unloader control system arrangements.

Step unloading

- Unloading is usually three step or five step
  - *Three step unloading* provides:
    - 0% rated compressor capacity
    - 50% rated compressor capacity
    - 100% rated compressor capacity
  - *Five step unloading* allows:
    - 0% rated compressor capacity
    - 25% rated compressor capacity
    - 50% rated compressor capacity
    - 75% rated compressor capacity
    - 100% rated compressor capacity

Figure 27.6 Step unloading

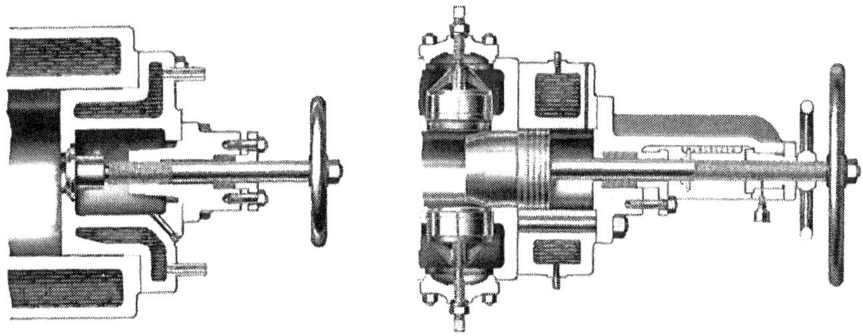

Function: increases the 'dead' (clearance) volume in the cylinder thus reducing compressor cylinder flow

Figure 27.7 Volume (clearance) pockets. Left: fixed volume. Right: variable volume (Courtesy of Dresser Rand)

It must be remembered that a cylinder volume flow is reduced by unloading the cylinder. That is, the unloaded cylinder does not move flow in the unloaded condition. If liquid becomes trapped in the cylinder by virtue of a cooling system leak, condensation or cylinder lubrication accumulation, serious damage to the cylinder can occur. Since any liquid is incompressible, if the volume of the accumulated liquid exceeds the free volume of the cylinder, the piston's action can fracture the cylinder head, bend the piston rod, destroy the cross head, connecting rod or crank shaft bearings. For this reason, most unloader systems are controlled by a timer and will load up to pump any accumulated liquid out of the cylinder. Typical cycle times are 30 minutes to one (1) hour.

Another method of varying the cylinder flow rate is to employ volume or clearance pockets. Figure 27.7 presents schematics and the function of clearance pockets.

Clearance pockets can be designed for a specific cylinder flow rate as shown on the left hand side of Figure 27.7 or can provide a variable cylinder flow rate as shown on the right hand side of this figure. A closeup view of a constant flow 'head end' volume pocket is shown in Figure 27.8.

Figure 27.9 summarizes the need for unloading 'cycle times' if valve, port or clearance pocket unloading is used in lubricated cylinders.

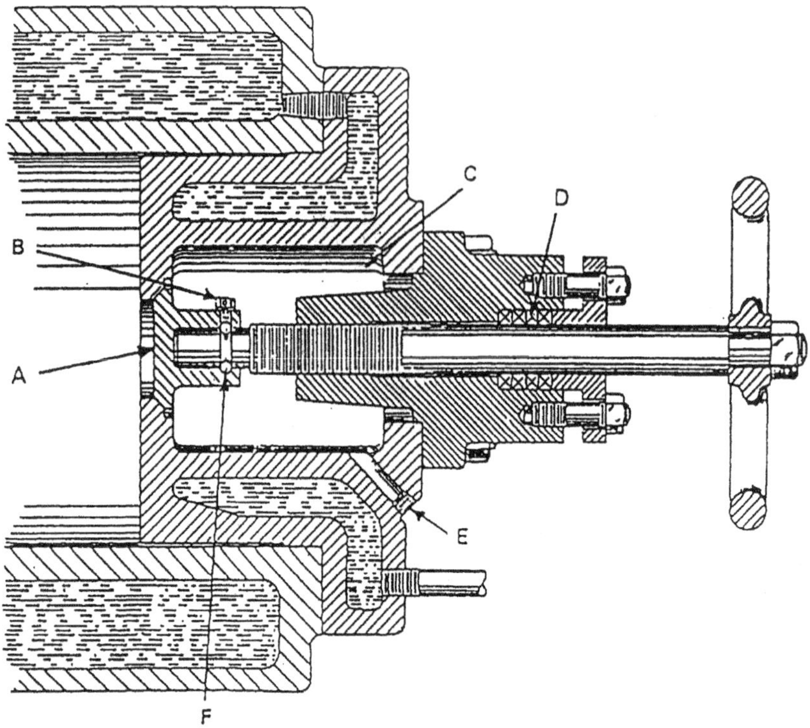

Figure 27.8 HDS off-gas-fixed volume clearance pocket (Courtesy of Dresser Rand)

Unloading the limitations

- If compressor volume is changed by valve unloading, port unloading or clearance pockets
  - Additional head is generated
  - Lubrication can accumulate in cylinders (lubricated compressors only)
- As a result, unloading time is limited

Figure 27.9 Unloading the limitations

The last option available for cylinder volume control is to use a cylinder bypass valve. This valve, which is external to the cylinder or compressor can be manually operated or automatically actuated. The bypass control function is explained in Figure 27.10.

Bypass control

- Controls delivered volume to the process by recycling ('bypassing') excess flow to the suction
- It is exactly the same control scheme as:
  - Surge control (centrifugal compressors)
  - Min flow bypass (centrifugal pumps)

Figure 27.10 Bypass control

Usually, a manually controlled bypass valve will be used only for compressor start-up duty while an automatic control valve will be used for modulating (infinite step) control during operation. Regardless of the type of bypass valve actuation, this method of control is inefficient when compared to other volume control alternatives. However, for low power applications, it is a viable option and increases overall compressor reliability since both suction unloaders and volume pockets require more maintenance.

Start-up methods

A reciprocating compressor, like any type of driven rotating equipment, requires a reduced load start-up. This requirement is based on reducing all driver and compressor component loads (piston rod, bearings, etc.) during start-up for maximum unit reliability. Figure 27.11 presents these facts.

Reciprocating compressor start-up requirements

- As for any pump or compressor, the objective is to reduce driver and component loads during start-up
- Since power and component loads are *reduced* with *reduced flow*, it is necessary to reduce flow by:
  - Unloading
  - Volume (clearance) pockets
  - Bypass

Figure 27.11 Reciprocating compressor start-up requirements

In summary, any of the control methods discussed in this chapter can be used individually or combined with one or more methods for volume control during start-up and normal operation. Most modern reciprocating control systems employ a PLC to modulate the compressor's flow. This is accomplished by using an input signal from a flow or pressure controller and sending an output signal to the appropriate volume control device. Figure 27.12 presents the 'total control scheme' options.

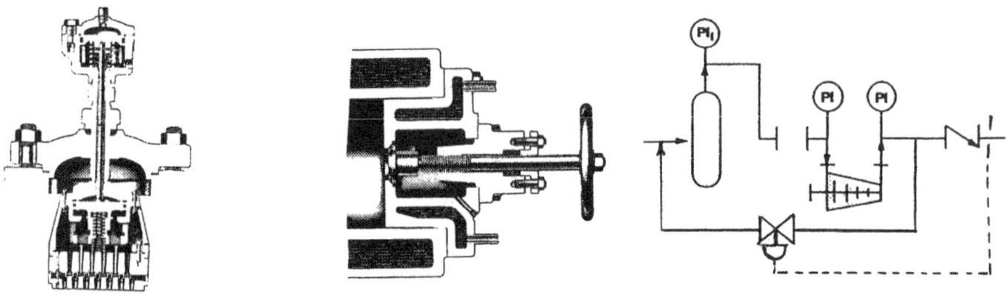

A reciprocating control scheme will use one or all of the mthods described in this section

Figure 27.12 The total control scheme. Left: valve unloading. Middle: clearance pockets. Right: bypass

28 Screw compressors

- Introduction
- Principles of operation
- Screw compressor types
- Performance relationships
- Mechanical components
- Selection guidelines
- Condition monitoring

Introduction

Screw compressors are the newest type of compressors. The dry screw compressor was developed in the late 1940's and did not experience wide use until the 1960's for low to medium flow plant air services. On the other hand, reciprocating compressors were developed 100 years before (circa 1850) and centrifugal compressors at the turn of the last century. In the 1980's the concept of the dry screw compressor was modified by continuously injecting a liquid (usually lube oil), which enabled much higher compression ratios and simplified the mechanical design by eliminating the timing gears. Figure 28.1 and 28.2 illustrate the two major screw compressor designs.

Screw compressors have many inherent advantages over their competition in the low to medium flow range. Since they are positive displacement compressors, like the reciprocating type, they will draw a constant inlet volume (assuming constant speed), can meet the varying differential pressure requirements of the process and are not significantly affected by gas density changes.

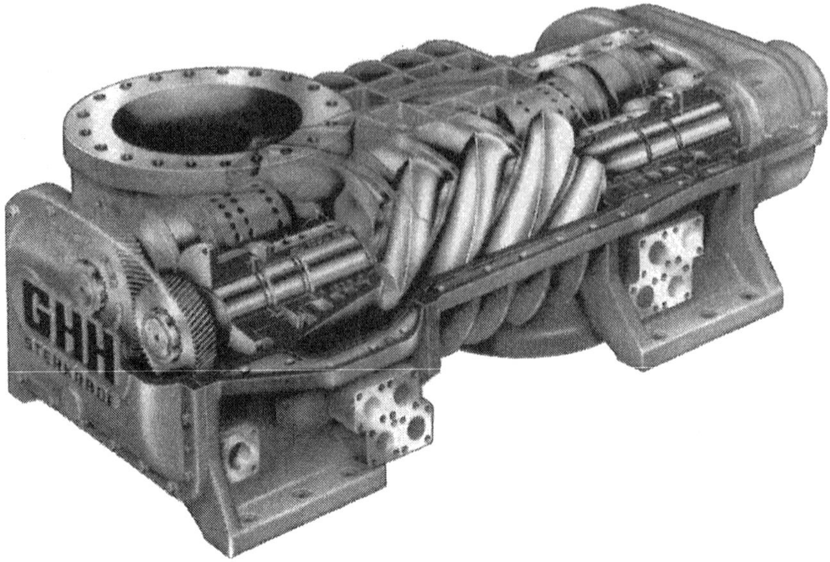

Figure 28.1 Dry twin screw compressor (Courtesy of Man/GHH)

However, unlike their reciprocating cousins, they can accomplish the above tasks by drawing a continuous, non-pulsating volume. As a result, pulsations are minimized, suction and discharge valves and troublesome unloaders are not required as well as high maintenance packing. Since there is only rotary motion, all of the conventional

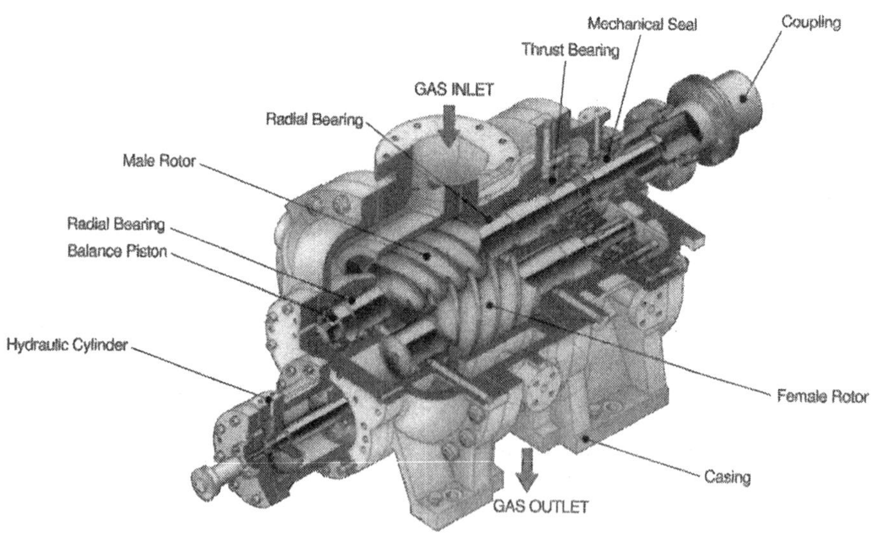

Figure 28.2 Oil injected twin screw compressor (Courtesy of Kobelco-Kobe Steel Ltd)

sealing alternatives are available (including well proven dry gas mechanical seals).

In addition, the rotary motion significantly reduces the 'footprint' of the screw compressor unit compared to other positive displacement alternatives. The result is an efficient, reliable compressor type that is very competitive from an initial cost and installed cost standpoint. The advantages of screw type compressors are presented in Figure 28.3.

Screw compressors exhibit the following advantages over other positive displacement compressor types:

- Constant inlet volume flow
- Can meet variable process differential pressure requirements
- Can handle varying density and dirty gases
- Pulsation free
- Valves, unloaders and volume pockets not required
- Mechanical seals used, packing not required
- High reliability (close to centrifugal types)
- Small 'footprint'
- Low installed cost

Figure 28.3 Screw compressor advantages

As a result of their many advantages, dry screw and flooded screw compressors, have become a dominant force in low to medium flow process applications. They have also continued to grow as the preferred type of plant and instrument air compressor in this flow range.

In the 'upstream' exploration and production industry and in gas plants, the screw compressor has become 'the type to use'. For gas gathering, the depleted fields of North America have utilized the advantages of the screw compressor to provide highly reliable, cost effective service. Many 'upstream' applications utilize gas turbines as prime movers for power generation, large pump and compressor drives. The screw is rapidly becoming the compressor of choice in gas turbine fuel gas booster applications, which require medium flows, pressure ratios and the ability to handle varying gas densities. As the characteristics and advantages of screw compressors become more widely appreciated, their use in all types of applications will increase.

The objective of this chapter is to familiarize the reader with the design, principle of operation, application and condition monitoring of the different types of screw compressors available. Figure 28.4 defines the objectives of this chapter.

Screw compressor chapter objectives:

- Define design features and component function
- Present selection and application guidelines
- Inform regarding field condition monitoring requirements

Figure 28.4 Screw compressor chapter objectives

Throughout this section, the reader must remember that all screw compressors are positive displacement compressors.

Positive displacement – dynamic pump comparison

	Positive displacement	Dynamic
Definition	Increases pressure by operating on fixed volume in a confined space	Increases pressure by using rotary blades to increase fluid velocity
Types	Screw, gear, reciprocating	Centrifugal, axial
Characteristics	■ Constant volume ■ Variable differential head ■ Relatively insensitive to liquid properties ■ Relatively insensitive to system changes ■ Not self-limiting	■ Variable volume ■ Constant differential head ■ Sensitive to liquid properties ■ Sensitive to system changes ■ Self-limiting
Characteristic flow vs. differential head curves	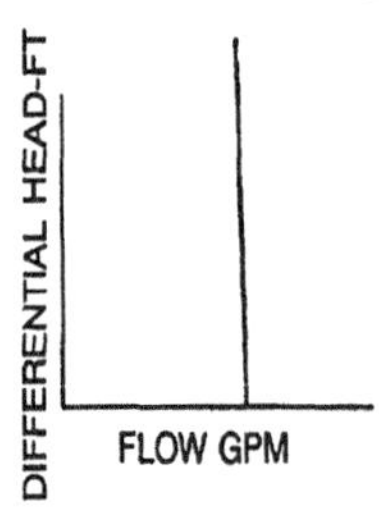	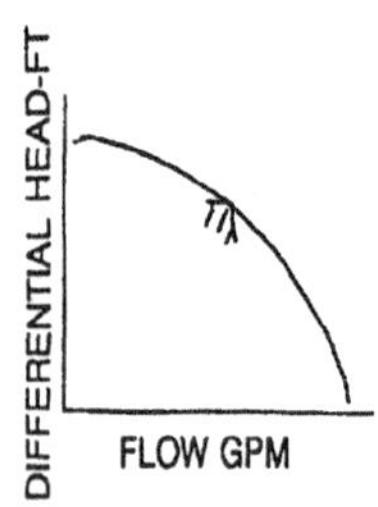

Figure 28.5 Positive displacement and dynamic characateristics

As will be discussed in this section, all types of positive displacement compressors present the designer with a challenge. This challenge is to provide varying flow requirements to the process system in a safe and reliable manner. You will discover that the screw compressor industry has met the challenge in a safe, reliable and most efficient way.

Principles of operation

Figure 28.6 shows the basic principles of operation for a twin screw compressor. Regardless of type, dry or flooded, the principle is the same.

As the screws separate at the suction end, the volume between the male (drive) and female (idle) screw is filled with gas until the outlet screw flute passes out of the suction volute, the suction volume is a function of the mating screw volume and the speed of the compressor.

Once the screw flute passes through the section volute, the compression phase begins and continues until the screw flute enters the discharge volute. The designer determines the length of the compression phase by the machine specified requirements. The compressor ratio is a function of the volume reduction ratio and the gas characteristics (specific heat ratio K=cp/cv). Details concerning performance will be discussed later in this chapter.

Figure 28.7 shows the effects of operating any screw compressor on input power at greater and lower than specified compressor ratio. This situation is always the actual operating mode in the field. Since power required is a function of volumetric efficiency (related to internal leakage through the screws) the effects of off design operation (over or under pressurizing) are minimal (usually less than 3%).

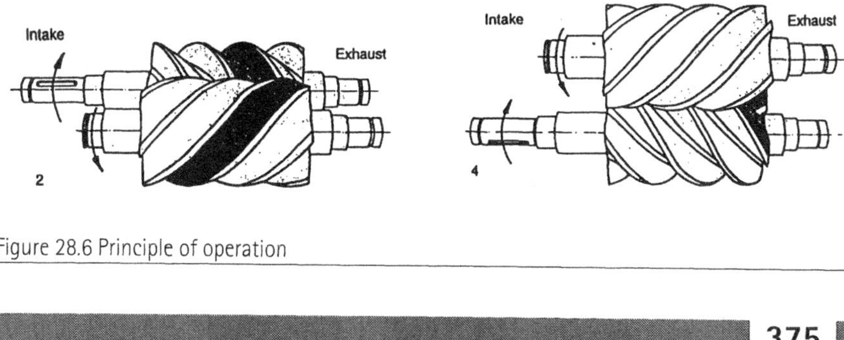

Figure 28.6 Principle of operation

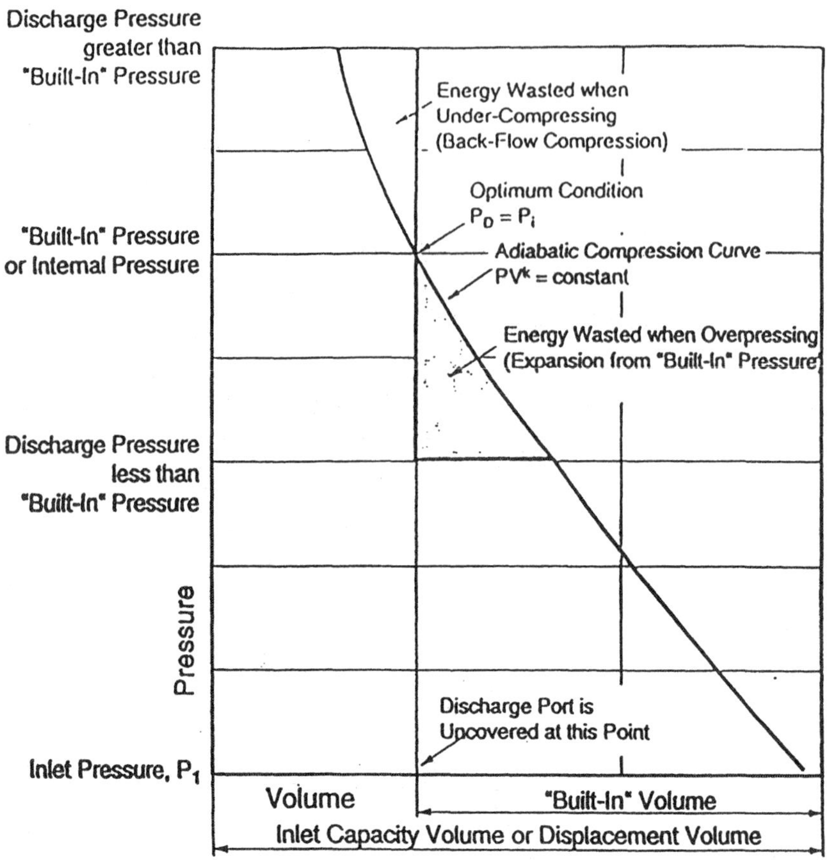

Figure 28.7 Screw pv diagram

Screw compressor types

Twin screw – oil free (non oil injected)

Figure 28.8 shows a single stage, oil free twin screw compressor with major components noted.

The oil free screw compressor was the first type developed for use as a plant and instrument air compressor. Since no lubrication is introduced in the gas stream, an external, close tolerance timing gear is used to separate the rotors. Consequently, both screws must be sealed on each end to prevent oil from entering the gas stream.

To maintain high volumetric efficiency by minimizing internal leakage (slip) losses, the backlash of the timing gears (clearance between teeth) control the screw rotor clearance to small valves. A major reliability

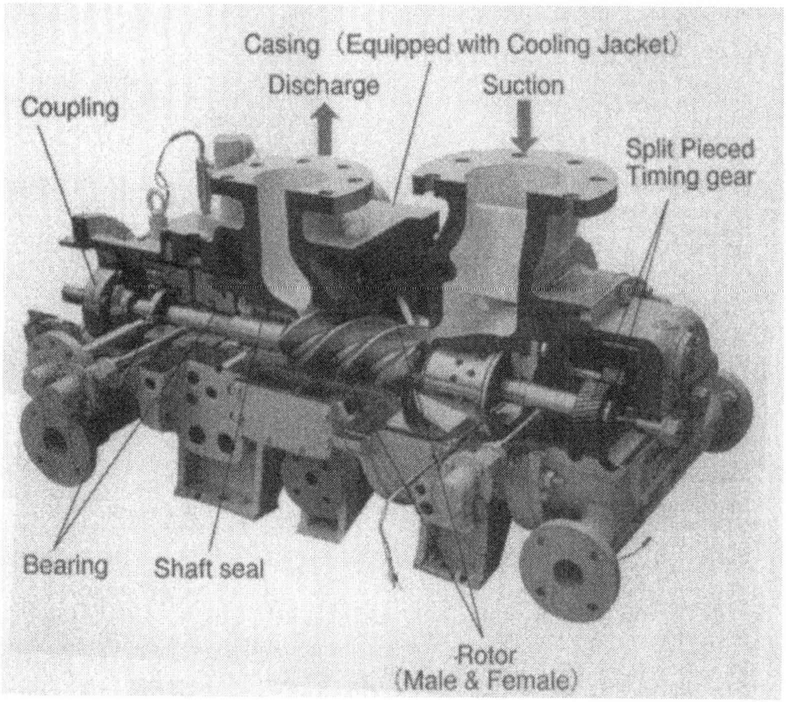

Figure 28.8 Single stage twin oil free compressor (Courtesy of Kobelco – Kobe Steel Ltd)

factor in dry screw type compressors is rotor deflection at high compression ratio's. Typically, the compression ratio for dry screw compressors is limited to around 5:1. With special screw rotor profiles, external cooling jackets and liquid injection (as great as 20% of inlet mass flow) higher compression ratio's can be attained in one casing (approx. 10:1) with discharge gas temperatures as high as 550°F. Flow range for dry screw compressors is 300–40,000 ACFM. Please refer to Figure 28.9 to view the various screw compressor rotor profiles and their applications.

Dry screw compressors were the preferred type for plant and instrument air services. However, in recent years, oil flooded screw compressors have gained popularity in this field of application. This has been due to the development of high efficiency oil separation coalessers and the lower cost for a flooded screw type compressor unit. Dry screw compressors should be considered for sour process services since sour gas can cause oil quality deterioration, which can result in frequent oil changes and possible component damage. Facts concerning dry screw compressors are noted in Figure 28.10.

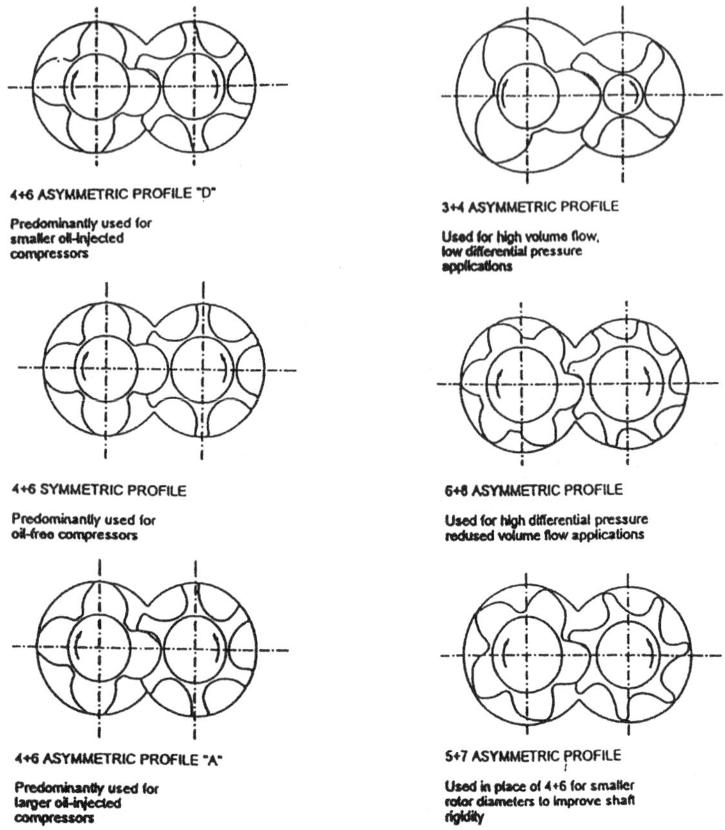

Figure 28.9 Screw compressor rotor profiles

Dry screw facts

- External timing gears required
- Normal compression ratio 5:1
- Jacket cooling and/or rotor cooling and liquid injection allows operation up to 550° F. max ratio = 10:1 Δ p max = 375 psi with special screw profile
- Multi staging may be required to prevent high discharge temperature and rotor deflection. 300–40,000 ACFM
- Approximately 1/3 speed of centrifugal compressors and 3x speed of flooded screw compressors
- Seals required at each end of screws.

Figure 28.10 Dry screw compressor facts

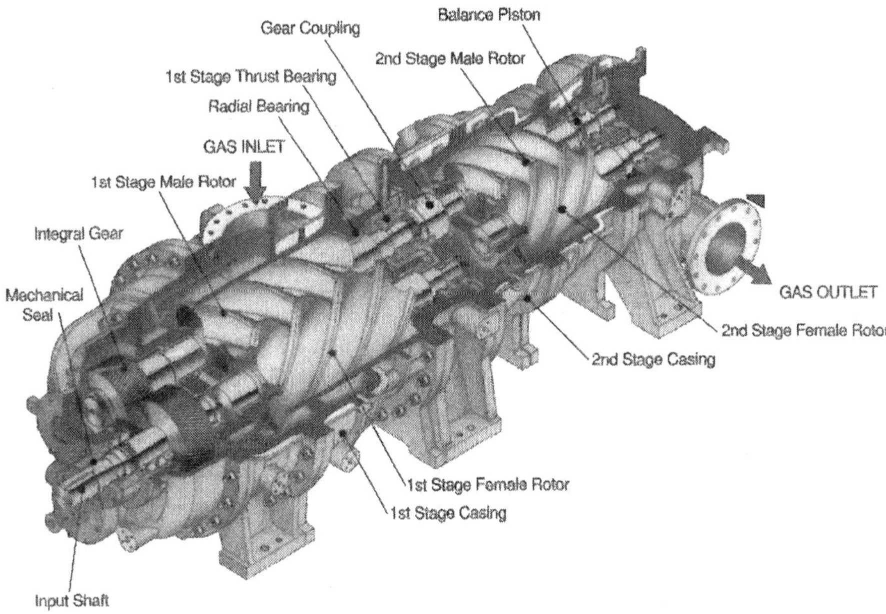

Figure 28.11 Two-stage, tandem screw oil flooded compressor. Note: integral gear used for speed increase of male rotors (Courtesy of Kobelco - Kobe Steel Ltd)

Twin screw – oil flooded

A twin screw, oil flooded compressor is shown in Figure 28.11 with major components noted.

The oil flooded twin screw compressor, developed in the 1960's has become the most widely used variety. The primary reason for its success is the ability to handle very high compression ratios without external cooling. Compression ratios as high as 25:1 are possible and Δ P's of 800 PSI can be attained with special screw profiles (6+8). Flow range for this type is 100 – 8,500 ACFM. Oil injection enables operation at high compression ratios but limits discharge temperature to 210° F. Use of synthetic oils allows discharges temperature to reach 250° F. The injection of oil also eliminates the necessity of timing gears and reduces the number of shaft end seals required to one.

Another advantage of oil flooded screw compressors is the efficiency of its capacity control system. The use of oil internal to the compressor allows the use of a slide valve in the compressor section of the compressor. By varying the stroke of the valve, the volume ratio can be varied. The result is that flow can be adjusted between 10–100%.

Finally, flooded screw compressors have proven that they can tolerate dirty, difficult to handle gases. The continuous oil injection serves as an 'anti foulant liquid' and prevents fouling build up.

Therefore, it can be seen that the screw compressor has many advantages over the other types and can usually be purchased for a lower cost. The facts concerning flooded screw compressors are detailed in Figure 28.12.

Flooded screw facts

- External timing gears not required
- Compression ratios as high as 25:1
- Discharge temperature limited to 210°F (mineral oil) 250°F (synthetic oil)
- 100-8,500 ACFM
- 1/3 speed of dry screw compressors
- Only one shaft end seal required
- Rotor cooling not required
- Compatability of oil with gas must be confirmed

Figure 28.12 Flooded screw facts

However, the compatibility of the injected oil with the process gas must be examined before purchasing a flooded type. There are two primary concerns:

- Process system deterioration
- Oil degradation

Prior to purchase of a flooded screw compressor, the entire downstream process system must be examined to confirm small quantities of oil will not affect coolers, reactors etc. The efficiency of the oil separation systems used is high 99.9%. However, upsets can cause oil to be transferred downstream. A typical oil separation system used for a flooded screw compressor is shown in Figure 28.13.

The quoting screw compressor vendors should be consulted regarding modification to the standard system that can be made to meet requirements. Available options are:

- Separate lube and seal oil injection systems
- Increased separator vessel retention time
- Self cleaning separator vessels upstream of the compressor

However, this activity must take place early in the project (phase 1) since the cost of the unit will increase.

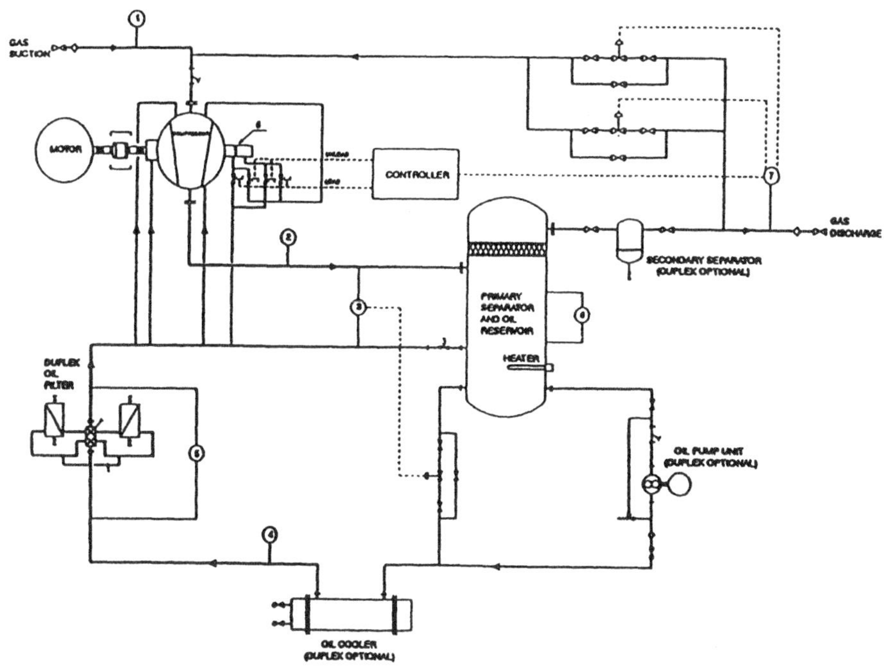

1. Suction instrumentation
2. Discharge instrumentation
3. Oil/gas differential pressure instrumentation
4. Oil temperature/pressure instrumentation
5. Oil filter differential pressure instrumentation
6. Oil level instrumentation
7. Skid discharge instrumentation
8. Capacity control slide valve

Figure 28.13 Oil separation system (Courtesy of Man/GHH)

The other concern with flooded screw compressors is oil degradation caused by interaction of the process gas with the injected oil. Please refer again to Figure 28.13. The retention time of the oil in the separator is typically 1–1 ½ minutes. Therefore, any contaminant in the process gas may not have sufficient time for removal. There have been some bad experiences, particularly in gas recovery applications where a sour, dirty gas (asphaltines) has required frequent oil and oil filter change out. Our recommendation is to discuss these facts in detail with the vendor prior to the order (Phase 1 of project) and require references of where the equipment is operating. It has been our experience that this action, unfortunately, is taken after installation of the compressor. These facts are presented in Figure 28.14.

Flooded screw reliability factors

In phase 1 of project

- Confirm compatability of oil with process
- Confirm oil will not be degraded by process gas
- Visit installations to confirm vendor claims

Figure 28.14 Flooded screw reliability

Performance relationships

All screw compressors are positive displacement compressors. Performance of screw compressors is calculated using the adiabiatic process. An adiabiatic compression process assumes that no heat is lost in the compression process. The path of compression is defined by:

$$\frac{T_2}{T_1} = \left(\frac{P_2}{P_1}\right)^{\left(\frac{k-1}{k}\right)}$$

Figure 28.15 defines the performance relationships used for screw compressors.

Screw compressor performance relationships

- CFM = (RPM) (DM^3) (L/DM) (C_0) where:

  DM = male rotor diameter – inches

  L/DM = length to diameter ratio of male rotor

  C_0 = rotor profile constant
  4 + 6 profile = 15.853
  3 + 4 profile = 18.082
  4 + 6 'A' profile = 17.17
  4 + 6 'D' profile = 16.325
  3 + 4 'A' profile = 19.372
  6 + 8 'D' profile = 10.364

Figure 28.15 Performance relationships

- $P_2 = P_1 \dfrac{(T_2)}{(T_1)}^{(K/K-1)}$
- $V_2/V_1 = \dfrac{(P_2)}{(P_1)}^{(1/K)}$

where: P = PSIA
T = °R
°R = °F + 460°
V = volume – cubic FT/min

- $\text{Adiabatic H.P.} = \dfrac{P_1 Q_1 (P2/P1)}{229 \frac{(K-1)}{K}}$
- $\text{BHP} = \dfrac{\text{Adiabatic H.P.}}{\text{Adiabatic efficiency (TOTAL)}}$

Where: adiabatic efficiency total =
(adiabatic efficiency) (mechanical efficiency)

- $T_2 = T_1 \left(\dfrac{P_2}{P_1}\right)^{\left(\frac{k-1}{k}\right)}$
- $\text{Adiabatic efficiency} = \dfrac{\Delta T \text{ Adiabatic}}{\Delta T \text{ actual}}$

Figure 28.15 Continued – performance relationships

The relationships in Figure 28.15 can be used to either estimate screw compressor performance for selection purposes or to determine field performance. When determining screw compressor field performance, a trend of delivered flow is most useful since all screw compressors are positive displacement compressors and will deliver a constant volume flow. Reduction of flow is an indication of an increase of slip (internal leakage). Care must be used in trending flow however. Rotor speed, suction throttle valve, slide valve and bypass valve (if supplied) position must be constant.

Trending adiabatic efficiency is another useful indicator of performances. Again, care must be taken to assure oil injection (flooded screws), jacket water temperatures and rotor cooling flows (dry screws), in addition to valve positions are constant. These facts are presented in Figure 28.16.

Screw compressor field performance check guidelines

- Trend volume flow

  Note: assure speed and control devices are at constant values.

- Trend adiabiatic efficiency

  Note: assure oil injection rates, jacket cooling temperature and rotor cooling flows and valve positions are at constant values.

Figure 28.16 Field performance checks

Mechanical components

It is useful to always remember that any type of rotating machine contains five major components. It is important to know their function and monitor their condition. Screw compressors are no exceptions. The five major component systems are;

- Rotor
- Journal bearing
- Thrust bearing
- Seals
- Auxiliary systems

Please refer to Figure 28.17 which shows a typical dry screw compressor and Figure 28.18 which shows a typical flooded screw compressor. Both views are top (plan) views. These figures will be used to describe the function of the major mechanical components.

Rotor

As previously shown, there are various screw profiles available. Their use depends on the process condition and type of screw compressor (dry or flooded). The most common profile is the asymmetric 4 + 6 (4 lobe male rotor and 6 lobe female rotor). The major mechanical concern is rotor deflection. Therefore, the L/D (rotor supported length divided by diameter) and number of lobes will vary directly with capacity and pressure ratios. High flow rates may use only 3 male lobes and high pressure ratio applications may employ 6 male lobes.

We recommend that the contractor (or end user) require vendor rotor experience references during the bidding phase. These references

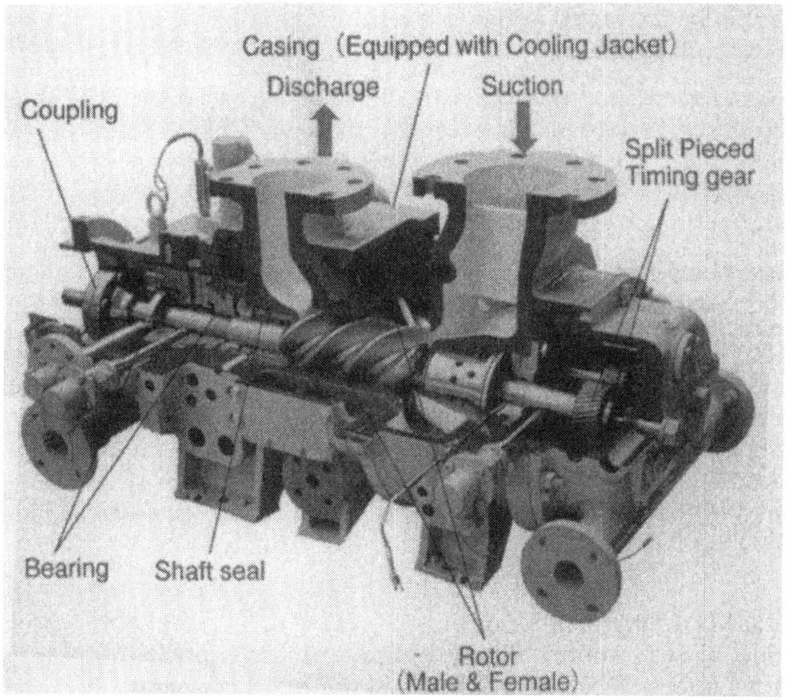

Figure 28.17 Dry screw compressor (Courtesy of Kobelco – Kobe Steel Ltd)

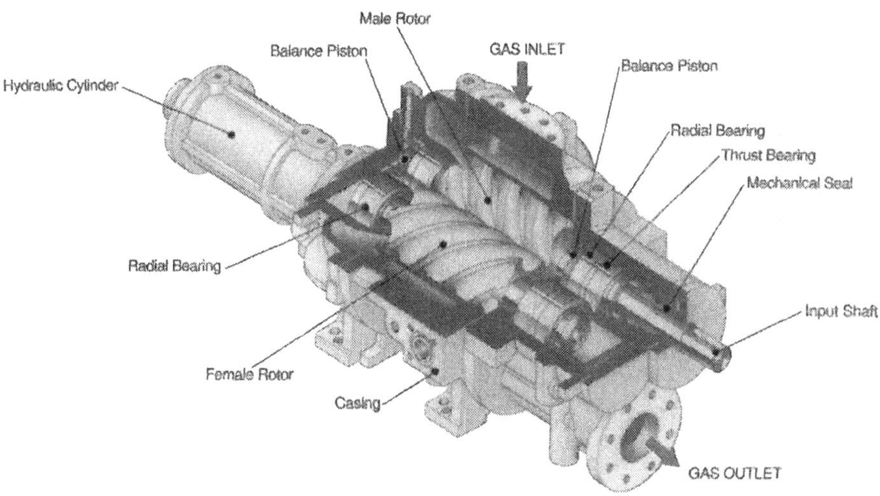

Figure 28.18 Flooded screw compressor (Courtesy of Kobelco – Kobe Steel Ltd)

should be contacted to confirm reliable field operation for the proposed rotor configuration.

Rotor speeds are low relative to centrifugal compressors. Consequently, screw compressor rotors normally are rigid (they operate below their first critical speed). Dry screw compressor tip speeds are 350 ft/sec and flooded screw speeds are 100–175 ft/sec.

Journal bearings

The usual Journal bearing selected are roller type anti-friction bearings. Larger compressors (above 300 BHP) use sleeve or tilt pad bearings.

It is important to confirm proper bearing size and selection during the bidding process. Anti-friction bearings should be checked to confirm that DN number (diameter of bearing bore in mm multiplied by shaft speed) is within acceptable limits and the L-10 life is a minimum of 25,000 hours. Sleeve bearing loads (based on projected area) should be less than 250 PSI.

Thrust bearings

Small screw compressors use angular contact anti-friction bearings. As stated above, DN and L-10 life should be confirmed to be acceptable during the bid phase.

Larger screw compressors above 300 BHP may use plan, tapered land or even tilt pad thrust bearings for larger size (above 1000 BHP). Regardless of type, bearing loads should be less than 250 PSI. Larger screw compressors use balancing devices to control the thrust load (balance pistons).

Timing gears

As previously mentioned, dry screw compressors require external timing gears to assure proper rotor clearances. An increase in rotor clearance of 0.0004" can reduce efficiency 1%. The timing gears are precision ground gears that require continuous lubrication. Small compressors (less than 300 BHP) can use self contained ring oil lubrication. Larger compressors will require a pressurized lubrication system.

Flooded screw compressors do not require external timing gears but do require significant amounts of oil that is injected into the screws (usually in the bottom of the casing). As an example, a typical oil injection rate is 7 GPM per 100 ACFM flow rate for air compressors.

Sealing devices

Dry screw compressors require that sealing devices be installed at each end of each rotor. This is necessary to prevent timing gear oil from entering the process stream and process gas entering the lube oil system.

Flooded screw compressors only require a sealing device on the drive screw coupling end.

Regardless of type of screw compressor, there are many sealing alternatives available. Some of the common types are:

- Labyrinth
- Restrictive carbon ring type
- Mechanical (liquid) seals
- Dry gas seals

Auxiliary systems

Small screw compressors can be supplied with self-contained lubrication (ring oil) and sealing systems. Depending on operating conditions, a jacket cooling system (dry screws) may be required.

Larger screw compressors will employ a pressurized lubrication system and perhaps a liquid or gas seal buffer system.

All flooded screw compressors will require an oil separation system which is usually combined with the lubrication system. A typical P and ID for such a system is shown in Figure 28.19.

Experience has shown that bearing and seal component reliability is a direct function of auxiliary system component selection and design. The auxiliary system(s) are the only major source of potential cost reduction for screw compressor vendors. Since screw compressors are relatively new, the specifications have not reached the sophistication of

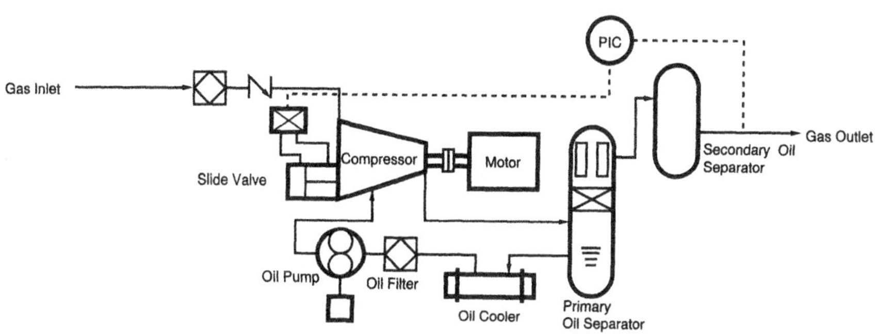

Figure 28.19 Typical oil flooded screw oil separation system (Courtesy of Kobelco – Kobe Steel Ltd)

reciprocating and centrifugal compressors. We recommend that all auxiliary systems be thoroughly reviewed in the bidding phase, with referneces required and during the co-ordination meeting for large compressors. Special care should be given in the pre-order phase to gas/oil separator retention time and vendor experience with this item in similar applications.

Capacity control

Since screw compressors are positive displacement compressors, inlet volume flow is constant. If the speed of the rotor remains constant, like reciprocating compressors, various methods are used to allow capacity to be varied. The four control methods used are presented in Figure 28.20.

Screw compressor capacity control methods

- Variable speed
- Suction throttling
- Slide valve
- Bypass

Figure 28.20 Capacity control methods

Variable speed

As shown in the previous section, the capacity of the compressor is a function of rotor profile and rotor speed. Variable speed is the preferred method of capacity control for oil free compressors. It is not the preferred method for oil flooded compressors since injection oil would have to be varied with speed and the slide valve method is available.

Suction throttling

Suction throttling enables the mass flow of the screw compressor to be controlled by changing the inlet gas density. Care must be used to assure that the maximum compressor ratio is not exceeded which could cause rotor deflection and/or excessive discharge gas temperature.

Slide valve

Oil injected screw compressors are usually fitted with a slide valve that allows the suction volume to be varied between 10–100%. A slide valve inside the screw case housing is shown in Figure 28.21.

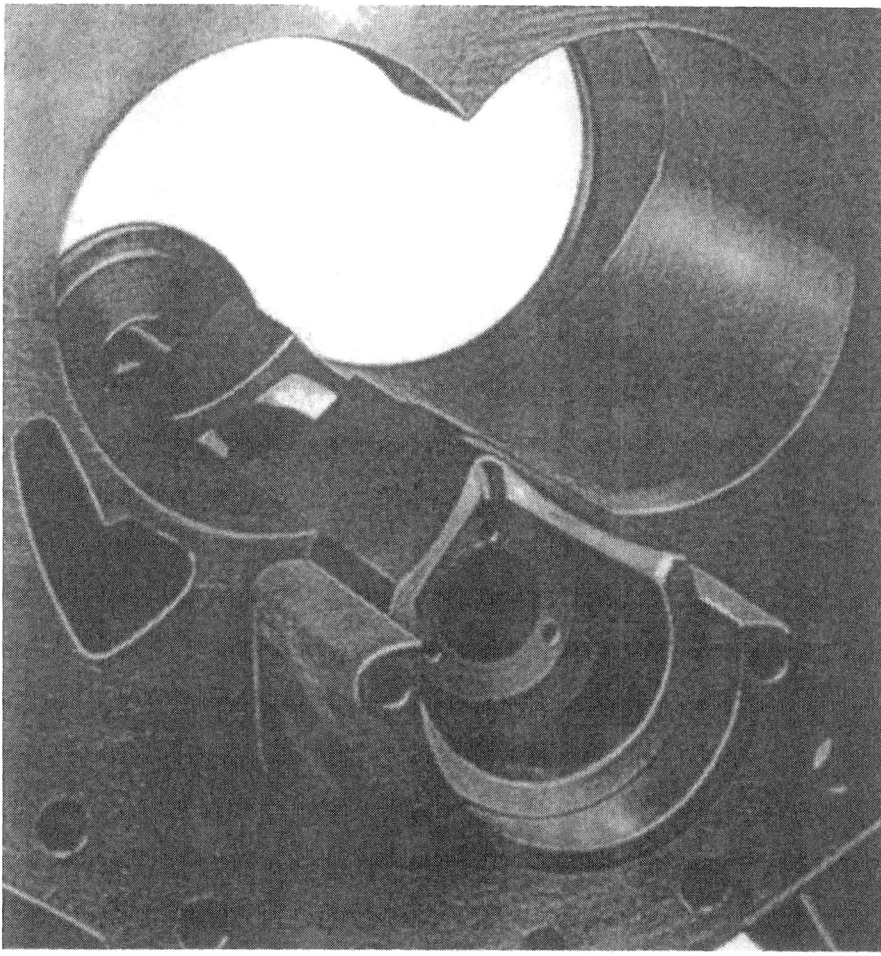

Figure 28.21 Slide valve control

This stepless method of capacity control maintains high efficiency and is the main advantage of oil injected screw compressors. The slide valve method has recently been employed in oil free compressors for a limited number of applications.

The operation of a slide valve is shown in Figure 28.22. The slide valve is an axial moveable segment of the casing cylinder wall. As it is moved from the suction end towards the discharge end, flow is bypassed to the suction. Oil is injected through ports in the valve cooling the recycled flow. This method of control is the most efficient after the variable speed method.

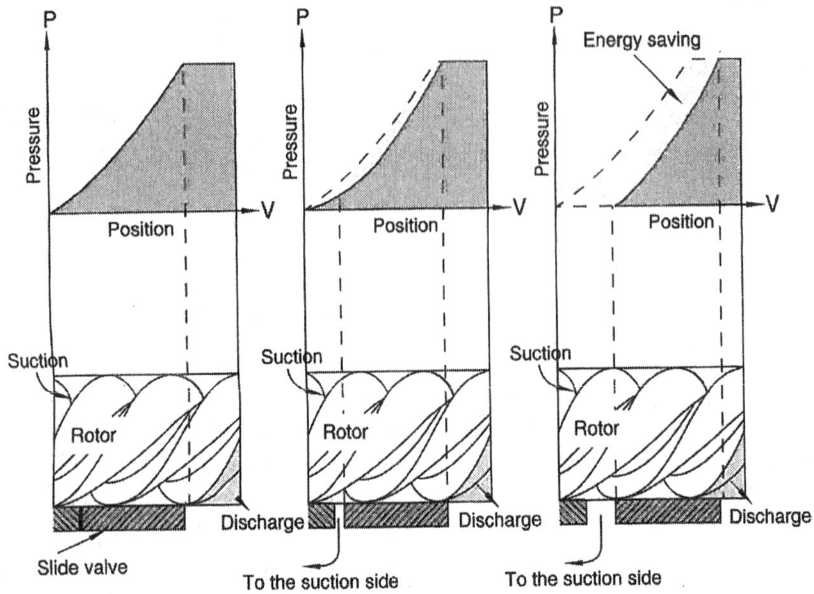

Figure 28.22 Slide valve operation (Courtesy of Kobelco – Kobe Steel Ltd)

Bypass control

This method utilizes an external control valve to bypass excess gas back to the suction. The recycled gas must be cooled. The control of the bypass valve can be either pressure or flow. Bypass control is the most inefficient method of capacity control.

Selection guidelines

As previously mentioned, screw compressors have a flow range of 100–40,000 ACFM and can produce compression ratios as high as 25:1. In order to optimize compressor efficiency, accurate process data must be provided to the quoting vendors. The required input data is the same as for other types of compressors and is shown in Figure 28.23.

Required process data

- P_1, P_2
- T_1
- Gas analysis (for each case)
- Gas contaminents
- Flow rate
- Off design flow requirements

Figure 28.23 Required process data

It is most important to accurately define gas contaminants (asphaltines, etc.) and the percent per unit volume. Sour gas (H_2 S) must be accurately defined and will determine material types and if a flooded screw can be used.

Dry or Wet screw?

As discussed in this chapter, the technical and cost advantages of flooded screw compressors, has made them the compressor of choice. However, special modifications may be required for use of flooded screws in sour gas service. The low retention time of oil in the separation vessel (60–90 seconds) do not allow sour gas components to be vented from the reservoir oil. This can lead to oil contamination, screw component damage and frequent oil changes. Reservoir capacity can be increased, but at an extra cost. The use of dry screw compressors should be considered in sour gas service and evaluated against properly designed oil flooded compressors. Vendor experience lists should be required and checked.

Condition monitoring

Condition monitoring requirements for screw compressors should follow the principle of component condition monitoring. The following major components should be monitored:

- Rotor
- Journal bearing
- Thrust bearing
- Seals

The suggested minimal condition monitoring requirements are presented in Figure 28.24.

Srew compressor minimum condition monitoring requirements

- Performance (rotor)
  - Flow trend
  - Efficiency check
- Manual vibration readings (accelerometers/velocity)
- Thrust bearing temperature measurement (rtd's)

Figure 28.24 Minimum condition monitoring required

29 Conversion to metric system

Unit Nomenclature

Symbol	Name	Quantity
a	annum (year)	time
bar	bar	pressure
°C	degree Celsius	temperature
°	degree	plane angle
d	day	time
g	gram	mass
h	hour	time
Hz	hertz	frequency
J	joule	work, energy
K	kelvin	temperature
kg	kilogram	mass
litre	litre	volume
m	metre	length
min.	minute	time
N	newton	force
Pa	pascal	pressure
rad	radian	plane angle
s	second	time
t	tonne	mass
V	volt	electric potential
W	watt	power

Conversion tables – Customary to metric (SI) units

Quantity and SI unit	Customary unit	Metric unit	Conversion factor Multiply customary unit by factor to get metric unit	
Space, time				
Length – m	mi	km	1.609 344	E+00
	m	m	1	
	yd	m	9.144	E–01
	ft	m	3.048	E–01
	in	mm	2.54	E+01
	cm	mm	1.0	E+01
	mm	mm	1	
	mil	µm	2.54	E+01
	micron (µ)	µm	1	
Area – m^2	mi^2	km^2	2.589 988	E+00
	yd^2	m^2	8.361 274	E–01
	ft^2	m^2	9.290 304	E–02
	in^2	mm^2	6.451 6	E+02
	cm^2	mm^2	1.0	E+02
	mm^2	mm^2	1	
Volume, capacity – m^3	m^3	m^3	1	
	yd^3	m^3	7.645 549	E–01
	bbl (42 US gal)	m^3	1.589 873	E–01
	ft^3	m^3	2.831 685	E–02
	UK gal	m^3	4.546 092	E–03
	US gal	m^3	3.785 412	E–03
	litre	dm^3	1	
	UK qt	dm^3	1.136 523	E+00
	US qt	dm^3	9.463 529	E–01
	UK pt	dm^3	5.682 609	E–01
	US pt	dm^3	4.731 765	E–01
	US fl oz	cm^3	2.957 353	E+01
	UK fl oz	cm^3	2.841 305	E+01
	in^3	cm^3	1.638 706	E+01
	ml	cm^3	1	

Quantity and SI unit	Customary unit	Metric unit	Conversion factor Multiply customary unit by factor to get metric unit	
Plane angle – rad	rad	rad	1	
	deg (°)	rad	1.745 329	E–02
	min (′)	rad	2.908 882	E–04
	sec (″)	rad	4.848 137	E–06

Conversion to the metric (SI) system

Quantity and SI unit	Customary unit	Metric unit	Conversion factor Multiply customary unit by factor to get metric unit	
Mass, amount of substance				
Mass – kg	UK ton	t	1.016 047	E+00
	US ton	t	9.071 847	E–01
	UK cwt	kg	5.080 234	E+01
	US cwt	kg	4.535 924	E+01
	kg	kg	1	
	lb	kg	4.535 924	E–01
	oz (troy)	g	3.110 348	E+01
	oz (av)	g	2.834 952	E+01
	g	g	1	
	grain	mg	6.479 891	E+01
	mg	mg	1	
	µg	µg	1	
Amount of substance – mol	lb mol	kmol	4.535 924	E–01
	g mol	kmol	1.0	E–03
	std m^3 (0°C, 1atm)	kmol	4.461 58	E–02
	std m^3 (15°C, 1atm)	kmol	4.229 32	E–02
	std ft^3 (60°C, 1 atm)	kmol	1.195 30	E–03

Quantity and SI unit	Customary unit	Metric unit	Conversion factor Multiply customary unit by factor to get metric unit	
Calorific value, heat, entropy, heat capacity				
Calorific value – J/kg (Mass basis)	Btu/lb	MJ/kg	2.326 000	E–03
		kJ/kg	2.326 000	E+00
	cal/g	kJ/kg	4.184	E+00
	cal/lb	J/kg	9.224 141	E+00
Calorific Value –J/mol (Mole basis)	kcal/g mol	kJ/kmol	4.184	E+03
	Btu/lb mol	MJ/kmol	2.326 000	E–03
		kJ/kmol	2.326 000	E+00
Calorific value – J/m³ (Volume basis – solids and liquids)	therm/UK gal	MJ/m³	2.320 800	E+04
		kJ/m³	2.320 800	E+07
	Btu/US gal	MJ/m³	2.787 163	E–01
		kJ/m³	2.787 163	E+02
	Btu/UK gal	MJ/m³	2.320 800	E–01
		kJ/m³	2.320 800	E+02
	Btu/ft³	MJ/m³	3.725 895	E–02
		kJ/m³	3.725 895	E+01
	kcal/m³	MJ/m³	4.184	E–03
		kJ/m³	4.184	E+00
	cal/ml	MJ/m³	4.184	E+00
	ft•lb$_f$/US gal	kJ/m³	3.581 692	E–01
Calorific value – J/m³ (Volume basis – gases)	cal/ml	kJ/m³	4.184	E+03
	kcal/m³	kJ/m³	4.184	E+00
	Btu/ft³	kJ/m³	3.725 895	E+01
Specific entropy – J/kg•K	Btu/lb•°R	kJ/kg•K	4.186 8	E+00
	cal/g•°K	kJ/kg•K	4.184	E+00
	kcal/kg•°C	kJ/kg•K	4.184	E+00
Specific heat capacity – J/kg•K (Mass basis)	kW·h/kg•°C	kJ/kg•°C	3.6	E+03
	Btu/lb•°F	kJ/kg•°C	4.186 8	E+00
	kcal/kg•°C	kJ/kg•°C	4.184	E+00
Specific heat capacity – J/mol•K (Mole basis)	Btu/lb mol•°F	kJ/kmol•°C	4.186 8	E+00
	cal/g mol•°C	kJ/kmol•°C	4.184	E+00

Quantity and SI unit	Customary unit	Metric unit	Conversion factor Multiply customary unit by factor to get metric unit	
Temperature, pressure, vacuum				
Temperature – K	°R	K	5/9	
(Absolute)	°K	K	1	
Temperature – K	°F	°C	5/9 (°F – 32)	
(Traditional)	°C	°C	1	
Temperature – K	°F	°C	5/9	
(Difference)	°C	°C	1	
Temperature/Length – K/m (Geothermal gradient)	°F per 100 ft	mK/m	1.822 689	E+01
Pressure – Pa	atm	MPa	1.013 250	E–01
		kPa	1.013 250	E+02
	bar	MPa	1.0	E–01
		kPa	1.0	E+02
	at (kg_f/cm^2)	MPa	9.806 650	E–02
		kPa	9.806 650	E+01
	lb_f/in^2 (psi)	MPa	6.894 757	E–03
		kPa	6.894 757	E+00
	in Hg (60°F)	kPa	3.376 85	E+00
	in H_20 (39.2°F)	kPa	2.490 82	E–01
	in H_20 (60°F)	kPa	2.488 4	E–01
	mm Hg = torr (0°C)	kPa	1.333 224	E–01
	cm H_20 (4°C)	kPa	9.806 38	E–02
	lb_f/ft^2 (psf)	kPa	4.788 026	E–02
	μm Hg (0°C)	Pa	1.333 224	E–01
	μbar	Pa	1.0	E–01
	dyn/cm²	Pa	1.0	E–01
Vacuum, draft – Pa	in Hg (60°F)	kPa	3.376 85	E+00
	in H_20 (39.2°F)	kPa	2.490 82	E–01
	in H_20 (60°F)	kPa	2.488 4	E–01
	mm Hg = torr (0°C)	kPa	1.33 224	E–01
	cm H_20 (4°C)	kPa	9.806 38	E–02
Liquid head – m	ft	m	3.048	E–01
	in	mm	2.54	E+01
Pressure drop/length – Pa/m	psi/ft	kPa/m	2.262 059	E+01
	psi/100 ft	kPa/m	2.262 059	E–01

Quantity and SI unit	Customary unit	Metric unit	Conversion factor Multiply customary unit by factor to get metric unit	
Density, specific volume, concentration, dosage				
Density (gases) – kg/m^3	lb/ft^3	kg/m^3	1.601 846	E+01
		g/m^3	1.601 846	E+04
Density (liquids) – kg/m^3	lb/US gal	kg/dm^3	1.198 264	E–01
	lb/UK gal	kg/dm^3	9.977 644	E–02
	lb/ft^3	kg/dm^3	1.601 846	E–02
	g/cm^3	kg/dm^3	1	
Density (solids) – kg/m^3	lb/ft^3	kg/dm^3	1.601 846	E–02
Specific volume – m^3/kg (gases)	ft^3/lb	m^3/kg	6.242 796	E–02
		m^3/g	6.242 796	E–05
Specific volume – m^3/kg (liquids)	ft^3/lb	dm^3/kg	6.242 796	E+01
	UK gal/lb	dm^3/kg	1.022 241	E+01
	US gal/lb	dm^3/kg	8.345 404	E+00
Specific volume – m^3/mol (mole basis)	litre/g mol	$m^3/kmol$	1	
	ft^3/lb mol	$m^3/kmol$	6.242 796	E–02
Concentration – kg/kg (mass/mass)	wt %	kg/kg	1.0	E–02
		g/kg	1.0	E–05
	wt ppm	mg/kg	1	
Concentration – kg/m^3 (mass/volume)	lb/bbl	kg/m^3	2.853 010	E+00
	g/US gal	kg/m^3	2.641 720	E–01
	g/UK gal	kg/m^3	2.199 692	E–01

Quantity and SI unit	Customary unit	Metric unit	Conversion factor Multiply customary unit by factor to get metric unit	
Facility throughput, capacity				
Throughput – kg/s (mass basis)	million lb/yr	t/a	4.535 924	E+02
	UK ton/yr	t/a	1.016 047	E+00
	US ton/yr	t/a	9.071 847	E–01
	UK ton/d	kg/h	1.016 047	E+00
	US ton/d	t/d	9.071 847	E–01
	UK ton/h	t/h	1.016 047	E+00
	US ton/h	t/h	9.071 847	E–01
	lb/h	t/d	4.535 924	E–01
Throughput – m^3/s (volume basis)	bbl/d	t/a	5.803 036	E+01
		m^3/h	6.624 471	E–03
	ft^3/d	m^3/h	1.179 869	E–03
	bbl/h	m^3/h	1.589 873	E–01
	ft^3/h	m^3/h	2.831 685	E–02
	UK gal/h	m^3/h	4.546 092	E–03
	US gal/h	m^3/h	3.785 412	E–03
	UK gal/min	m^3/h	2.727 655	E–01
	US gal/min	m^3/h	2.271 247	E–01
Throughput – mol/s (mole basis)	lb mol/h	kmol/h	4.535 924	E–01

Quantity and SI unit	Customary unit	Metric unit	Conversion factor Multiply customary unit by factor to get metric unit	
Flow rate				
Flow rate – kg/s (mass basis)	UK ton/min	kg/s	1.693 412	E+01
	US ton/min	kg/s	1.511 974	E+01
	UK ton/h	kg/s	2.822 353	E–01
	US ton/h	kg/s	2.519 958	E–01
	UK ton/d	kg/s	1.175 980	E–02
	US ton/d	kg/s	1.049 982	E–02
	million lb/yr	kg/s	5.249 912	E+00
	UK ton/yr	kg/s	3.221 864	E–05
	US ton/yr	kg/s	2.876 664	E–05
	lb/s	kg/s	4.535 924	E–01
	lb/min	kg/s	7.559 873	E–03
	lb/h	kg/s	1.259 979	E–04
Flow rate – m^3/s (volume basis)	bbl/d	dm^3/s	1.840 131	E–03
	ft^3/d	dm^3/s	3.277 413	E–04
	bbl/h	dm^3/s	4.416 314	E–02
	ft^3/h	dm^3/s	7.865 791	E–03
	UK gal/h	dm^3/s	1.262 803	E–03
	US gal/h	dm^3/s	1.051 503	E–03
	UK gal/min	dm^3/s	7.576 820	E–02
	US gal/min	dm^3/s	6.309 020	E–02
	ft^3/min	dm^3/s	4.719 474	E–01
	ft^3/s	dm^3/s	2.831 685	E+01
Flow rate – mol/s (mole basis)	lb mol/s	kmol/s	4.535 924	E–01
	lb mol/h	kmol/s	1.259 979	E–04

Quantity and SI unit	Customary unit	Metric unit	Conversion factor Multiply customary unit by factor to get metric unit	
Energy, work, power				
Energy, work – J	therm	MJ	1.055 056	E+02
		kJ	1.055 056	E+05
		kJ	3.6	E+03
	Btu	kJ	1.055 056	E+00
	kcal	kJ	4.184	E+00
	cal	kJ	4.184	E–03
	ft•lb_f	kJ	1.355 818	E–03
	lb•ft	kJ	1.355 818	E–03
	J	kJ	1.0	E–03
	lb•ft^2/s^2	kJ	4.214 011	E–05
	erg	J	1.0	E–07
Power – W	million Btu/h	MW	2.930 711	E–01
	ton of refrigeration	kW	3.516 853	E+00
	Btu/s	kW	1.055 056	E+00
	kW	kW	1	
	hydraulic horsepower–hhp	kW	7.460 43	E–01
	hp (electric)	kW	7.46	E–01
	Btu/min	kW	1.758 427	E–02
	ft•lb_f/s	kW	1.355 818	E–03
	kcal/h	W	1.162 222	E+00
	Btu/h	W	2.930 711	E–01
	ft•lb_f/min	W	2.259 697	E–02
Power/area – W/m^2	Btu/s•ft^2	kW/m^2	1.135 653	E+01
	Cal/h•cm^2	kW/m^2	1.162 222	E–02
	Btu/h•ft^2	kW/m^2	3.154 591	E–03
Cooling duty – W/W (machinery)	Btu/bhp•h	W/kW	3.930 148	E–01
Specific fuel – kg/J consumption (mass basis)	lb/hp•h	mg/J	1.689 659	E–01
Specific fuel – m^3/J consumption (volume basis)	m^3/kW•h	dm^3/MJ	2.777 778	E+02
	US gal/hp•h	dm^3/MJ	1.410 089	E+00

Quantity and SI unit	Customary unit	Metric unit	Conversion factor Multiply customary unit by factor to get metric unit	
Transport properties continued				
Viscosity – m^2/s (kinematic)	ft^2/s	mm^2/s	9.290 304	E+04
	in^2/s	mm^2/s	6.451 6	E+02
	m^2/h	mm^2/s	2.777 778	E+02
	cm^2/s	mm^2/s	1.0	E+02
	ft^2/h	mm^2/s	2.580 64	E+01
	cSt	mm^2/s	1	

Index

FSC
www.fsc.org
MIX
Papier | Fördert
gute Waldnutzung
FSC® C083411

Kolibri 360 GmbH
Ferdinand-Jühlke-Straße 7
99095 Erfurt, Deutschland
produktsicherheit@kolibri360.de